History of Mechanism and Machine Science

Volume 43

This bookseries establishes a well-defined forum for Monographs and Proceedings on the History of Mechanism and Machine Science (MMS). The series publishes works that give an overview of the historical developments, from the earliest times up to and including the recent past, of MMS in all its technical aspects.

This technical approach is an essential characteristic of the series. By discussing technical details and formulations and even reformulating those in terms of modern formalisms the possibility is created not only to track the historical technical developments but also to use past experiences in technical teaching and research today. In order to do so, the emphasis must be on technical aspects rather than a purely historical focus, although the latter has its place too.

Furthermore, the series will consider the republication of out-of-print older works with English translation and comments.

The book series is intended to collect technical views on historical developments of the broad field of MMS in a unique frame that can be seen in its totality as an Encyclopaedia of the History of MMS but with the additional purpose of archiving and teaching the History of MMS. Therefore. the book series is intended not only for researchers of the History of Engineering but also for professionals and students who are interested in obtaining a clear perspective of the past for their future technical works. The books will be written in general by engineers but not only for engineers. The series is promoted under the auspices of International Federation for the Promotion of Mechanism and Machine Science (IFToMM).

Prospective authors and editors can contact Mr. Pierpaolo Riva (publishing editor, Springer) at: pierpaolo.riva@springer.com

Indexed by SCOPUS and Google Scholar.

Rafael López-García · Marco Ceccarelli
Editors

Distinguished Figures in Mechanical Engineering in Spain and Ibero-America

Springer

Editors
Rafael López-García
Department of Mechanical and Mining Engineering
University of Jaén
Jaén, Spain

Marco Ceccarelli
Department of Industrial Engineering
University of Rome Tor Vergata
Rome, Italy

ISSN 1875-3442 ISSN 1875-3426 (electronic)
History of Mechanism and Machine Science
ISBN 978-3-031-31074-4 ISBN 978-3-031-31075-1 (eBook)
https://doi.org/10.1007/978-3-031-31075-1

This Springer imprint is published by the registered company Springer Nature Switzerland AG
The registered company address is: Gewerbestrasse 11, 6330 Cham, Switzerland

Preface

This is the sixth volume in a series of books whose aim is to compile contributions by distinguished figures to Mechanism and Machine Science (MMS). The last volume in the series follows the five already published and is dedicated specifically to distinguished figures from the Spanish and Latin American community in the field of the science of machines and mechanisms and mechanical engineering.

This series of books represents a who's who in MMS dictionary project with the encyclopaedic character of the entire book series by highlighting the importance of MMS through time. The unifying feature of the volumes is that all articles recognize individuals whose scientific work resulted in relevant technical-scientific achievements with an impact on technology and science in the historical evolution of the fields of MMS and mechanical engineering. The attention is focused not only within the IFToMM or other associations such as the Spanish Association of Mechanical Engineering AEIM or the Ibero-American Federation of Mechanical Engineering FEIBIM, but also looking at outstanding influence on the development of the society of their time. Biographical notes describing the efforts, training, experiences, and achievements of these individuals are also included, although the fundamental core of each chapter is on their technical-scientific achievements and results.

This is the sixth volume in a series of edited books that started in 2007 with the aim to collect contributed papers on distinguished figures in Mechanism and Machine Science (MMS) and Mechanical Engineering. It is a continuation of the first volume which was published in 2007 (ISBN 978-1-402-06365-7), the second in 2010 (ISBN 978-9-048-12345-2), the third in 2014 (ISBN 978-9-401-78946-2), the fourth in 2020 (ISBN: 978-3-030-32398-1), and the fifth in 2023 (ISBN: 978-3-031-18288-4), all combining ancient and recent scholars in order to give not only an encyclopaedic character to this project but also to emphasize the significance of the MMS over time.

The sixth volume of the series project has been made possible, thanks to the invited authors, who have enthusiastically shared this initiative and devoted their time and effort to prepare the chapters well in advance. The chapters cover the wide field of Mechanical Engineering History with a specific focus on MMS by specific discussions of the distinguished figures and the specific activities they have carried out during their lifetime with impacts also on the following developments with fame also today.

In this book, special attention is devoted to the distinguished figures who contributed to the history of Mechanical Engineering and MMS in the Spanish and Ibero-American community. The aim is to record and give merit to all those figures that may not yet be well known in the international community, also with the aim to show that the evolution of science and technology is built day by day even with very recent contributions that need to be recognized as soon as possible not only for historical credits but also to trace the future and to attract more works and attention to those topics and aggregation of associations. The contributions in this volume are selected from the books published in Spanish, in revised and expanded versions, as "Figuras Ilustres de la Ingeniería Mecánica en España" in 2018 and "Figuras Ilustres de la Ingeniería Mecánica en España e IberoAmérica" in 2020.

We believe that readers will take advantage of each of the chapters in this book and future ones by getting further satisfaction and motivation for her or his work (historical or not).

We also wish to acknowledge the professional assistance of the Springer staff, and especially Mr. Pierpaolo Riva, who has enthusiastically supported this project with his help and advice in the preparation of the sixth volume.

We would especially like to thank the hard work and effort of the invited authors of each of the excellent chapters that compose this book.

We are grateful to our families and our friends and colleagues for their patience and understanding, which have made it possible for us to work on this book and the book-series project of distinguished figures on MMS.

Jaén, Spain
Rome, Italy
March 2023

Rafael López-García
Marco Ceccarelli

Contents

Eduardo Barreiros (1919–1992)

M. Lage Marco

Abstract E. Barreiros developed his extraordinary automotive activity from 1951, when he began converting petrol engines to diesel, through to 1969, when he sold his big company to Chrysler. In the 80s, he initiated a second industrial venture producing diesel engines in Cuba. This article is an extract from the book *Barreiros. Historia completa* by the same author, which will be published this year.

Keywords Spanish automotive history · Barreiros engines · Trucks · Tractors and buses

M. Lage Marco (✉)
ASEPA, Madrid, Spain
e-mail: manuel.lage@asepa.es

R. López-García and M. Ceccarelli (eds.), *Distinguished Figures in Mechanical Engineering in Spain and Ibero-America*, History of Mechanism and Machine Science 43, https://doi.org/10.1007/978-3-031-31075-1_1

1 Biographical Notes

2 1927–1952. From Bus Transport to Engines Conversion

Eduardo Barreiros Rodríguez was eight years old when his father was awarded the licence to operate a bus line between Orense and Los Peares in 1927. Eduardo would often accompany his father as he drove the bus, serving as conductor and seating passengers inside the bus and on the open-top level, which was euphemistically and rather pompously known as 'the imperial section'. He fitted this real work around his attendance at school. Before long, the young Eduardo had learned to drive the bus. Every Sunday, he would carry out maintenance tasks that were essential in those days: washing and oiling the vehicle.

The problem was that if the bus broke down, which was a frequent occurrence back then, the passengers on the line would be left without transport. The only feasible solution was to buy another bus, which was an old truck chassis fitted out for passengers. The buses were repaired at a garage in Orense. At that time, maintenance and repair required ongoing attention; when a part broke it was repaired by hand, or another similar scrap part was adapted instead. This system of replacing parts, as well as larger components such as engines, axles, gearboxes, etc., gave rise to what was known as a *recastado* (crossbreed) or a vehicle whose original brand could not easily be identified.

In 1929, the family moved to Orense because they wanted their youngest sons, Valeriano and Graciliano, to get a good education that was only possible in the regional capital. Eduardo, meanwhile, was keen to spend more time surrounded by engines and chassis in the garage, which was what he most enjoyed doing. In autumn 1931, the 12-year-old Eduardo informed his parents that he no longer wished to pursue his education and wanted instead to work at the garage and learn everything he could about the mechanics of trucks and their engines. Back then, these types of vehicles all had petrol engines.

The family's passenger transport business prospered and a year after arriving in Orense, they moved to a new house with a garden and a garage and workshop on the ground floor. In 1933, Empresa Barreiros applied to Orense City Council for a licence to extend the bus line to Ferreira de Pantón in Lugo province, which was granted. This allowed the company to grow steadily as it was the only public transport service connecting the different villages in the area. The bus transported goods between villages as well as passengers, which proved particularly profitable.

The young Eduardo worked half days in the garage and provided attentive service to the salespersons and brand representatives who visited it. He was particularly taken by the Dodge chassis that the brand's dealer had come to offer the garage. The chassis were assembled in Spain by the Seida company, which was the sole importer of cars and trucks from the American Chrysler Corporation Group.

In 1935, Seida built a large workshop hall in Zorroza (Bilbao), which had been designed by Chrysler itself, and began to assemble Dodge Brothers trucks. The proportion of locally sourced parts in the vehicles quickly reached 35% thanks to the large auxiliary industry in the area. The registration plate on the Dodge belonging to Empresa Barreiros, OR-1824, indicates that it was registered in 1934 and was most likely assembled in Spain. The new bus was a modern, powerful vehicle with a six-cylinder 70 CV petrol engine, making it far superior to the average vehicle on the road at that time.

Eduardo had convinced his father to buy the new chassis so that they could offer a better service to passengers. It could also be rented out on Sundays to transport football fans to Vigo, more than 100 kms away, and back to Orense on the same day. Ultimately, Eduardo Sr. accepted his son's suggestion and purchased the new chassis, using one of his old buses as part payment.

The second chassis, which his bargain-spotting son also urged him to buy, was a four-cylinder 40 CV Ford A. The plan was to attach the bodywork from an older bus to the chassis to create a *recastado* (crossbreed) vehicle. Empresa Barreiros, as the company's name appeared on the buses, now had two very modern vehicles and another, older spare vehicle.

In October 1934, the Asturias uprising broke out and the army requisitioned the Dodge bus, which was used to transport troops with its owner E.B.N. at the wheel. A few days later, the Ford was also requisitioned, along with its driver Arturo. It was fortunate that the vehicles were taken to provide this compulsory service with their usual drivers, as they were familiar with the maintenance and care required. As noted above, vehicles in the 1930s needed to be checked and oiled almost every week.

The young E.B. followed every aspect of the transport business closely, taking more and more of an interest in the mechanics of cars and their engines as time went on. The temporary requisitioning of the bus brought a halt to the passenger transport and mail service, as well as the parcel and delivery service, which had become a healthy source of income for the family. Trying to address the situation, Eduardo, who was now 15, requested a loan vehicle from the military fleet so that service could resume, albeit on a smaller scale. The director of the fleet understood the need for a replacement vehicle and loaned him a Buick convertible to which E.B. added two extra bench seats to increase its capacity from 6 to 12 seats. With this, Empresa Barreiros resumed its services.

In 1936, shortly after the Spanish Civil War began, E.B. enlisted as a volunteer on one of the first expeditions, despite not yet being 17 years old. The following year, just after his 18th birthday (the minimum legal age for drivers), he passed his driving test and was issued with a driving licence. He joined the army in April 1938 and was assigned to drive a Ford V-8 truck towing a 37 mm anti-tank gun. His skill and daring at the wheel earned him the nickname 'the thrills driver'.

E.B. was 19 years old when the war ended but he had already completed three years of military service and was discharged in October 1939, a few days before his 20th birthday. The young E.B.'s return to the family business as an experienced driver breathed new life into Empresa Barreiros, which purchased an old 1925 Chevrolet bus with a four-cylinder 35 CV engine. E.B. rebuilt the front of the rundown four-cylinder 40 CV Ford, fitting it with a single-pane windscreen and a larger radiator in a longer engine boot to give it a more modern appearance.

The new bus, OR-1513, had been registered as a van in February 1931. In 1939, it was recognised as a bus with 20 seats and another 8 on the roof. The engine was replaced by a six-cylinder Chevrolet in 1943, later by a Buick in 1944 and finally by a Fiat. It served the company for more than 15 years and was eventually fitted with a converted diesel engine. This bus evolution perfectly exemplifies the life of an industrial vehicle during an era of hardship and supply shortages.

E.B. carried out this important refurbishing and rebuilding work on the buses on his father's behalf, although it was, he who decided on the purchases to be made. Meanwhile, he also bought his own vehicles as his positive experience of purchasing and rebuilding vehicles had shown him the potential profits to be made from a business that he felt perfectly equipped to pursue and expand. One of E.B.'s more significant purchases was a 1930 Mercedes saloon that he bought in Orense with the intention of dismantling it and selling all the components except the engine. These pre-war saloon cars consumed large amounts of fuel and became unaffordable for their owners during the 1940s, when petrol was scarce. This made it possible to purchase them for an economical price.

E.B. fitted the six-cylinder 70 CV Mercedes engine to a truck chassis purchased as useless, built new bus bodywork for it and installed a wood gas generator as a solution to the petrol shortage. In these devices, wood or charcoal were burned in a vertical chimney and the incomplete combustion produced gases that passed over charcoal embers, resulting in a gas containing high levels of carbon monoxide enriched with hydrogen from the water in the wood. Engines fuelled in this way were 30% less powerful due to the different calorific values of gas and petrol, but they freed drivers from their dependence on scarce supplies of petrol.

There was not enough scrap in Orense to satisfy Eduardo's urge to build vehicles, so he travelled to Madrid in search of parts. He found everything he was looking for at the stalls selling scrap and spare parts in the Rastro street market. He purchased used engines, gearboxes, differentials, and wheels, which he was already imagining on the new vehicles he intended to create. On behalf of his father, he also acquired a Reo bus registered in 1929, which he rebuilt at the garage in Orense until only the original engine and number plate remained.

Besides rebuilding and repairing the buses belonging to the family business, E.B. had become an experienced mechanic capable of building a new vehicle from parts from different brands and sources and he decided to extend this activity by selling his vehicles to third parties. This was the only way to obtain means of transport that could provide a reasonable service, and one that was very necessary at that time.

Another vehicle assembled by E.B. was a stylish Hispano Suiza truck with a 1945 new registration plate reading OR-2198, with a payload of 5 tons. It had an engine and gearbox from a 1930 HS 56bis luxury car purchased near Orense in 1940, featuring six cylinders in-line, overhead camshaft and 160 CV. The chassis, of unknown origin, was finished with a Dodge front axle and a Krupp rear axle, while the cab was newly built. At the age of 26, E.B. sold the truck for the sizeable sum of 150,000 pesetas, making a generous profit.

During this time, E.B. continued to work for Empresa Barreiros as a driver in the afternoons and as a mechanic at the garage in the mornings, when he would also work on his own projects. In 1945, Eduardo was the only sibling not to pursue their studies: Valeriano was now 21 years old and was about to qualify as a business expert, while Graciliano was 19 and had left the seminary the year before after deciding to study Mechanical Engineering. He moved to Madrid, where he enrolled at a prestigious academy to prepare the access to the Engineering High School. His decision to become an engineer was most likely influenced by his brother Eduardo, who is sure to have regretted missing out on more comprehensive technical training than he was able to obtain from working at the garage, despite his success as a mechanic and buyer and seller of vehicles. Eduardo was already considered the leader of the family, who had a very harmonious relationship, as we will see in subsequent years.

3 1943. Building a Pier at Castellón. The Petrol Engine Trucks

As he drove the bus along the route from Monforte to Parada do Sil, the young E.B. noticed that numerous road repair and construction works were taking place. Seeing that the names of the companies awarded the contracts all belonged to rich people, he concluded that public works offered a prosperous future with large profits. He discussed this matter with his father and his brother Valeriano, who had graduated as a business expert.

Despite the company belonging to both brothers, with Valeriano keeping the accounts as its legal representative, they had agreed that Eduardo Barreiros would be the name and image of the company. He often referred to the company in the first person. In November 1946, the company was renamed 'Eduardo Barreiros. Public works contractor. General construction' and given a round logo containing the initials E.B.

E.B. decided to start bidding for public works contracts, which were being auctioned all over Spain at that time. The new company, which employed around 10 people, won a series of contracts to repair secondary roads in the province of Orense, fulfilling all the requirements and making a healthy profit. E.B. proved highly capable of managing his workers, who felt comfortable with him after seeing him work even harder and longer than them.

E.B. was a man of new business ventures and ideas, which he was able to implement quickly and develop effectively and innovatively thanks to his strong personality and extraordinary energy. His extensive experience and expertise in mechanical matters of all kinds enabled him to purchase and adapt machines for use in the company's work, often applying original techniques. Valeriano, who was five years his junior, handled the accounting and monitored expenditure, ensuring that E.B.'s initiatives became real businesses that yielded the anticipated results and profits.

In August 1946, E.B. married Dorinda Ramos, a girl from Cerreda, where E.B. passed by every day on his bus. They travelled to La Coruña for their honeymoon in a 1933 Standard Special sports car, which E.B. had purchased with a broken cylinder head. It had to be sent to foundry La Industriosa in Vigo, where they designed, cast, and machined a new part in a process now known as reverse engineering. The Standard Special cylinder head, followed by another for the Fiat engine in the OR-1513 bus, both petrol engines, were the first on the list of cylinder heads designed and built for Empresa Barreiros. Five years later, La Industriosa designed and cast special cylinder heads for engines converted to diesel.

In 1947, the company won more contracts: one of the most important was to repair and surface 80 km of provincial roads, which gave E.B. the opportunity to display his creativity and efficacy once again. In Bilbao, he bought mixers to prepare the bitumen with water so that it could be applied in the form of an emulsion; usually, the bitumen was heated to over 100 °C before having water and emulsifier poured over it to produce a sudden burst of steam. E.B. thought about the process and observed after several tests that doing it the other way round and pouring the bitumen over the boiling water produced the same result without any problems. He was extremely satisfied with his new procedure.

Before the roads could be surfaced, they had to be meticulously cleaned to remove any dust, soil or stones so that the asphalt would stick. This work was usually carried out by a team of sweepers with steel-spiked brushes, who moved along the road removing the mud stuck to it. Another team of women carrying twig brooms then swept the road to leave the surface clean. This was slow, physically demanding work so Eduardo decided that it could be mechanised to increase output. In less than a month, he had designed and manufactured a large steel-spiked roller, which was 150 cm wide with a 40 cm diameter and was powered by a four-cylinder Chevrolet engine that rotated it at 250 rpm.

The invention was a great success: the mechanical roller removed all the mud and stones stuck to the road in a very short space of time, leaving the original macadam surface uncovered, but it still had to be swept thoroughly before being covered in asphalt. Sweeping the dust was a simpler operation, which was mechanised by attaching a large fan to the roller that was also driven by the petrol engine. The machine, dubbed in Galician language *a máquina do demo* (the devil's machine) by one of the company's employees, may be considered E.B.'s first major invention. It allowed the company to fulfil the contract by the agreed deadline and cemented Eduardo's reputation as a great mechanic and inventor in his hometown, Orense.

In January 1947, E.B. heard of a tender to extend the pier in Garrucha (Almería) but he did not submit a bid because he needed funds of one million pesetas, which he was unable to obtain. A month later, he learned of a similar project to extend the pier at the port in Castellón de la Plana. E.B. travelled there to find out more about the project, which entailed building a new pier four kilometres long and four metres wide, with several breakwaters. The cost would exceed four million pesetas and was considerably higher than the project in Garrucha, but E.B. and Valeriano believed that they would be able to borrow the funds. Empresa Barreiros submitted a bid and won the contract. Once again, E.B. as the person in command, he would demonstrate his inventive approach to finding original, effective solutions to any problems that arose.

The port authority had assumed that the small railway at the port would have to be used to move the large amounts of earth required to carry out the works, as it would not be possible for trucks to turn on the narrow pier. E.B. considered it perfectly viable to use trucks if a detachable side platform could be built to allow them to turn around at the end of the wall and empty their cargo in reverse. The platform would be moved each week as the works progressed. He used this simple, novel solution to convince the port authority to leave the contractor the option to transport the rock in the manner they considered most efficient in the specifications. Barreiros won the contract.

E.B. travelled to Castellón and spent two months there planning the work, with a particular focus on transporting the stone and setting up a garage to maintain his trucks and machinery. Once work on the sea pier had begun, E.B. returned to Orense and the company's legal representative, Valeriano, replaced him as supervisor at the site in Castellón (Fig. 1).

4 1951. Krupp, The First Petrol Engine Converted to Diesel

Diesel trucks were not a new development in Spanish industry, as the Naval-Somua factory in Bilbao launched a chassis with an English Gardner diesel engine in 1935 and Hispano Suiza sold several four- and six-cylinder models with a Hungarian Ganz

Fig. 1 Krupp/Barreiros Diesel truck on the maneuvering platform at the pier in Castellón, 1951. Courtesy of Fundación E. Barreiros

diesel engine. Following the end of the Civil War, Hispano Suiza was developing its own 66-D diesel engine, which began to be manufactured in 1945; the engine formed the basis of the first Pegaso Diesel in 1949, which was a direct injection diesel engine.

When the new 125 CV Pegaso Diesel engine reached the market, it was fitted to an eight-ton truck, which was a very large tonnage at the time: most of the trucks in the existing fleet had a payload of no more than 3 or 4 tons, so they required far less power. Except for the small numbers of HS 66-D manufactured between 1945 and 1946, until Enasa was founded, the few diesel trucks seen in Spain were imported, mostly from Germany.

E.B. quickly realised that diesel engines offered a huge competitive advantage due to their lower consumption and the unlimited availability of diesel, which was far cheaper than petrol at 6.25 pesetas per litre for petrol and 1.80 for diesel. Even though for a similar power output, diesel engines had a larger displacement because they turned at a lower rpm. Diesel or heavy oil engines, as they were called at the time, continued to offer far greater benefits.

E.B. was 29 years old by this time and his brother Graciliano, who was studying Mechanical Engineering in Madrid, was 22. There can be no doubt that E.B. asked his brother to find technical literature on diesel engines. The Bible of diesel engines in those years was *High-Speed Diesel Engines* by American author P.M. Heldt, which was published in 1940 and translated by Professor R. Marqués at the Escuela Especial de Ingenieros de Barcelona. The 460-page volume explored almost all the diesel engines available in 1940 in detail: engines from the USA, Germany, Italy, England, etc. in all their variants, including direct injection, swirl chamber, air chamber and pre-chamber. The description of the different types of combustion chambers and pre-chambers in E.B.'s patent, which we will see later, is practically the same as the one in the book.

In 1948, E.B. had two trucks for construction work but he was very worried about the scarcity and high price of petrol. Despite this, it was not feasible to purchase diesel trucks as there were none available second-hand and new trucks were prohibitively expensive. He began to ponder a new idea: would it be possible to convert a petrol engine to a diesel engine? Any expert in engines would have said that it could not be done but it all depended on the components that you were willing to modify or replace.

This is where Graciliano, a Mechanical Engineering student, came in. At his brother's request, he began to study the question in a very pragmatic manner, drawing on the excellent information in *High-Speed Diesel Engines*. The high-quality illustrations showing cross sections and longitudinal sections of many different engines allowed measurements to be taken directly from the drawing.

Graciliano's first step was to take measurements from the drawings and study and tabulate the main dimensions of the diesel engines described in the book, seeking the basic relationships between the main dimensions: cylinder bore to crankshaft crankpin area, connecting rod length and piston stroke, etc. He also studied the usual compression ratios in direct injection engines and pre-chamber engines.

The company E.B. had purchased two Krupp trucks with twin drive axles from an Ejército del Aire auction. The engines were six cylinders 100 × 160 mm, 7,540 cm^3, with 95 CV at 1,400 rpm and 100 CV at 1,800 rpm. At that time, power for continuous use and maximum power, which was not compatible with intensive use, were both stated.

The first conversion to take place was the Krupp. The side housing of the spark plug in the upper part of the cylinder block was used to install the horizontal injector, but this arrangement made it necessary to build intermediate parts to serve as water-cooled pre-chambers, including a resistor to help with cold starts. The system is perfectly described in the patent:

> In some engines, the holes on the spark plugs can be used to install the combustion pre-chambers or air chambers. When this is not possible, they are installed on the cylinder head or block. The chambers or pre-chambers are cooled by air or water, as appropriate. The injectors are assembled on the air chambers in the same manner as in well-known diesel engines.

The injection pump, purchased from the Bosch agency in La Coruña, was installed on the right-hand side of the engine, which had the timing gears behind it so a front gear train had to be designed from the camshaft on the left, with a large intermediate gear to achieve the required side elevation. The entire system was encased in a new, very visible front timing case.

The engines were assembled at a workshop in Orense where there was little more than a drawing board. This was where the company's designs were drawn up and the first ЗИС would be converted until operations were transferred to Madrid in 1952.

The new pistons, which were higher for greater compression and had a smaller diameter than the original pistons to compensate for the extra stress on the crankshaft and connecting rods, were manufactured by Tarabusi, a company based in Bilbao that was the main supplier of pistons for series production and spare parts in Spain. After two months of work to modify the block, install the pre-chambers and injectors, fit the injection pump, and change the pistons, the Krupp engine finally began to run with its new diesel cycle. According to E.B., "it sounded great". Once the truck was ready, it was driven just over 900 km from Orense to Castellón, laden with 10 tons of material required for the work on the pier. Years later, E.B. said: "I made a wonderful truck with the first Krupp engine, which provided excellent service in Castellón."

The cost savings offered by the engines that he had converted for his own use were so significant and attractive that he began to think about converting engines for other users. The truck's first use was as a vehicle for exhibiting the converted engines. Its engine was exposed and another converted engine, the first ЗИС (ZIS), was mounted on the back of the vehicle and could start up and run idle. E.B. drove it around several cities in Galicia to promote his converted engines. The ЗИС (ZIS) would be the most used engine in E.B.'s new activity converting engines to diesel for third parties, which will be explored later on.

The advertised fuel consumption for the Krupp petrol engine was 60 l/100 km and the converted engine is likely to have reduced it to around 30 L. In 1950, the price of diesel was 1.80 pesetas per litre compared to 6.25 pesetas for petrol, which was also rationed. Of the two Krupp engines, one was rebuilt with 90 mm bore and the other with 94 mm and higher compression so that it would deliver greater power.

Given the sheer number of changes made to the original petrol engine, it is apparent that what was termed 'conversion' entailed building a diesel engine using most of the components from a basic petrol engine. E.B. had shown that it was possible to convert a petrol engine to a diesel engine and had done so pragmatically and effectively by changing the necessary components for new, purpose-built designs. The technical side of the process was overseen by his brother Graciliano, still a Mechanical Engineering student.

It is interesting to note that, from the very beginning of his motoring activities, Barreiros took great care to protect his ideas and achievements with the corresponding patents.

5 Converting the Soviet ЗИС (ZIS) Engines to Diesel

Interestingly, the book *High-Speed Diesel Engines* features the Hercules diesel engine; the Soviet ЗИС was originally a licence from Hercules in the USA and the fact that a diesel version had now appeared on the market with the same piston stroke and crankshaft was enough for Graciliano and E.B. to decide that the architecture of the ЗИС would be a good base for conversion. The book gave two possible cylinder bores, 89 and 94 mm, which were adopted by Barreiros for the company's conversions.

In Spain there was a large running park of Soviet ЗИС trucks, which had been purchased by the Republican Government during the Civil War and were gradually auctioned by the army. The brand was ZIS, or ЗИС in Cyrillic letters, which the Spanish soldiers read as the initials of '3 Hermanos Comunistas' [3 Communist Brothers] as the trucks were manufactured in Moscow. The main problem with the trucks were the side valves, making it necessary to design and cast a new cylinder head and replace the original valve set up with an overhead valve arrangement actuated by push rods. In practice, the aim was to build a newly designed diesel engine using ЗИС components.

After their positive experience with Krupp, E.B. identified ЗИС engines as particularly suitable for conversion to diesel and used them for their first job for a customer. In October 1951, it was reported that the company had delivered a diesel ЗИС engine to its first customer, Autocares Soler.

Diesel conversions: This afternoon, E.B. visited us with the newly converted engines that we have all been so enthused about. Then Mr Soler came to visit us and when he saw how quickly it started up and how well it worked, he was amazed, and the same happened to several mechanics and gentlemen who are interested in having their engines converted.

The ЗИС engine was robust with a 7-bearing crankshaft but it had side valves in the cylinder block, making it necessary to manufacture a new, purpose-built cylinder head for the engine with new overhead valves. In subsequent versions, the single cylinder head was replaced by two-cylinder heads with three cylinders each. This was possible because there was a larger gap between the central cylinders in the ЗИС block and four screws between them.

Although this may appear to be a purely semantic concern, we must insist that the conversions performed by Barreiros involved building a new engine using the main components from an older one. The successive designs for the new cylinder heads with direct injection or pre-chambers, the different position of the intake and exhaust manifolds, the adding of an injection pump, injectors and preheating spark plugs and the modifications to the water circuit, new pistons, etc. were equivalent to designing and building a whole new engine.

In 1951, Barreiros launched a first advertising campaign to promote its converted engines, displaying a Krupp truck chassis without a cabin so that the converted engine was visible, with another ЗИС engine mounted on the back of the same chassis, which could be used as a test bench when the vehicle was stationary. An advertisement published in *El Ideal Gallego* in August 1951 read "You too can convert your truck to diesel for just 25% of the factory cost. Guaranteed performance equal to the original petrol engine. The converted engine delivers the same power and speed. Delivery time: six weeks. Patent from E. Barreiros. P.O. Box 24. Orense (Fig. 2)."

Fig. 2 Display chassis with Krupp/Barreiros Diesel engine, with another ЗИС engine mounted on the chassis, 1951. Courtesy of Fundación E. Barreiros

6 1951–1952. Engine Conversion Develops at the Orense Workshop

At the end of 1951, Barreiros announced the delivery of 20 converted engines in their first year of operations. Studying the charts kept in the workshop for each of the engines, it gives us a more precise idea of the scope and detail of the work carried out by the company. Conversion entailed dismantling the whole engine, repairing any worn components such as the crankshaft, the camshaft, and their housings in the engine block by replacing the bearings; the cylinders were rebored with liners to match the new pistons, new connecting rods were manufactured when necessary and the injection system was adapted and assembled. Through this process, a petrol engine could be rebuilt from scratch by designing and manufacturing all the components needed for it to operate with diesel.

For the first year, the company sought out trucks and engines from different brands and then studied, calculated, designed, and manufactured the parts needed to convert each model: each case was approached as a standalone project. Although the most widespread brands in the Spanish market were Ford and Chevrolet from the USA, their engines could not be converted as the components were not robust enough.

E.B. often had to purchase components and equipment from regional and national suppliers. Not all the operations could be completed at the company's garage so for special machining tasks such as repairing the crankshafts, designing the blocks and cylinder heads, etc., the parts were sent to external workshops in Galicia and beyond. Engine blocks were sent to be repaired at Mintegui in Bilbao, for example, which was one of the best equipped truck garage and workshops in Spain at the time.

The workshops in Orense had no power test bench, but they were fitted with simple benches where each finished engine was set to rotate for several hours, powered by an electric engine. The diesel engine rotated cold and on idle, with no combustion, with the sole aim of 'smoothing' all the surfaces working in contact with other: pistons and jackets, crankshaft supports, connecting rods, etc. Of course, the oil pump was working, and the engine turned with oil pressure in the circuit. This running in method was not very effective as there was no combustion so the components did not expand as they would normally.

Between 1951 and 1952, Barreiros converted 57 engines, 37 of which were ЗИС and 20 were from a variety of other brands. Besides those mentioned above, we find: Dodge, International, Waukesha, Wisconsin (all American); Ferguson (tractor engine), Standard Vanguard (English) and even a petrol Pegaso, despite the brand already offering its own diesel engine. The converted engines were all in-line six-cylinder engines, except for the first Leyland, the Ferguson and the Standard Vanguard, which had four cylinders.

The engines converted in 1951 were all sold in Galicia, except for the engine for Autocares Soler in Castellón and another two in Asturias. In November 1951, the advertisement was published in the national press for the first time, appearing in *ABC* (Figs. 3, 4, 5, 6, 7, 8, 9 and 10).

Fig. 3 Advertisement for the EB-0 engine in national newspaper ABC. The company continued to operate from Orense, 1951. Author's archive

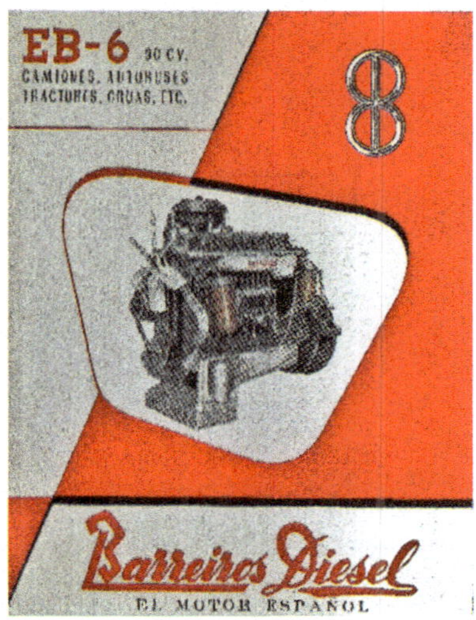

Fig. 4 Publicity for the EB-6 engine, 1956. Author's archive

7 1952. Exhibition at the Engineering High School of Madrid

As we already know, Graciliano Barreiros, the third of the brothers, had left the seminary in 1945 to study Mechanical Engineering. In 1946, he moved to Madrid to prepare his admission in the Escuela Especial de Ingenieros Industriales, where he was accepted in 1949.

Fig. 5 E.B. poses proudly with one of his first trucks, September 1957. © Regional Archive of the Community of Madrid. Cristóbal Portillo Fund

Fig. 6 Publicity of the Barreiros engine range, including the 150 CV, 1957. Author's archive

Fig. 7 The military prototype with ellipsoid wheels, 1958. Courtesy of Fundación E. Barreiros

Fig. 8 Full-page publicity in *Auto-Revista* showing the definitive cab, December 1959. Author's archive

Fig. 9 The first Hanomag-Barreiros tractor: the R-545, 1959. Author's archive

Graciliano followed his father and brothers' activities in the family bus and construction ventures with keen interest. He played an important role in the company's engine conversions; despite still being a student, he oversaw the technical side of the first conversions in Orense. Comparative study of the dimensions and characteristics of the modern diesel engines featured in Heldt's book was used to calculate and determine the compression ratios and turbulence chambers for the different engines converted to diesel.

In February 1952, a series of events were scheduled to mark the Centenary of the Engineering High School of Madrid, including a large technical exhibition open from 20 January to 20 March. The First Official National Exhibition of Metalworking and Electricity was held at indoor and outdoor venues at the Escuela Especial, where Graciliano was in the second year of his studies.

For Graciliano, the technical exhibition being prepared at the school was a unique opportunity to present the engine conversions from petrol to diesel that he had overseen to a high-level technical and political audience in the Spanish capital. He informed his brothers of the event; they enthusiastically began to prepare the material

Fig. 10 Publicity for the C-14 engines targeting the taxi sector, 1961. Author's archive

they wished to exhibit. One of the first steps they took to prepare for the exhibition was to publish the first national advertisement for the engine conversions in the newspaper *ABC*. Barreiros prepared an eye-catching stand featuring two converted engines: a six-cylinder ЗИС and a four-cylinder Leyland. Both engines were prepared to run as a demonstration of their functioning with diesel.

From the outset, E.B. was always keen to publicise his engine conversion work as widely as possible. The consumption of an original ЗИС was 50 L of petrol per 100 km and the diesel engine reduced this to 25 L. Given that the price of diesel at the time was 2 pesetas per litre and the price of petrol was 5.50, the 80% saving per kilometre between the original engines and the converted Barreiros Diesel engines that was advertised appears to be accurate (Figs. 11, 12, 13, 14, 15, 16 and 17).

8 Moving to Madrid. The New EB-1 and EB-2 Engines

1952 was a very important year for Barreiros as the company transferred from Orense to Madrid, occupying a new workshop at km 7 on the Carretera de Andalucía (Santos et al. 1952). By this point, they employed around 50 staff. Engine conversion quickly became their main activity, overshadowing bus transport and construction.

The rented facility, which was later bought by Barreiros, had a surface area of 2,000 m^2 and was well equipped for repairing engines as it had a power test bench.

Fig. 11 First range of Barreiros trucks with specific names, 1961. Author's archive

Fig. 12 Barreiros-AEC bus, 1965. Author's archive

Fig. 13 Centauro cement mixer truck with the big cab, 1964. Courtesy of Fundación E. Barreiros

Fig. 14 The BH-38 flat engine was the forerunner of the six-in-line B-36. Courtesy of Fundación E. Barreiros

Fig. 15 When the new heavy trucks were launched, E.B. was no longer in the company, 1969. Author's archive

Fig. 16 Cuban Taíno truck with Barreiros/Taíno engine. Author's archive

Fig. 17 One of the last photos taken of E. Barreiros. Havana, 1991. Courtesy of Fundación E. Barreiros

The machines were all manual as they were designed for repairs rather than manufacturing, even in small series. The workshop featured a block crankshaft and camshaft line boring machines, a cylinder boring machine, a crankshaft grinding machine, etc.

The cylinder heads continued to be cast in Vigo, where the basic machining of the contact face and holes for fixing them to the block took place; they were then sent to Madrid for the rest of the machining, especially the valve seats and guides. Still a lack of most appropriated machinery is illustrated by the fact that the contact face of cylinder heads was machined on a lathe with a large-diameter plate, to which the cylinder head was attached and faced.

By July 1953, 124 engines had been converted at the new garage in Madrid, 117 of which were ЗИС. By that time, the company focused less on converting customers' engines and more on purchasing all the ЗИС engines they could find to convert them to diesel and sell them as fully rebuilt engines.

The availability of a power test bench at the new garage enabled tests to be carried out with different configurations and led to the decision to develop a new prechamber with indirect injection. Ultimately, the company opted for a Lanova-type combustion chamber, where the injector was positioned horizontally and directed the pulverised fuel towards an air chamber opposite.

This brought about a major change to the design of the cylinder heads, which now required more elaborate machining. In mechanical terms, these new engines were 'flat piston' (flat-headed) engines. Since the company was no longer converting engines from other brands and the ЗИС engines had undergone extensive development work on the test bench to fine-tune the new combustion chamber designed by Barreiros, they were considered original engines and assigned the nomenclature EB-1.

With the declining availability of original ЗИС engines and the company's extensive experience in manufacturing engine components, E.B. decided that the next step would be to manufacture all the components, starting with the casting and machining of new blocks to take advantage of the stock of crankshafts that the Republican Army had received as spare parts and that E.B. had purchased. The crossflow cylinder heads were also designed from scratch, keeping the horizontal injectors on the right-hand side. Barreiros no longer converted engines; the company had become a manufacturer of new engines using components based on the ЗИС design.

The engine, with all its new components, became known as the EB-2. The block was modified and reinforced, with two main suppliers: Clúa and Fundiciones Padró, both from Barcelona. With these modifications, the engine speed and power increased from 65 CV at 2,000 rpm to 73 CV at 2,300 rpm. This version was named and began to be manufactured in April 1954. A total of 486 EB-2 engines were produced by the company, dramatically exceeding the 297 engines it had converted. By 1954, its fourth year of motoring operations, Barreiros Diesel was a fully-fledged engine manufacturer.

At the end of 1953, an advertisement referred to 223 trucks "travelling Spain with our diesel converted engines". This figure is consistent with the offer made that same year by E.B. himself to convert military ЗИС engines that were still in service, stating that "we have already converted more than 200 engines". Of the 223 units, 57 were produced between 1951 and 1952 and 166 in 1953.

Publicity from 1955 indicates that the number of engines in service was 700. In 1954, another 237 engines were made, followed by 314 in 1955. Approximately 40 engines were manufactured each month until the EB-6 model was launched, with a total of 783 units produced. Except for around 20 units from different brands in 1951 and 1952, the remaining engines were rebuilt ЗИСs or new builds.

E.B. also created his own engine oil brand: EBROIL, based on his initials.

9 1954. Barreiros Diesel, Fábrica Española de Motores y Tractores

The first engine developed from scratch by Barreiros was the EB-1bis, a two-stroke, twin-cylinder engine with a scavenging blower. Only two units were built despite ample publicity in 1953. It had a displacement of 1,998 cm^3, a maximum power of 44.5 CV at 2,000 rpm and a tractor power of 26 CV at 900 rpm. The engine was advertised for use in tractors, industry, and agriculture. Images indicate that it was a two-stroke engine with exhaust valves, a modern solution generating greater specific power.

In February 1952, Barreiros applied for a licence to manufacture tractors using the company's own technology in response to a government decree to organise agricultural tractor production. E.B. was confident of obtaining the licence as he developed a specific two-stroke engine for use in tractors, as well as an entire tractor. When it proved infeasible to build complete tractors, E.B. fitted the two-stroke engine to a Ferguson tractor (most probably the one he had converted the previous year). Advertising from 1954 suggests that this was an attempt to tap the petrol engine replacement market, but prospects for the two-stroke engine must have been discouraging as it was ultimately abandoned.

Tractors were the first vehicles that Barreiros sought to manufacture, and the full name given to the limited company when it was registered in March 1954 was Barreiros Diesel, Fábrica Española de Motores y Tractores SA [Barreiros Diesel, Spanish Engine and Tractor Manufacturing]. The two-stroke engine was intended for use in the tractors that E.B. hoped to manufacture as it offered lower production costs and required less maintenance. That same year, the ministry rejected the company's application to manufacture tractors and the two-stroke engine went unused. The licences were awarded to Ford Motor Ibérica in Barcelona, which also manufactured trucks and the recently created Lanz Ibérica.

The public register for the new company was signed on August 25th, 1955, with a share capital of 10 million pesetas. The company began operating on December 1st, 1954. The Barreiros family owned 64% of the new company's shares. Graciliano was not among the shareholders as he was still a student, although he was already working part-time at the company.

The activities declared by the company were:

- Conversion of petrol engines to diesel
- Construction of engines and vehicles.
- Financial, agricultural, and industrial activities.
- Procurement and execution of public works and services.
- Promotion of domestic and international trade.

It is interesting that public works and services were among these declared activities, as the company sought to continue the transport service delivered by the original Empresa Barreiros and the construction work it had carried out subsequently.

In 1954, 237 diesel engines had been produced and the expansion of the premises in Villaverde (near Madrid) began at a pace of one engine per day. Barreiros now had a 4,000 m^2 factory and 150 employees. At the end of the year, the site was extended again when the company purchased the adjoining plots, which covered a total area of 16,000 m^2.

10 1955. Developing the EB-6 and EB-4 Engines

In the 1950s, the booming engine replacement market encompassed the entire prewar truck and bus fleet and even some more modern vehicles, all of which had petrol engines. Most of the trucks in Spain at that time were American: Chevrolet, Ford, GMC, Dodge, Autocar, Federal, Reo, Diamond-T, Studebaker, Stewart and International. There were also Italian trucks brought over during the war: OM, SPA and Fiat, along with Mercedes, Renault, Citroën, Bedford, a significant number of Spanish Hispano Suiza and the soviet ЗИС.

The most famous diesel engine in Spain was the Perkins from England and Barreiros contacted the company to request a licence to manufacture it in Spain. The request was denied, partly because Perkins Hispania had recently been set up to import and sell the company's engines. This was when the adventure really began: Barreiros sought information on the legal status of Perkins' designs in Spain and found out that the English company had patented its engines in 1932 but that the patent had expired after 20 years. The company quickly applied for a 'patent of introduction' for two major improvements: the first was the Perkins combustion pre-chamber, no. 216,240, dated 30 June 1954, which was granted for the 10 years stipulated by law, occupying the legal ground when it came to the use of the English engine's combustion system in Spain. The next day, Barreiros registered another patent of importation to cover the flame cold start system in the intake manifold, no. 216,263, also from Perkins.

Both patents of importation were registered under the name of Mr Soto Rodríguez, Barreiros right-hand man who oversaw the work in Castellón and was also a shareholder of Barreiros Diesel SA. E.B. immediately requested a ministerial licence to manufacture diesel engines, as his activity until then had been considered repair rather than manufacture of new engines. The new company was granted ministerial permission to produce 2,500 six-cylinder engines and 1,200 four-cylinder engines each year: these were the EB-6 and EB-4, which were copies of the Perkins P-6 and P-4. The engine was such an exact copy that many of the parts were interchangeable, which made it possible to use English parts mostly smuggled from Portugal to overcome the supply shortage in Spain.

In 1954, J. Merino joined Barreiros as workshop manager. He was the first external qualified technician to work at the company and, in his own words, he provided a first degree of technical supervision on the work taking place. In November 1954, drawings for the future EB-6 engine were ready after being drawn up by a draughtsman supervised by Graciliano Barreiros. However, the plans were based on sketches of the English Perkins P-6 engine and were purely geometrical; they did not indicate tolerances, surface finishings or material for each part. These characteristics were clarified later in the manufacturing process, which was the only engineering activity at Barreiros at that time. These days, this approach would be unthinkable, but it demonstrates the company's determination to progress and learn as it went along.

Despite the EB-6 being a direct copy of the Perkins P-6, it featured several improvements on the original. The most important of these was the decision to incorporate the oil circuit into the engine block and cylinder head; on the Perkins engine, it was external, and the pipes were at risk of damage, potentially causing the engine to seize. Not having the right, and very expensive machine to drill such long. E.B. had an excellent idea, which would be used in all the company's engines from then on: inserting a steel tube into the sand moulds at the foundry process, so that the longitudinal oil gallery in the block was automatically created in the casting. Many technical aspects were overseen by E.B. himself, who was very intelligent despite having completed little formal education.

The system was hard to clean from shavings and burrs formed during the machining process risking scratching the crankshaft bearings. It finally was fine-tuned and used in all Barreiros engines until production ceased in 1985. Another major improvement was the addition of a larger capacity oil pump. This initial development took place at the factory on the Andalusia Road.

Initially, the crankshafts and camshafts were original Perkins parts purchased as spares and smuggled into Spain, most often from Britain's Perkins Ltda. in Portugal. Purchasing the parts legally with an import licence would have been impossible, as the INI/Ministry of Industry opposed the creation of an industry that would overshadow the state-owned enterprise Pegaso.

The first EB-6 engine was manufactured in July 1955; production of the four-cylinder EB-4 began a year later in 1956.

11 Series Production. Creating Cabsa and Ceesa

In 1956, Barreiros Diesel supplied a certificate of origin with its engines, which stated that Spanish materials had been used to manufacture them and that import duties had been paid for any foreign parts. The warranty covered any manufacturing defects and premature wear and tear for six months, including replacement parts where necessary.

The new engines were serviced twice under warranty, once after 1,000 km and again after 3,000 km. Each service included an oil change, general lubrication, filter and injector cleaning and cylinder head nut tightening, as well as checks on leaks, oil pressure, tappet clearance, injection pump level, dynamo charge, voltage regulator, batteries, and starter. The company recommended using "very clean diesel, filtered if possible" and checking the oil level (ideally EBROIL or E.B.R. oil), radiator water level and fan belt tension daily.

During his trips to England, where he discovered the Perkins P-6 engine, E.B. contacted engine design company Ricardo & Co., one of the leading companies in the sector both then and now. Ricardo worked with Barreiros to fine-tune and manufacture the EB-6, which was known at Ricardo as the pseudo-Perkins. The visit by the engineers from Ricardo served as motivation for E.B., who immediately contacted the ministry to ask for the licence to manufacture six-cylinder engines to be increased to 25,000 units or ten times the number permitted under the existing licence. His request was met with vehement opposition from the other Spanish manufacturers, Enasa, Motor Ibérica and Babcock (Historia de la industria española de automoción 2005). Following a series of appeals and commitments to export, the official licence was increased to 5,000 engines up to 100 CV in the first year, rising to 10,000 over the following five years.

In February 1956, the company produced 10 engines per day, reaching a total of 3,494 engines that year; by September 1956, 216 of those engines were the new EB-6 model and 29 were EB-4. Between 1957 and 1959, the company's production rate rose from 14 to 35 engines per day. In 1957, a total of 4,416 engines were manufactured. Demand for diesel engines was very high as almost all existing vehicles ran on petrol, which was considerably more expensive than diesel.

Now that Barreiros had become an engine manufacturer, he began to think about manufacturing entire trucks. In 1955, he requested a licence from the Ministry of Industry to produce 1,500 units, which was rejected despite positive feedback from the official technicians. The main obstacle was a political one: that same year, the state-owned Pegaso factory in Barajas (Un millón de camiones y buses españoles 2008) had been completed and it was inconceivable to issue a licence to another truck manufacturer that would compete with the national brand.

In the minutes of the Enasa board meeting in October 1955, a brief note alludes to an issue that would be very concerning for Barreiros:

> The managing director voiced objections to the establishment of an industry in Madrid assembling medium-sized trucks, with an anticipated production of 1,000 units per year. He has duly informed the INI (National Institute of Industry) of this matter. Managing director Mr Ricart commented on the continual requests for authorisation to set up motor vehicle factories.
>
> In his eyes, it is easy to obtain a partnership contract with a foreign company and begin manufacturing motor vehicles using imported parts to the obvious detriment of an industry such as our own, which, in times of isolation, had to meet such an important need as motorised road transport. He considers it necessary to implement protective measures like those in force in other countries with a long industrial tradition and strong automotive industry. If comparisons are made with a company based on imported parts, rudimentary assembly and the name of a large foreign firm, the price and market issues for genuine national production will worsen and will be inadequately understood if there is no intention to protect Spanish labour as we have been requesting.

The company in question was Barreiros Diesel, which had only been manufacturing engines for the replacement market until then but had now decided to embark on producing entire trucks. There is a clear arrogance in Ricart explanations as he refuses to accept that a private company be allowed to produce trucks based on its own design. It is also true that Barreiros industrial activity in 1955 was limited to manufacturing indirect injection diesel engines with a prechamber, copied from the English Perkins engines to the extent that the parts were interchangeable. Only three years had passed since Ricart had seen the converted engines at the Engineering High School exhibition and, based on his limited knowledge of E.B., it did not seem credible that the same company was now equipped to manufacture entire trucks.

One of the main issues concerned the procurement of injection pumps as they had to be imported, which was difficult at the time because of the lack of foreign currency. E.B. could see only one possible solution: start manufacturing the pumps locally to keep up with the rising demand forecast. Barreiros knew that any request for a licence to manufacture injection pumps would be immediately rejected by the ministry as it would be viewed as competition for Pegaso; instead, he opted to apply for a licence to make unspecified industrial pumps, which an inexpert administrator would no doubt interpret as water pumps for irrigation or something similar. Using this strategy, he was able to obtain a licence to manufacture his own injection pumps, which were one of the hardest parts to procure on the market at that time. Within four months, the new Compañía Anónima de Bombas (Cabsa) was already producing the pumps and injectors needed to meet Barreiros needs.

In April 1956, shortly after operations began, Cabsa applied to the Ministry of Industry for permission to double its production from 3,600 to 7,200 pumps per year, which required 18,000 injection elements and an equal number of valves and nozzles to be imported at a total cost of 8.5 million pesetas. As always, the application was rejected by Enasa, which alleged that its contract with Bosch "tends to meet the country's needs". This was a rather devious way of phrasing it, as there were no plans to supply other engine manufacturers.

Cabsa replied by pointing out that Enasa had not implemented the agreement with the German company and had no right to "monopolise all the industries in Spain"; the company also explained that it believed it necessary to shake off any sense of inferiority and manufacture products using Spanish technology, while acknowledging the technical advice being sought from Italian company Spica. The application was approved in June 1957, with technical support ultimately provided by American company Roosa-Master.

Given the urgent need for injection equipment to keep up with the rise in engine production, Cabsa production capacity was expanded in 1958 without submitting a request for authorisation. The licence was obtained in July 1959 once the corresponding fine had been paid. In 1960 and 1961, the company had no choice but to apply for authorisation to import new machinery. This was challenged by the Spanish Association of Machine Tool Manufacturers, which suggested that it should instead purchase equipment manufactured in Spain. The case was decided in Barreiros favour, indicating that the machines in question were high-precision machines required to manufacture components for the pumps that were not available in Spain.

Besides mechanical elements, the engines also contained the basic electrical equipment for the trucks, including the starter, dynamo, and regulator. Horns, windscreen wiper motors and batteries were also offered and were all provided by an external supplier that was soon unable to keep up with the rising numbers of Barreiros engines being produced. In response to the situation, Ceesa (Constructora Eléctrica Española SA) was set up in January 1956 under licence from the French company Ducellier-Bendix, which held 25% of the share capital, to manufacture electrical starters, dynamos, and regulators. Later, the company's range of products was expanded to include generators, welding equipment and electrical brakes for trucks. To bypass the need for a licence, the new company set up the facilities at an existing workshop that was already authorised to manufacture electrical components. Production began within the record time of three months.

As we can see, the growth of Barreiros and the company's strategies for overcoming any supply problems they encountered were unstoppable and were becoming an increasing cause for concern for the state-owned enterprise Pegaso.

Manufacturing crankshafts was a particular problem for the company. In Spain, Patricio Echevarría from Bilbao was the main supplier of forged crankshafts but, for some reason, Barreiros did not get on with him and never used his products. Another major company was Forjas y Aceros de Reinosa but there was a three year wait for crankshafts while the company was installing its new manufacturing machinery. Despite this, Reinosa provided invaluable technical assistance to design the required tooling, and this allowed all the designs to be drawn up at Barreiros. The parts were then manufactured at La Farga Casanova. This company in Campdevánol (Gerona) became Barreiros main crankshaft supplier but it was obliged to purchase a larger hammer press than the one it had so that it could stamp the new parts for Barreiros.

In the meantime, a free fall drop hammer was designed and built in Villaverde to allow forging to take place on the factory premises. The problem was that the foundation slab beneath the factory extended across such a large area that when the machine started up and began to hit the stamping steel, the vibrations affected the entire neighbourhood, making the houses shake and breaking windowpanes. Once everyone had got over the shock and the cause of the vibrations had been identified, the mayor of Villaverde angrily complained and no more crankshafts were forged at the factory.

Besides the difficulties involved in procuring blanks, which were delivered with the stamping burrs still intact, hindering the turning process, there were also issues with the crankshafts breaking while in use. Initially, they were not heat-treated as the company did not have the necessary facilities. The National Institute of Aerospace Technology was tasked with analysing the breakages and its director H. Pérez Vázquez wrote a detailed report on the problem, which he personally presented to Barreiros. Not long after, he began to work with Barreiros to create the company's materials laboratory, which served numerous other suppliers.

The help received from Swiss company Brown Boveri was key when it came to fine-tuning the surface hardening heat treatment on the main bearings and crankpins, but the process led to deformations and micro-cracks that could cause breakages. The problem with the machine was that the induction coils had a limited lifespan and had to be replaced but the bureaucratic procedure for requesting an import licence was so lengthy that the licences were not granted in time to maintain production if they were granted at all. The most practical solution was to talk to Brown Boveri in Geneva and collect the big diameter coils in person. On several occasions, J.M. Antoñanzas, chief engineer in the engine department at the time, flew to Switzerland with his pockets full of dollar notes and a large hatbox, which passed all the security controls without arousing suspicion, and returned with the induction coils so that production of the crankshafts could continue in Villaverde.

12 1958. First Buses and Trucks. Barreiros Undertakes the Development of Heavy-Duty Engines

In 1958, Barreiros built his first complete buses with an EB-6 engine, with different coachbuilders providing the bodywork. Two more units were bodied as trucks with a bus-style cab. The chassis used were Hispano Suiza Type 202 with their original axles. The first was bodied with a nose and with the steering wheel on the right, reflecting the model's original configuration. These chassis, which came with a four-cylinder petrol engine, had been manufactured by HS in the early 1940s to fulfil an order for several hundred units from the Ejército del Aire.

After the first bus was made with a nose and an extended rear overhang to accommodate more passengers, the company decided to modify the front of the chassis as well to add a flat cab by using angle brackets to extend the side members forwards, without touching the leaf spring supports. When this major change to the steering system was made, the steering wheel was positioned on the left in some chassis. Barreiros, who had already been awarded the Portuguese army contract by that time, had a licence to build 1,500 trucks so these Hispano Suiza original chassis were the first units registered under the new Barreiros brand.

Like the last engine conversions to be carried out by the company, these first truck and bus chassis were essentially new products using modified components from other brands. By extending the Hispano Suiza chassis, changing the suspension system and the driving position, and fitting them with a new diesel engine, the company produced a new Barreiros-branded truck.

As we have seen, the number of engines manufactured by the company was rising rapidly. Producing 3,500 units in 1956 and 4,400 in 1957, its production capacity continued to rise from 14 units per day in 1957 to 35 in 1959. The business was growing and E.B. was already embarking on new ventures.

In 1956, a major programme to develop larger displacement, direct injection engines were launched: the six-cylinder, 9.35 L EB-150 and the four-cylinder, 6.23 L EB-100. They featured chrome-plated steel wet liners that were very resistant to wear, which became a typical feature of Barreiros engines. Although they were clearly inspired by the similar Swedish Scania engines released in 1954, these were the first engines to be entirely designed by Barreiros. Once again, Eduardo and his brother Graciliano had come up with the idea themselves.

The engines were designed from scratch at Barreiros with the corresponding tolerances in each dimension and specifications for the materials to be used. By this time, the company had extensive machining and assembly experience. In April 1957, it was announced that the machinery and tools for the engines would be ready in time to start production by the end of the year. Advertising for the EB-150 began in 1957 and continued in 1958 and 1959, but the lengthy development and fine-tuning process meant that production could not start until 1960.

13 1957. The Big Contract for the Portuguese Army (Barreiros: Historia completa. To be Published)

In 1955 E.B. requested a licence from the Ministry of Industry to manufacture 1,500 trucks, which was rejected despite positive feedback from official technicians. There is no doubt that this decision came from the INI, which could not allow any other truck manufacturer to compete with Enasa, declared as being of national interest.

Barreiros argued that Enasa was unable to meet all national demand, which was estimated at 25,000 trucks and 1,300 buses per year, but the licence was still not forthcoming.

In March 1956, a group of military officials from Portugal, accompanied by the Portuguese ambassador and the Embassy's military attaché, visited the new Pegaso factory in Barajas. During the visit, the Portuguese authorities are sure to have discussed their army's rolling stock needs because in 1957, Portugal launched a major tender for the purchase of 300 military trucks for use in Angola. E.B. had followed the Portuguese visit with interest and for sure he knew about this tender from his many contacts in Orense province, on the border with Portugal.

With his engine production business thriving and providing a solid foundation for the company, E.B. continued to think about manufacturing trucks, which he had not been authorised to do by the Ministry of Industry. Now, however, a unique opportunity had arisen, and he could not let it pass him by: the construction of a four-wheel drive prototype for the Portuguese army tender. In a way, the Portuguese visit to the new Pegaso factory in Barajas may be viewed as the catalyst for the major Portuguese army contract that was awarded to Barreiros, transforming him into a truck manufacturer from one day to the next.

E.B. commissioned M. Gamarra to design and build an off-road military truck. The only component that had already been decided was the EB-6 engine, although the chassis and cab were to be designed from scratch. Given the shortage of components suppliers, the drive shafts and transfer case had to be taken from a scrapped Douglas crane and adapted to the new vehicle. The steering came from a Ford truck and the winch from a military GMC.

E.B. worked on building the prototype himself. The open cab was designed by the company and built by hand by an external panel beater. The cargo box was made by Manufacturas Metálicas Madrileñas, which would go on to supply other components. The 90 CV EB-6 engine was coupled to a four-speed gearbox with a reduction gear in the transfer case, which made it an eight-speed. A main advantage of the engine was to be diesel, but having a set of parts that, with two skilled mechanics, could be converted to petrol in 90 min. The rear suspension was a conventional semi-elliptical leaf spring system with a rigid axle, but an original system was designed for the front with a transverse rocker linking the front bearings of the springs, with high vertical displacement capacity to each of the wheels, and a rigid axle. This original suspension system was covered by a patent registered in December 1957.

Named TT 90.22 (Todo Terreno [off-road], 90 CV, 2 axle, 2-axle drive), the prototype was presented in Portugal, driven by E.B. himself, who showed its advantages against the other trucks in service in the Portughese Army, and finally won a contract to produce 300 units. These Portuguese trucks were very difficult to manufacture as the prototype had been made by hand without any drawings; it was only after winning the tender that the company began to produce designs and component specifications. After winning the tender, Barreiros paid a visit to the Minister of Industry, Mr Planell, to obtain authorization to build 1,500 trucks. The licence was granted on the condition that the company did not exceed this number.

From the very beginning of his business venture, E.B. had had a clear vision for publicising and promoting his products and for the company's public relations. In August 1957, the San Sebastián Sea Fair took place. Franco spent part of his summer vacation in the city, so it was common knowledge that he would be attending. E.B. took advantage of the opportunity to exhibit his marine engines and the prototype for the military truck that won the Portuguese contract, despite it having nothing to do with the sea. As he had anticipated, Franco visited his stand and no doubt heard E.B.'s explanations about the truck and the contract it had won in Portugal first-hand.

As a result of this first conversation with the head of state, E.B. was invited to give a demonstration of his truck's performance to the highest government authorities in the land surrounding El Pardo in February 1958. He drove the truck himself. Franco was very impressed by the Spanish-built truck that had won a foreign tender. The demonstration was also attended by Suanzes, who was the president of the INI and opposed Barreiros activity, and the Minister of Industry, Planell. It undoubtedly challenged the INI's arrogance regarding its state-owned enterprise Pegaso, as it saw how a private industrialist had won a major truck export contract that Pegaso should have been involved in.

Not one to miss an opportunity, E.B. took advantage of the occasion to inform Franco of the bureaucratic problems he was experiencing (which were caused by Suanzes and Planell). Franco listened to him and congratulated him on the demonstration, saying "keep up the good work and everything will work out; onward, Barreiros, onward!". Official authorisations began to come through from that point on, although the licence for the mass production of trucks for the domestic market took another two years to arrive and was finally issued in 1960.

In October 1958, Barreiros was the only Spanish company to attend the Paris Salon. It also attended the event in 1960, 1963 and 1966.

14 1958. The First Full Barreiros Trucks. David Brown Engranajes S.A.

Another smart move by E.B. was to sign an agreement in 1958 with Moto Import, the Polish state-owned agency for the automotive industry. Poland manufactured Star 21 trucks with petrol engines; no diesel engines were produced in the country and there was no available currency to import them. E.B. came up with the idea of exchanging engines for trucks without an engine. For each Star 21 chassis, the Polish factory would receive two Barreiros engines so no currency would be exchanged, which was one of the country's main concerns at that time. In Spain, Barreiros founded Comercial Star SA to sell the Polish trucks with the company's own diesel engine, but the poor performance of the chassis limited the scope of the operation.

In 1959, the company Sava based in Valladolid obtained the ministry's authorisation to build 1,000 light four-wheel trucks with either a petrol or diesel engine. That same year, it revealed its new four-wheel P-58 model, which had a load capacity of

2.5 tons and featured a 55 CV Barreiros EB-4 engine. It was the first series produced Spanish truck to be fitted with a Barreiros Diesel engine from factory. In 1960, Sava attended the first International Fair of Lisbon with its P-58 trucks equipped with Barreiros engines. Sava continued to use Barreiros engines in its trucks until 1962, when the company began to produce its own engines with a licence from English company BMC.

In early 1958, E.B. travelled to England to speak to the directors of David Brown. He signed an agreement to create a new company with a capital of 100 million pesetas. Barreiros owned 75% of the company, while David Brown owned 25%. In just six months, the gearbox factory that was so essential for the production process was up and running.

In 1958, the company received a licence to manufacture 1,500 TT 90 trucks without any issues or delay. The fact that it was to produce military trucks for export helped its case, although it was already planning to harness the off-road capabilities of the TT 90 for civil applications within Spain, such as mining, construction, forestry, etc. In April 1959, the Ministry of Industry, Planell, and his entire team visited the Villaverde factory. This was quite a triumph for Eduardo Barreiros, who had been unable to convince the ministry to grant him a manufacturing licence just a few years previously.

The first TT 90.22 truck for Portugal was delivered in January 1959 and the first batch of 30 units was shipped in March. In late 1959, the prototype for the first civilian chassis was completed. It was named the TT 90.21: 21 stood for two axles and one drive shaft, like the trucks normally used for road transport. The gearbox and differential were already being manufactured in Villaverde. The chassis had been reinforced to carry a load of up to 6,000 kg. EB-6 engines were used; a large initial series of these engines featured original Perkins crankshafts purchased in Portugal and smuggled into Spain.

Following the important demonstrations of the Portuguese military truck to Franco in 1958 and to Carrero Blanco, the Minister of the Presidency, a few months later Barreiros began to be advertised as a new truck brand offering two-wheel and four-wheel drive models with the same open military cab. In November 1958, after receiving a licence to manufacture 1,500 TT 90 trucks, the company published its first catalogue featuring the open military cab, which was the only model available at that time. The description of the two-wheel drive TT 90.21 model stated that it would be delivered without a cab. The first advertisement for the new truck occupied a full page of the magazine *Auto-Revista* in December 1959.

15 Hanomag–Barreiros Tractors and A Diesel Engine for Cars

In 1958, a new department was taking shape at the company, which was initially named Tests and Experimentation, then Research and finally Product Engineering. It focused on improving the performance of the EB-4 and EB-6 engines to bring them up to 60 and 90 CV respectively.

The EB-150 had begun to be advertised in 1957 and in 1958 it continued to be promoted as a high-output diesel engine to replace petrol engines in large trucks such as military Federal trucks, Leyland trucks and petrol Pegaso trucks. However, series production of this model had not yet begun.

At the end of 1958, Barreiros signed a contract with the German Rheinstahl-Hanomag to create a joint venture to manufacture agricultural and industrial tractors. Like David Brown Engranajes SA, the new company Rheinstahl Hanomag Barreiros, founded in July 1959 with a capital of 100 million pesetas, was controlled by Barreiros, which had contributed 75% of the capital; the other 25% came from Hanomag in the form of manufacturing machinery. Initially, 40% of the components were imported from Germany but within a year they were all manufactured in Spain.

In February 1959, the long-awaited licence to manufacture wheeled and tracked tractors with 50–150 CV was issued. However, only wheeled models were produced as Pegaso had begun to make tracked VCC-5 tractors in 1957. Official documents show that the company had a total of 1,860 employees at the time of the application: almost 10% (174) were technicians, revealing the extent of the design and development work taking place at Barreiros Diesel SA at that time. In terms of annual production, the company manufactured 1,500 EB-6 engines, 1.200 EB-4 engines and 1,500 off-road vehicles. There were plans to increase this to 3,000 tractors (wheeled and tracked) with 50–150 CV.

The first Hanomag-Barreiros R-545 tractors came onto the market in 1959, with around 50% of the components made in Spain. The tractor was the first vehicle to use the new four-cylinder EB-100 direct injection engine. For many years, its 70 CV engine made it the most powerful wheeled agricultural tractor to be manufactured in Spain. This application was the determining factor in the development of the four-cylinder version of the new B-series engine, which would later also be used in trucks.

A diesel engine for cars. Alongside its work to develop engines for trucks, in 1959 Barreiros decided to break into the market for light diesel engines for cars, where the only domestic product was the English BMC engine, which was assembled using some local components by its importer Conde Medín (2007) in La Coruña.

Peugeot in France had developed a light diesel engine for use in cars, working with world experts in fast diesel engines with a combustion pre-chamber, Ricardo & Co. In 1927, the new engine had been fitted to 20 Paris taxis to assess its performance and durability on the ground over a two-year period. In 1959, following the positive results of the assessment, 2,500 taxis in Paris were fitted with the Peugeot Diesel engine.

This major innovation did not go unnoticed by E.B., who quickly realised how important this type of engine would be in the future for professional car use, especially taxis. J. Merino was dispatched to Paris to find out more about the novel Peugeot engine, where he met with specialist magazines to sound out the opinions of experts and the market. The outcome of the mission was highly positive and E.B. immediately decided to embark on developing a new engine, which would be one of Barreiros biggest success stories.

In 1959, the company had around 4,000 employees. Miguel Aldecoa was appointed director of Product Engineering, leading a team made up of A. Pérez Rodríguez, F. Gutiérrez Nogales, J. Gayá and J.M. Castaño. The new engine was named Series C, although it had initially been called EB-55. It featured Ricardo's basic design commissioned in July 1958, which covered the combustion chamber, connecting rods and crankshaft, as well as the overall dimensions. However, the execution design was carried out by Barreiros.

The EB-55 engine started up for the first time on 7 March 1959 at 10 pm under the watchful eye of Don Eduardo, who demanded to attend every time a new engine was started. It was intended to replace the petrol engines in Seat 1400 taxis, where it was used on a large scale throughout the 1960s and the early 1970s. It was a four-cylinder indirect injection engine like all fast diesel engines at that time, with a displacement of 1.9 L.

Changing the engine in a Seat 1400 offered huge cost savings, reducing consumption by around 33%. Throughout the 1960s, diesel cost half as much as petrol so the Barreiros Diesel engine reduced the cost per kilometre to one-third of the original cost with the petrol engine. The operation to switch engines took just five hours.

16 The New 100 and 150 CV Engines

Thanks to its solid track record, Barreiros finally obtained its long-awaited licence to manufacture trucks in 1960, when the company had evolved into a major manufacturer of engines, tractors, and trucks with 5,000 employees. That year, the company produced 1,120 trucks and 12,000 engines.

The new Series B engine, inspired by the Scania, proved highly innovative as it was the first to feature wet cylinder liners, three-cylinder heads and a crankshaft with counterweights. It was also the first direct injection engine, with a turbulence chamber in the piston. It generated 150 CV at 2,200 rpm. The block and cylinder heads were cast at the new own foundry in Villaverde. The injection equipment, pump and injectors were also manufactured at Cabsa, while the electrical equipment was made by Ceesa.

The EB-100 engine, which had the same design as the EB-150 but was a four-cylinder model, began to be manufactured in 1959 to equip the R-545 tractors. Its use in the tractors gave it an extraordinary reputation for power and reliability, as we will see later. Once the new 150 CV B-16 engine was available, the company received an order from the Spanish army to mount it to 100 three-axle Continental military trucks with petrol engines. That same year, tests began on the new 170 CV B-26 version.

In 1960, the factory's extension reached 80,000 m^2. The number of employees rose to 8,000, with 6,000 working in the main factory and 2,000 working at the components factories. In 1961, once it had obtained all the necessary licences, Barreiros released a wide range of trucks featuring the two available engines: the six-cylinder 83 CV A-16 engine and the new four-cylinder 115 CV B-24 engine, which was based on the B-14 model fitted to the R-545 tractors.

Valeriano Barreiros, who represented the company at the Lisbon Fair in 1961, had the opportunity to speak with the president of the Portuguese Republic, General Américo Thomaz, and inform him of the creation of the Companhia Portuguesa de Motores e Camiões, which would manufacture engines and trucks at a new factory in Setúbal scheduled to open in 1962. President Thomaz visited Villaverde in November 1961.

Following the agreements for the gearboxes and tractors, another weak point in the national components industry was the production of springs. Once again, Barreiros took the initiative and signed an agreement with English company Ratcliffe, obtaining official authorisation in February 1961 with manufacturing to begin within a maximum period of 12 months. As always, the new company Ratcliffe Ibérica SA would be 25% owned by the technological partner.

At the end of 1963, the new six-cylinder 170 CV B-26 engine would be used in the new heavyweight range comprising the new tandem drive axle Centauro and the Super Azor tractor. The market for the long-chassis, on-road version of the tandem drive Centauro was reduced, even being the company's first trucks with a 10-speed gearbox, as well as the first to feature an electric retarder as standard. The short-chassis version for construction tippers was the most successful.

Barreiros experienced a dramatic rise in sales of its trucks from 2,850 units in 1961 to 6,000 in 1962, compared to the 6,500 units manufactured by Pegaso.

17 Barreiros-AEC Buses, Forklifts and Other

The most important event to take place in 1961 was the signing of an agreement with A.E.C. (The Associated Equipment Company Ltd.) from England to create Barreiros AEC SA with a capital of 50 million pesetas (75% Barreiros and 25% AEC), which would manufacture chassis for urban and intercity buses.

In May 1961, it was announced that production at the new Barreiros factory in Toledo would rise from 250 buses in the first year to 1,000 standard buses and 275 double-decker buses in the fourth year. In August, the ministry approved the plans submitted. The buses were put on the market in the first half of 1962.

The new buses were equipped with top power B-26 and B-24 engines. During the late 1950s, the only bus chassis on the Spanish market were made by Pegaso with a front engine that left no space for the front door needed to allow drivers to act as conductors at the same time. At that time, the trend in Europe was for urban buses with a horizontal engine under the floor, which allowed for an open internal platform suitable for installing a front door.

Barreiros AEC SA was founded in Madrid on 13 November 1961 with E.B. as the chair of the board of directors. Barreiros was aware that Leyland had become a shareholder of Pegaso, but he never imagined that a year after signing the agreement, AEC would also be bought out by Leyland, with serious consequences for the Barreiros AEC partnership.

Forklifts. The tractor manufacturing plant was equipped with a section for producing forklifts, which were fitted with the same A-24 engine as the tractors. They were designed by Barreiros, based on a copy of a Clark unit.

Barreiros was also keen to break into the medium power marine engine sector, contacted the French company Société Alsacienne de Constructions Mécaniques, which manufactured MGO engines. Although Barreiros was awarded a licence to build engines from 200 to 1,400 CV in 1961, this activity never got off the ground.

18 The New Challenge of Car Manufacturing. The Borgward Operation

Once again, Barreiros was quick to take the initiative and came to an agreement with a European manufacturer to produce high-end passenger cars in Spain. In May 1960 E.B. submitted a plan to the ministry to manufacture 5-to-7-seater passenger cars with a displacement between 1,900 and 2,200 cm^3, which could be used as taxis. In the application, the company stated that it planned to produce 2,000 units in the first year, rising to 10,000 from the fourth year. Half of the vehicles manufactured would be fitted with the company's diesel engines.

The application was rejected as expected, with fierce opposition coming from Seat president, Ortiz Echagüe. Barreiros did not give up and continued to explore the possibilities of obtaining a licence from a foreign car brand, entering contact with Aston Martin-Lagonda, Simca, Jaguar, Rootes Group and General Motors. Ultimately, the operation went ahead when the company purchased premises belonging to German company Borgward.

In 1961 E.B. learned that Borgward had declared bankruptcy and quickly set to work to see if it would be feasible to purchase the company. In the early days of 1962, a delegation from Barreiros Diesel visited the company in Bremen, which was in insolvency proceedings and offered an excellent opportunity for Barreiros to enter the passenger car market quickly and with little risk. The good purchase prospects allowed Barreiros to envisage manufacturing the Borgward 2300 in Spain, a high-end model with only 4,000 units produced by that time. The company planned to export 50% of the units manufactured to Mexico, in agreement with the other interested party. The discussions were fruitful, and Barreiros purchased Borgward to allow him to move into the car manufacturing sector that he had so longed to join. The machinery was shipped to Toledo, where the new car factory was to open.

However, the economic difficulties experienced by Barreiros Diesel SA throughout 1962 and the entering of Chrysler into the company in 1963, stopped the Borgward program and the machinery was never even unpacked and was finally sold to a Mexican group, which founded the Fábrica Nacional de Automóviles SA in Escobedo, Nuevo León, in 1962.

19 1961–1963. Financial Difficulties. Chrysler Enters Barreiros Diesel SA

As we have seen, Barreiros had grown and grown since it transferred its operations to the factory in Madrid in 1952 and set up as a limited company in 1954. At the end of 1961, the company's capital stood at 1.03 billion pesetas, with reserves of 850 million. Investment in manufacturing equipment, buildings and land exceeded 1.3 billion pesetas. The Barreiros group was manufacturing engines, tractors, and trucks, as well as important components such as gearboxes, electrical equipment, and injection elements. Sales had reached 12,700 engines, 2,800 trucks and 2,300 tractors. Exports to Portugal, Poland, Egypt, Tunisia, Brazil, and Uruguay continued apace.

The financial problem now emerging was that the company itself was funding half of the instalment sales made by its dealers and distributors. To maintain normal activity, a loan of 500 million pesetas was now needed. In 1963, a series of major events at Barreiros Diesel SA would determine the company's future. E.B. had been a pioneer in instalment sales as he was convinced that the system would prevail in the future and created his own finance company Fibasa in 1960. The possibility of paying for purchases in instalments had shaken up the market and multiplied sales but in early 1963, Barreiros experienced a severe lack of liquidity with no option for additional loans from the banks.

As a result of the personal letter that E.B. had sent to several ministers in May 1962, an inter-ministerial committee was set up to study the situation of the company, which was at risk of having to suspend its activities and draw up a downsizing plan that would affect 7,800 workers. The study took three months and had a positive outcome: the company was granted a 300 million pesetas loan.

This was the first traumatic episode in the company's history, but it would not be the last. Despite its cash flow problems, E.B. remained determined to establish Barreiros as a car manufacturer. His efforts were no longer focused on obtaining a licence from a prestigious brand but on building a partnership with a large automotive group that would be able to provide the resources that Barreiros lacked for the project.

Chrysler Corporation, the third largest American automotive company, acquired a majority share in Simca in France in January 1963, renaming it Chrysler France. In 1964, it purchased 30% of the English Rootes Group, acquiring a majority share in 1967 and transforming it into Chrysler UK. Given the company's European expansion strategy, it was clear that it was the best option for Barreiros. The agreement with

Chrysler was reached in October 1963, when the American company acquired 40% of the capital of Barreiros Diesel SA.

E.B. had been warned by I. Cavero, head of the legal department at the time, that being in partnership with a company as large as Chrysler meant that sooner or later it would take over the majority shareholding and control the company, as the Barreiros family would be unable to participate in the capital increases that would undoubtedly take place.

In spring 1963, Barreiros main financial backer, the Banco de Vizcaya, decided not to increase the existing risk so the negotiations with Chrysler took place under pressure and with little room for manoeuvre. The final contract was signed on October 31st, 1963. Chrysler acquired 40% of the company's capital, paying more than 1.1 billion pesetas. The contract stipulated that Chrysler vehicles would be manufactured in Spain, including light Dodge trucks up to five tons as well as cars. The possibility of manufacturing 20,000 Dodge vans was also considered.

20 1964. High Power Engines. The Flat-Eights and the 12-L B-36

As we have seen, by the end of 1963, Barreiros had finally developed a 170 CV engine with a displacement of 10 L capable of competing with the 165 CV Pegaso engines. The new engine was being fitted in the tractor version of the Super Azor and the Centauro with a maximum power of 170 CV. According to the 1964 report, "demand for the 18-ton Centauro truck currently exceeds production several times over and we hope to be able to meet demand as quickly as possible". This was referring to the short-chassis version for tippers.

In 1962, the first prototype of the new D-26 engine, indirect injection, dry liners and a greater displacement was built. However, it was not fitted to any of the trucks as standard either that year or in 1963, when a new configuration of the engine with eight opposed cylinders began to be developed. Given that the D-24 engine was generating 90 CV at 2,800 rpm, a horizontally opposed eight-cylinder version was predicted to reach 180 CV, which was an extremely high power at that time.

The prototype began to be built in late 1963 and by 1964 it was already running on the test bench. There is a rather revealing anecdote to be told about E.B. and the tests for this engine: he always demanded to be present when the prototype of a new engine was started up on the test bench and he was very excited about the smooth, vibration-free functioning expected from the DH-28's eight horizontal cylinders.

The DH-28 engine continued to be tested on the Azor truck throughout 1964 and 1965 until the project was abandoned. Although the engine was not successful, the experiments with the flat-eight engine were used to develop the new BH-28/BH-38 engine in late 1964. Once again, E.B. was keen to explore all the technologies available to ensure that he had a wide range of powers for his trucks.

In 1963, the company began to explore alternative solutions to reach 200 CV and designed an initial flat-eight version with components from the B-24, which was named the BH-28. During the development process for the engine, a major change was made by increasing the cylinder bore from 120 to 130 mm. This meant that the entire engine had to be redesigned and the outcome was the BH-38 with a unitary displacement of two litres and up to 270 CV at 2,200 rpm in the aspirated version.

The idea behind the BH-38 engine was to obtain greater power which, in the absence of a turbo, could only be achieved with a larger displacement. The difficulties involved in fitting an opposed-piston engine beneath the chassis, behind the cab, were like those experienced with a horizontal six-cylinder engine.

Meanwhile, turbocharged engines were being launched on the market, which delivered greater power in a simpler, less cumbersome manner. Barreiros was already testing turbocharging with the new generations of KKK and Holset turbos, which performed far better than the first CAVs. Therefore, the BH-38 project was ultimately abandoned: a total of 10 units were manufactured up to May 1967, which were used in Ceesa generators.

In 1964, tests began on a new, larger diesel engine, the biggest and most powerful built by the company so far: the direct injection B-36 with a displacement of 12 L, 215 CV at 2,000 rpm in the aspirated version and 260 CV in the turbocharged BS-36 version. The 130 × 150 mm dimensions of the engine, with a total displacement of 12 L, remain the most common in the European industrial sector 50 years on. Barreiros introduced the aftercooled turbo engine to Spain, which is now used for all modern engines.

One unique feature of all Barreiros engines was the use of chrome-plated steel liners, which were extremely durable and offered effective protection against the external cavitation problems that had caused so many issues for Pegaso. In 1969, the B-36 engine with 216 CV at 2,200 rpm began to be series produced. It was fitted to all trucks in the new heavy range released at the end of the year. That same year, the BS-36 engine with a KKK turbo was completed and the first series-produced units were launched on the market in 1970.

21 Chrysler Acquires the Majority Share. E.B. Abandons the Company

1966 had been an acceptable year in terms of overall sales but the company was experiencing a lack of liquidity due to the large volume of financed purchases and found itself in a very difficult situation by the end of the year. Exports were not growing, and most African countries were affected by economic instability.

During the first few months of the year, E.B. continued to fight for the company's independence, which was proving increasingly difficult to maintain. He suggested to the Ministry of Industry that Barreiros could merge with Sava and state-owned company Enasa, with him as president. It was a brave offer that made a lot of sense.

The Minister of Industry, López Bravo, was quoted in the newspaper *Informaciones* on 14 February 1967 as saying that it would be preferable to have "just two manufacturers to mass produce passenger cars in Spain" and that a merger between Enasa and Barreiros to make industrial vehicles "was not essential but would not harm consumers".

In August 1967, E.B. wrote another letter to López Bravo stating that Barreiros Diesel's sales and cash flow had been satisfactory up to July 1966. This was when the Barreiros family had received a loan of 1 billion pesetas from the Banco de Crédito Industrial. The company's finance department believed that an increase of 2 billion would be adequate to enable it to deliver on its commitments, but this was based on an assumed minimum annual turnover of 12 billion, which was not going to be met.

On 6 October 1967, the inevitable occurred: the Council of Ministers approved Chrysler's acquisition of a majority share in Barreiros Diesel, rising from 45 to 77%. Franco had forbidden his ministers from taking notes during cabinet discussions, so none of their opinions on the matter were recorded. The operation was subject to three conditions:

- Chrysler's managers at the company would continue to produce diesel engines and the buses and trucks fitted with them. This was an attempt to ensure that the factory in Villaverde would continue to operate in a market where it was strong, drawing on Chrysler's global network to support exports.
- Chrysler would maintain a similar level of non-Spanish staffing to the existing situation and any partnerships with other foreign companies would be limited to what was strictly necessary.
- As well as purchasing the Barreiros family's shares, Chrysler would take note of the company's current financial needs and provide 35 million dollars in funds.

In the new chart, an administrative commission was created with four Americans and five Spaniards. Eduardo Barreiros continued as president and his brother Valeriano as vice president, but the position of executive vice president was occupied by an American man. From November 1967 to July 1969, E.B. went through quite an ordeal as a shareholder of the great company his family had founded. He remained president but the meetings of the board of directors were now led by the executive vice president.

In February 1969, it was reported that the first shipment of 200 Barreiros Panter III military trucks would be leaving for Saudi Arabia in an export operation negotiated directly by E.B. On 25 May 1969, upon his return from a trip to Saudi Arabia to negotiate the sale of another batch of 200 military trucks, E.B. announced that he would be standing down as president of the company and that his brothers would be resigning from all their managerial positions. The announcement read as follows:

> Mr. Eduardo Barreiros, President of the Board of Directors at Barreiros Diesel SA, has resigned. The founder of the company, Mr Barreiros, along with members of his family, was the majority shareholder until October 1967 when Chrysler Internacional SA acquired a major share. Since then, Mr Barreiros has continued as President of the company by mutual agreement.

The company has emphasised the progress it made from its creation in 1951 to 1963, when Chrysler made its first investment, which has continued to this day. Its capital has increased to 3.14 billion pesetas, and it currently produces a full range of Dodge and Simca cars, trucks, buses, tractors, and engines that are sold on the domestic and international market.

The number of employees at Barreiros Diesel SA and its subsidiary companies exceeded 13,000. The management team and senior executives are almost all Spanish, as there are only 14 North American employees in total. Chrysler's policy is for its companies to be managed by personnel from the country where they are located, as is the case in all its non-American partner companies.

Since it sold the majority share in its company to Chrysler, the Barreiros family have moved on to other activities, but they continue to hold a significant minority shareholding in Barreiros Diesel SA.

Resigning from the managerial positions they held at Barreiros Diesel SA, the Barreiros brothers—Eduardo, Valeriano, Graciliano and Celso— wrote a letter to J.L. Corral Sánchez and G. Sánchez Mayol, members of the works council, union representatives and employees of Barreiros Diesel SA, dated 24 May, which read:

> Dear friends, we never imagined this day would come. We founded Barreiros Diesel and only those who have worked alongside us are aware of our hard work, sacrifice and satisfaction; no words are sufficient to sing their praises as they deserve, nor our own. Today, we have resigned from the positions we held at the company. The reasons are too long to explain in this letter but suffice to say that there have been a series of causes that, as businessmen and Spaniards, we cannot accept. However, Mr Corral Sánchez and Mr Sánchez Mayol, the Directors representing the staff, will each receive a file detailing the causes, as will the other members of the board.
>
> What pains us most about this separation is the thought of losing touch with all those who have worked so tirelessly alongside us to make this company the most admired and prosperous in the country.
>
> As a token of our great affection for everyone with whom we have worked, we would like to gift them 25 million pesetas in company shares. To do so, we will meet shortly with the works council either to create a foundation on behalf of the company's personnel or to apply the sum in the manner we deem most appropriate.

This brought an end to 15 years of intense creative and industrial activity, which gave rise to the largest automotive company in Spain with products ranging from passenger and luxury cars—with hindsight, we can pinpoint the beginning of the end of Barreiros Diesel SA to this venture—to a variety of light, medium and heavy trucks, all with the company's custom-made engines and technology. The company had achieved all this despite its initial struggles to overcome official opposition to any challenge to the hegemony of the state-owned enterprise Pegaso.

22 1977. E.B. Returns to the Automotive Industry. Engines for Cuba

Under the contract governing the sale of the majority share in Barreiros Diesel SA to Chrysler International in 1967, Eduardo Barreiros was obliged to distance himself from any kind of automotive activity for a ten-year period. During this time out of the industry, E.B. launched several other ventures but engines remained as his main passion.

Barreiros Diesel had always had extensive, close commercial relations with Cuba, exporting a large number of trucks to the country in 1965. The company had a customer service office in Cuba to deliver assistance and supply spare parts. When Chrysler acquired the majority share in the company, the American embargo was applied, and exports of vehicles and spare parts ceased.

In December 1977, the Cuban ambassador in Madrid invited E.B. to a reception at the embassy in honour of the Cuban Vice president Carlos Rafael Rodríguez. The invitation was not entirely innocent, as C.R. Rodríguez was convinced of Ernesto Che Guevara's philosophy that "a country is never truly independent without its own automotive industry". Not long before, the Cuban government had asked Lloyd's of London to recommend a company capable of executing an ambitious engine manufacturing programme in Cuba: it appears that one of the names suggested was Eduardo Barreiros.

E.B. managed to persuade the embassy to screen the film *Barreiros 66* at the reception, which traced the origins of Barreiros Diesel and was a clever way of demonstrating his ability to create an engine manufacturing industry. The Cuban government also recalled the events of 1968, when E.B. had been forced by Chrysler, the company's majority shareholder, to cancel exports to Cuba and he had sent his brother Valeriano to explain the situation and convey his apologies.

After this initial contact with the Cuban Embassy, the project aroused a great deal of interest and a trip to Cuba was organised for E.B. to give a detailed presentation to Fidel Castro, who listened closely as he described the plans for the project. A few months later, two Cuban ministers and a group of technicians visited Madrid to study the proposal in detail and discuss it with Barreiros team.

E.B. was now free of his contractual obligations to Chrysler and was ready to repeat his own history. Twenty-five years after copying the Perkins engines, he was about to embark on a similar venture. Most European truck manufacturers at that time used supercharged in-line six-cylinder engines with between 280 and 320 CV, which were the most common engines in heavy road trucks. Meanwhile, Mercedes Benz and MAN had jointly developed a range of V-engines, allowing them to offer six, eight and ten-cylinder versions for trucks and twelve-cylinder versions for special applications, with a wide range of powers.

Observing the benefits of the modular configuration of V-engines in terms of construction and performance, E.B. brought together a group of trusted professionals to design and build prototypes for his own range of V-engines, using the MB-MAN range as a reference.

It is not hard to imagine what E.B. had in mind when he founded the new company Diesel Motores e Industrias SA (Dimisa) on 27 March 1980. Several activities were considered as part of the company's objectives, including the design and manufacture of earthmoving vehicles, but it eventually decided to develop a family of diesel engines with a V-cylinder configuration to cover a power range from 190 to 500 CV. The new company set up shop in a warehouse complex on the industrial estate in Pinto (Madrid). The new engine project had begun in 1978 and included three versions (6, 8 and 10 cylinders). To be noted that the first Cuban engine was tested in the engine laboratory of the Engineering High School of Madrid and this author, by then professor of engines practice at the school, oversaw the running in and power test of that engine, having the opportunity to know personally Eduardo Barreiros.

The second Congress of the Communist Party of Cuba was to be held at the end of 1980 and the party's vehicle fleet had been dominated until then by ZIL and Kamaz trucks, and Volga and Moskvitch cars, all of which were made in the Soviet Union using outdated, poor-quality technology. Part of the Cuban nomenklatura dreamed of having the national industry advocated by Che and saw Barreiros proposals as a way of obtaining trucks, engines for agricultural machinery, buses and other vehicles. The brand would be Taíno, named after an indigenous people living in Cuba prior to the discovery, which could be presented as a product of the Cuban Revolution.

Cuba provided a new area of operations where E.B. could repeat the adventure of creating a new diesel engine industry drawing on nothing more and nothing less than his extensive experience in the sector and a series of partners with whom he had worked from the very beginning of his career in Spain.

The Cuban government aspired to develop its own engine and truck industry, as well as a component manufacturing sector, to serve the local market and export to Latin America and the Soviet bloc. With this aim in mind, E.B. prepared the memorandum 'Industrial project for an engine factory in Guanajay' in the province of Pinar del Río, where there was already a Pegaso truck assembly factory. From that point on, the local industrialisation plan swung into action, requiring far more time than initially anticipated due to the lack of a local auxiliary industry capable of meeting supply needs.

After rigorous functional testing in Cuba, where it competed with other brands, the Barreiros engine was selected as the best solution and 10 full engines manufactured in Pinto were soon shipped: five V-6 engines and five V-8 engines. Production of the new engines began, and the next milestone was the local assembly of the first 100 engines, which were sent in full but completely disassembled. This took five months, which was quite a record given the relaxed working conditions and lack of experience among the Cuban workers.

The new six and eight-cylinder engines were also used to replace the engines in existing trucks and buses, as well as in generators, marine engines, loaders, and agricultural and industrial machinery, especially sugar cane harvesters.

From 1987 to 1990, Taíno trucks with six and eight-cylinder engines were exhibited at international fairs in countries across the Soviet bloc, including Leipzig (GDR), Plovdiv (Bulgaria) and Bucharest (Romania), where they received several awards. The V-6 and V-8 engines were manufactured from 1981 to 1996. The Cuban-made components were the block, cylinder heads, oil sump, manifolds and cast components in general, which were made by the 'Cuba-Soviet Friendship' factory. All the mobile components, including the crankshaft, connecting rods, camshaft, pistons, valves, tappets, etc., as well as the injection system, liners, and other parts, were shipped from Spain.

23 1983. Converting Soviet Engines Again. The ZIL V-8

Alongside the industrialisation plan for the new Taíno engines that was already underway, which would take a long time to materialise, E.B. set out to persuade the Cuban Ministry of Iron and Steel of the benefits and cost savings of converting ZIL V-8 engines from petrol to diesel. The ZIL 130 truck was one of the most popular in the Soviet Union and was exported to practically every country in its zone of influence. It had a payload of 5 to 6 tons, with a total laden weight of 10,500 kg. The truck had a petrol V-8 engine inspired by American designs from the 1950s, with a central camshaft in the block and overhead valves. It was fed by a twin-choke carburettor.

The initiative took E.B. back to his younger days converting petrol engines to diesel and it was a good opportunity given the number of ZIL trucks available in Cuba. By a strange twist of fate, ZIL (ЗИЛ in Cyrillic) was the same brand as the old ZIS (ЗИС) company, which was renamed after Stalin's death in 1953 from Z.I. Stalina (factory named after Stalin) to Z.I. Lijachova (factory named after Lijachov).

E.B. focused his attentions on this engine, which was too large for its limited performance, and decided that it would be a good candidate for conversion to diesel. Following initial exploratory talks, a contract was signed between Dimisa and the Ministry of Iron and Steel to carry out this conversion operation. To ensure its durability under the higher operating pressures of the diesel cycle, especially the crankshaft and its bearings, the cylinder bore was reduced from the original 100 mm to 92 mm. The engine had an indirect injection system with an in-line pump and the combustion pre-chamber was like that of the Barreiros C-24 engine, which had been such a success for the company.

At the start of 1984, road tests were carried out in Cuba to compare the new ZIL Diesel engine with the original petrol engine. The results were overwhelmingly positive as consumption fell from almost 35 L of petrol to 20 L of diesel, a reduction of 42%. It is relevant to note that the original ZIL petrol engines were far from optimal in terms of their fuel consumption. Based on these positive results, the Cuban government signed the new contract and the Taíno engine factory began to manufacture the new cylinder heads and other components for converting the ZIL trucks alongside its work to manufacture the V-6 and V-8 engines. 1983–1984 were certainly the ZIL truck's year.

Although 3,000 units were initially scheduled for conversion, only around 1,000 engines were converted in the end as they were not all usable and new crankshafts had to be imported from the USSR due to the poor condition of the engines in service.

24 Final Projects

E.B. continued to explore world business opportunities for manufacturing diesel engines in countries keen to boost their industrial sector. His team designed several new engines to be offered to different countries.

In early 1992, E.B. and his direct partner A. Guisasola (Eduardo Barreiros en la automoción 2002) were preparing a trip to Angola, from which Guisasola recalled: "We were both pushing 70, but we were filled with childlike enthusiasm by this new challenge".

Eduardo Barreiros death in Havana, on 19 February 1992 at the age of 73, brought an end to all these projects and initiatives.

References

Published Interviews

Conversation with Eduardo Barreiros, by M. Alcántara in *Ya*, July 1964
A unique interview with Eduardo Barreiros, by P. Fuentes in *Galería*, July-August 1965
What is Eduardo Barreiros doing now? by J. González-Cano in *La Gaceta Ilustrada*, May 1972
Interview with Eduardo Barreiros, by E. Daudet in *Reportaje ABC*, July 1972
Eduardo Barreiros, by G. Solana in *Coleccionable Los Españoles*, November. 1972
Eduardo Barreiros. The businessman who returned from Cuba, by M. Aznárez in *El País Semanal*, February 1989

Published Books

Barreiros Diesel. Santos, M. & García Ruiz, J.L. Fundación E. Barreiros. Madrid, 2002
Historia de la industria española de automoción, Lage M. (Fitsa, Madrid, 2005)
Un millón de camiones y buses españoles. Lage, M. Iveco. Madrid, 2008
Barreiros: Historia completa. Lage M. To be published
Barreiros. El motor de España. Thomas, H. Editorial Planeta. Barcelona, 2007
Eduardo Barreiros en la automoción. Su época cubana. Guisasola Berraondo, A. Unpublished manuscript. Madrid, 1997

Eduardo Barreiros at the top years of his industrial empire, 1964 [Courtesy of Fundación E. Barreiros] Note about the Barreiros name in the text.

The name Eduardo Barreiros Rodríguez, so often used, is replaced by the initials E.B. throughout the text for the sake of brevity.

His father's name Eduardo Barreiros Nespereira, is written E.B.N.

Rafael Escolá Gil (1919–1997)

Gabriel Vilallonga Elorza

Abstract A man of great ideals, during the first part of his life Rafael Escolá had to contain his powerful inner drive in the face of a series of adversities. He wanted to study engineering, but important events forced him to postpone his studies. When he wanted to become a researcher, his father stood in his way... Finally, in 1957, with the creation of IDOM, he was able to bring to fruition all his ambitions. He introduced the concept of free professional, breaking down the hierarchy of numerous bosses. He created a unique ownership structure, with the employees becoming the owners of the company. He saw the employees of the company as persons and considered himself just another one more. IDOM is now a prestigious company operating in 125 countries with 45 offices around the world. If Rafael were among us, he would say that he only planted the first seed because, during his lifetime, he did nothing else but attribute most of IDOM's success to the work and good deeds of his collaborators. But without that seed we would not have this tree. In IDOM, Escolá created a unique style of providing professional services. And the resonance of his work grows stronger and stronger with the passage of time.

Keywords Engineer · Consultant · IDOM · Employee-owned Company

G. V. Elorza (✉)
IDOM. Consulting Engineering Architecture, Madrid, Spain
e-mail: gve@idom.com

R. López-García and M. Ceccarelli (eds.), *Distinguished Figures in Mechanical Engineering in Spain and Ibero-America*, History of Mechanism and Machine Science 43, https://doi.org/10.1007/978-3-031-31075-1_2

"Pull the lever now!" shouted Rafael, visibly excited.[1]

"Are you sure it's not going to explode?", replied his brother Manolo from the other end of the hallway.

No, it's impossible, I've gone over the diagram a thousand times.

Manolo didn't see it so clearly. The two young men had built that machine with some junk found in the ruins of the neighborhood: pieces of an old clock, some wires.... And none of it seemed reliable. The paper reel, dry and crumpled, would rip apart every time they tried to unwind it.

Somewhere Rafael had read an article about Morse's machine and wanted to reproduce it at all costs. And Manolo knew how persistent his brother could be when he has a bee in his bonnet. All he needed was an awl, some gears, a socket and little else. He compensated for his lack of means with his enthusiasm.

Under pressure from his younger brother, Manolo finally pulled the lever. And what he feared happened. The invention gave off a spark, left the apartment in darkness and, worse still, set fire to the paper he was holding in his hands. Mercedes, the older sister, ran up the stairs, fearing for her brothers' lives. But she found them laughing, holding a torch in their hands:

"We have invented the electric lighter!".[2]

Student

Like his older brothers, Rafael Escolá (1919–1997) studied at the La Salle Brothers' school in Bonanova Street in Barcelona. He was a student with good grades who, at the end of high school in 1935, was not sure what profession to study. At school he was recommended to take a psycho-technical test that would take place during the summer vacations, creating a certain expectation among his siblings and friends in Caldetas, the town where the family spent the summer. On his return from Barcelona, Rafael showed everyone a piece of paper that was read with interest: "The student is qualified for any university studies, showing special suitability for technical subjects".

[1] This article is an edited version of the book "Rafael Escolá. Ingeniero", Ana Cardenal and Gabriel Vilallonga, Ed. Fundación Rafael Escolá, Madrid 2004.

[2] Fictitious dramatization based on the written memories of Mercedes Escolá. Cf. transcription of the interview made by Álvaro Chapa with Mercedes Escolá. Barcelona, 4.03.1996.

Fig. 1 Rafael was 16 years old when he studied the first year of the entrance exam for the School of Engineers (1935)

At that time, students who wanted to study engineering or architecture had to pass two entrance exams before enrolling in the corresponding school. The exams were usually prepared in private academies and Rafael prepared his first exam during the 1935–36 academic year in the Humet academy, linked to the School of Engineers of Barcelona,[3] where he passed in June.

Unfortunately, a month later the Spanish Civil War broke out (1936–1939). In the following three years Rafael lived countless hardships: he worked as a laborer in an aircraft factory, was imprisoned in the castle of Montjuïch,[4] almost lost his life on several occasions, managed to escape and was finally enlisted in the national army. All this, before he was 20 years old. Shortly before Christmas 1939 he was finally able to return to his parents.

His parents' home had been ransacked and the furniture burned, so the family stayed in the apartment of Teresa, one of the older married daughters, on Muntaner Street. Faced with the situation of hardship, Rafael suggested to his parents the possibility of abandoning his studies and looking for a job, but his father urged him to pursue his studies, because it was the best way to help the family (Fig. 1).

Engineer

At the age of twenty-one and having passed his first engineering course, Rafael began to prepare for his second course at the Humet Academy at the beginning of 1940, in a group of less than ten students. The course was accelerated and lasted barely one term because, after the war, the whole country wanted to make up for lost time and get back to normal as quickly as possible. In April 1940, the School of Engineers

[3] Francisco Prados de la Plaza, Fundación Rafael Escolá, pro manuscript, 1998, p. 45.

[4] Testimony written by Mercedes Escolá Gil in 1995. Rafael Escolá Foundation.

Fig. 2 **a** In 1943, Rafael was awarded a grant for research on photovoltaic cells, but family difficulties prevented him from pursuing the research. **b** Eighty years later, IDOM engineered the world's largest central tower photovoltaic plant. In the image below: Crescent Dunes solar thermal photovoltaic plant, United States

announced an extraordinary exam, which Rafael passed and finally, in June, he was able to enroll in the School of Engineering.[5]

The School of Barcelona, which was one of the few that conferred the degree of Industrial Engineering in Spain, had been provisionally located in an old mansion on Urgel Street. Rafael spent his five years of studies there.[6]

> I was attracted to the idea of thinking and doing things as tangible and as varied as those in the engineering profession. And I was lucky, because despite having missed three years of study as a result of the Civil War, I was able to pursue that field of study in Barcelona.

Escolá devoted himself to the task, obtaining grades that were a great source of admiration among his siblings. According to Mercedes, Rafael detested the limelight and therefore once threatened, jokingly, to lower his grades if he was praised in public. He obtained his engineering degree in 1945, along with about twenty of his classmates (Fig. 2).[7]

Researcher

While pursuing his academic studies, which were admittedly rather theoretical, Rafael sought to cultivate his practical curiosity through research and inventions. He involved his brother Manolo in this activity, with whom, in addition to sharing a bedroom, he carried out innumerable experiments.

The rest of the family was a reluctant witness to Rafael's inventions. On one occasion, for example, he built a Crystal radio, the novelty of which was that it could not only be heard through headphones—as in other Crystal radios—but also through a loudspeaker. There was only one small drawback: Rafael left the radio running day and night because he had found it very difficult to find the optimum tuning position...

---

[5] Joan Marqués Surinach (coord.), Testigos de la Fe durante la Guerra Civil (1936–1939), Ed. Palahí, Girona 1994, pp. 31–55.

[6] Francisco Prados de la Plaza, cit., p. 45.

[7] Francisco Prados de la Plaza, cit., p. 45.

Fig. 3 **a** As a contractor, Rafael carried out civil works such as the Vado de las Cabras Dam (1956), in the mountains of Madrid. **b** IDOM currently designs large infrastructures in the most remote places on the planet, such as the Lejana arch viaduct

1 Scholarship

Escolá was very drawn to what he called "applied science".[8] As soon as he started his studies, he began to look for opportunities to develop his research interests. And one fine day, according to his sister Mercedes:

> He came home excited because he had been granted a scholarship in Madrid to develop some invention that he had already worked on. I was there when he told Dad.[9]

This scholarship meant a lot to Rafael, because it would have allowed him to deepen the studies he had begun on the photovoltaic cell and also support himself financially, without being a burden on his family. But surprisingly, his father said "no", that he did not want him to go to Madrid. This decision baffled the brothers, who could not find an explanation:

> I don't know what my father's motives were because his character (…) was rather complacent. And besides, Rafael's scholarship was something to be proud of. I think he was simply ill, he was elderly, and everything seemed an uphill struggle.[10]

Mercedes was sorry to hear of her father's decision and Rafael's consequent disappointment. She always admired his respect for his father's wishes, putting aside his interests in the research.

[8] Rafael Escolá, Cómo nace y se hace…, cit., p. 3.

[9] Testimony of Mercedes Escolá Gil, cit.

[10] Testimony of Mercedes Escolá Gil, cit.

2 Philosopher

In those years, Rafael began to stand out as a "great talker",[11] who liked to theorize about everything that crossed his mind, not only about technical issues. Mercedes sometimes intervened to show her disagreement and Rafael gave in or, on the contrary, approached her reasoning from a new point of view. He liked to push his arguments and "exhaust" his listeners.

According to Mercedes, of all the civilizations that had passed through his Catalan land throughout history, the one with which Rafael, most identified himself was the Greek. He liked to delve into all matters, divine and human, elaborating them through a well-articulated discourse.

Of course, Raphael was not at all "Phoenician", if by "Phoenician" we mean a personality oriented to trade and the accumulation of goods. Nor was he a "Carthaginian", a historical figure that evokes the desire for domination and conquest, a common denominator of the entrepreneurial mentality. And of course, he felt alien to the caricature of "Roman" law, if by "Roman" we mean the legalistic mentality that subjects all aspects of life to regulation and norms.

As we shall see, this "Greek" naturalness, enhanced by a Christian vision of existence, will be the decisive force in the development of his entrepreneurial creation.

3 Change of Course

At the Humet Academy, Escolá had met a fellow student, Rafael Termes, with whom he became friends during his first year at the School of Engineering. Termes told him that he had joined a Catholic movement, recently founded, which proposed the search for Christian perfection in the midst of daily life. According to Rafael, one fine day, Termes called him aside and:

> (…) He explained to me the way of sanctifying work according to Opus Dei (…) His words opened up horizons I had never dreamed of before.[12]

Escolá decided to join *the Work* which, from that moment on, would be his family, and also the place where he would find the impetus to be a good professional and a good Christian.

> It was all very fast, almost instantaneous. That speed of decision defined his character. He was very expeditious, when he saw something, he went straight to it.[13]

Initially, this decision was not well received by the family:

[11] Testimony of Mercedes Escolá Gil, cit.

[12] Francisco Prados, cit., p. 47.

[13] Testimony of Rafael Termes, in Francisco Prados, cit., p. 47.

> My brothers tried to dissuade me from what they called an "my delusion" and very difficult years went by until, little by little, the truth came out.[14]

In the last years of his studies, Rafael went to live in the residence that Opus Dei had in Muntaner Street, not far from his family home. When he finished his studies, he decided to move to Madrid, where his new commitments demanded him.

During this period, the values learned in his family, as well as the adversities of the Civil War, strengthened some of Rafael's natural qualities: optimism, resolution, hardiness, perseverance, creativity and capacity for commitment.

Employee

In the early post-war years, Spain was immersed in a slow process of economic recovery that was initially without foreign support. The religious institution that Rafael had joined was a new and rapidly growing phenomenon. They needed financial means. And Rafael set to work to obtain them.

4 Manager

He participated in the creation of the company Edificios y Obras SA (EOSA), together with other engineers and architects. The company specialized in the renovation and reconstruction of buildings, a much-needed service at a time when it was not easy to find the means or financing to construct new buildings.

He started in 1945, working as a designer. In 1952 he became the company's manager. During these years, his ingenuity had to be applied, above all, to issues such as obtaining contracts of a certain importance that would nourish the Company's activity, obtaining resources, negotiating with trades and subcontractors, etc. Naturally, he also had to resolve technical issues raised by the restoration of deteriorated buildings. But this aspect of his work was very limited, and he felt that, little by little, he was becoming intellectually decapitalized:

> Twelve years is a long time: the knowledge gained during the course of study is quickly forgotten if not used in professional life.[15]

He often had to halt some of the projects due to lack of financial resources and, at other times, due to supply failures. During this stage, Rafael developed a modus operandi in which creativity and perseverance had to make up for the lack of means. That is why, later on, he demanded the same spirit from the people who worked with him:

---

[14] Andrés Vázquez de Prada, The Founder of Opus Dei, Volume II, edited by Rialp, 2002.

[15] Rafael Escolá, Cómo nace y se hace …, cit.

> It is necessary to carry out what one has to carry out, even if some resources are missing among the many that it would be desirable to have: people who do not wait for this resource or circumstance (which may never come) and manage without it are especially useful.[16]

Since Rafael had not only technical skills, but also commercial and financial skills, EOSA was growing, diversifying and expanding its radius of action. Some of the works took him to cities such as Segovia, Avila or Bilbao.

Although at the time he saw his job as a contractor as a constraint on the development of his engineering career, many years later he recognized the positive aspects of this work:

> I clearly realize that I would never have taken the initiative to offer engineering services (and perhaps no other), if I had not first acquired the confidence to offer them; and this confidence was given to me by twelve years of dedication to the activities of a building contractor: contracting and building strengthens and forces you to have self-confidence if you want to succeed.[17]

4.1 The Ideal of the "Independent Practice"

At the end of the 1950s, EOSA was awarded the contract for the El Espinar dam (Segovia). The construction type chosen for this project was called buttress dam, particularly suitable for river ravines with high quality rocky soil.

With the intention of presenting the particular characteristics of the future civil work,

Rafael organized a trip for the mayor and some councilmen of El Espinar to a dam with similar characteristics: the Ribadelago dam, which was being completed in 1956, in the region of Sanabria (Zamora).

In January 1959, the Ribadelago dam burst, and the water swept away houses, people and animals of the town, which was downstream. A total of 144 villagers died. The drama shocked the whole of Spain and the Minister of Public Works, Jorge Vigón, instructed the Spanish Committee of the International Commission on Large Dams to carry out a technical investigation into the dam rupture (Fig. 3).[18]

The expert reports concluded that the cause of the rupture was an accumulation of errors: design, execution and poor quality of the materials used. The contractor had built a concrete screen, which is not a very flexible material, together with masonry buttresses, which are more elastic. The masonry structure buckled under the pressure of the water, leaving the concrete screen to perform the resistance function, until it collapsed. The court sentenced those responsible for the works (the manager of the construction company, two engineers and a surveyor) to one year's imprisonment on a charge of recklessness.

---

[16] Rafael Escolá Gil, Deontología para ingenieros, published by the University of Navarra, 1987, p. 176.

[17] Cómo nace y se hace…, cit.

[18] "On the Fiftieth Anniversary of the Vega de Tera Dam Break and the Ribadelago Disaster of January 9, 1959", Ethnographic Museum of Castilla y León, p. 45.

While the causes of the malfunction were being investigated, suspicion spread to all works of a similar nature in Spain. At that time, the El Espinar dam was being completed and although Rafael was no longer working for EOSA, he was summoned to the meeting of specialists who were examining the case. Some of those present suggested that, like the Ribadelago dam, El Espinar might have been designed following unethical criteria. Escolá defended the project, explaining the design criteria and demonstrating that the safety coefficients recommended by European standards had been used.[19] Time has dispelled unfounded fears and today, sixty years later, the dam is still in perfect health.

The Ribadelago designers had clearly succumbed to the pressures of the company, which in turn had been beset by the shortage of cement, a material that was rationed at that time in Spain. The Ribadelago tragedy finally convinced Rafael of the need for the engineer to design independently, without being subject to economic, commercial or financial interests.

Financial independence became an indispensable condition for his professional development. A professional's judgment could not be subject to any conflict of interest. It was thanks to this independence that, over the years, in 1979, he was able to write:

> We have never had any catastrophes (which, among so many works, could have happened at some point).[20]

4.2 *New Change of Course*

It was clear to Rafael that independence of technical criteria was only possible in "free practice". Certainly, as the manager of EOSA, he had been able to maintain his independence in his judgment. But he had not managed to avoid other aspects that he naturally rejected: "the harshness that usually accompanies economic, commercial and financial environments", and their "implacable demands, such as dismissals, destitution, etc."[21]

His antipathy towards such matters had grown to the point that he felt uncomfortable with his work. In one of his later writings, Rafael refers to this period of his life as "dark, monotonous and heavy work".[22]

He wanted the freedom to design, to solve technical problems, to devise solutions: "it was always clear to me that I wanted to be an engineer",[23] an activity he described as "pleasurable and rewarding".[24] And he wanted to practice it without

[19] Deontología para ingenieros, cit., p. 87.

[20] Rafael Escolá, Letter to Luis Olaortúa, 1979. Rafael Escolá Foundation.

[21] Rafael Escolá, Principles of IDOM, 1976, n. 10. Rafael Escolá Foundation.

[22] Cómo nace y se hace…, cit.

[23] Cómo nace y se hace…, cit.

[24] Cómo nace y se hace…, cit.

being burdened with the aspects he disliked about working for others. In 1957, in Bilbao, he had his first opportunity to fulfill his desired ideal.

Consultant

When Rafael made his first trips to Bilbao with EOSA to take charge of the rehabilitation of buildings, in 1953, the Basque capital was in full industrial expansion. Altos Hornos de Vizcaya, supported by the protectionist policies of the State, was the main motor of the regional economy. Other metallurgical companies moved in its orbit, one of which was Basconia SA, founded in the nineteenth century as a supplier to the tinplate market and which, in 1957, was building in Etxebarri (Vizcaya)[25] Spain's first cold strip rolling mill using American technology.

Cold rolling is a high-speed continuous deformation process that, by maintaining the temperature below the crystallization point, produces steel sheets of controlled thickness with very fine tolerances. This process requires the construction of buildings of considerable height and length, as well as substantial investment in equipment and facilities.

Rafael had met the CEO of La Basconia, Fernando Gondra, in Madrid, precisely when the construction of the new factory (familiarly known as "Bandas") was beginning. From the first moment, the two understood each other perfectly and Gondra confided in him about a difficulty that threatened the future of his investment: the poor management of the project.[26]

Basconia had a number of engineers on staff, but they were specialists in process and plant maintenance. They needed someone with the ability to bring unity to the team and manage a large, new, greenfield project from scratch.

4.3 Independent Consultant

Gondra invited Rafael to visit the construction site and meet the half-dozen engineers working on it. He then offered him the sole, integral management of the project. Escolá realized that this was what he was looking for: engineering and coordination. But he also wanted to work from a position of independence, and that was incompatible with being part of La Basconia's staff. He proposed to Gondra to work in the client's offices and with the client's team, but as an external consultant, with the independence to select.

the different contractors and equipment suppliers as he saw fit.[27] He would have only one personal assistant:

---

[25] Conversation between Joaquín Mª Aguinaga and the journalist Francisco Prado, 1997. Francisco Parados, cit., p. 67.

[26] Conversation with Joaquín Mª Aguinaga, cit.

[27] Conversation with Joaquín Mª Aguinaga, cit., p. 67.

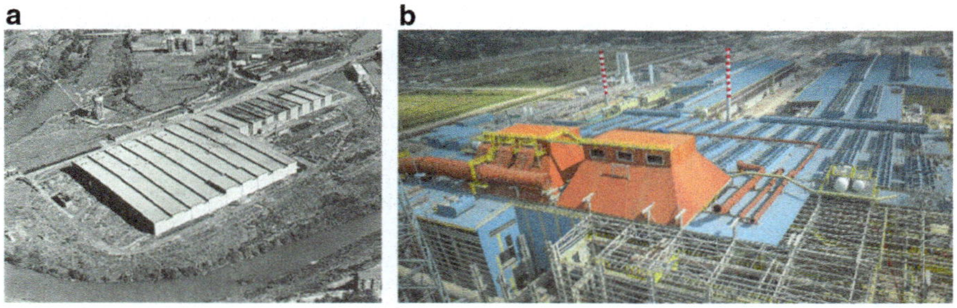

Fig. 4 **a**. Rafael's first project as an independent consultant was a cold steel rolling mill in 1957. **b**. IDOM is currently designing and managing the construction of the world's largest steel complex for Algerian Qatari Steel

Luis Olaortúa, a student who had just finished his fourth year of the Bilbao School of Engineering, whom he would pay out of his own pocket. Once the project was completed, Rafael would rejoin EOSA.

This approach was a break from the prevailing culture: at that time, it was normal for engineers, both civil and industrial, to seek the stability of a salary, whether in the company or the State.

It could be said, then, that Rafael proposed a new way of practicing engineering: external professional services. A modality of work that already existed in the Anglo-Saxon world, where it was known as "consulting engineer",[28] but of which he was not even aware (Fig. 4).

5 The "Free Practice"

By becoming an independent consultant, Rafael achieved "professional freedom", a situation in which the engineer has no "bosses" (except for the Client) and therefore:

> (...) can work professionally without taking orders from anyone. In such a situation of free practice, the person is a true "entrepreneur" in the exercise of their own profession; they do not work for others. (...) When there are no bosses, (...) the work is developed according to the ideas that are going through one's head and it is already seen that the great advantage of this way of working is that, in principle, the work is creative. It is also more personal, so it is more independent.[29]

[28] The formula of professional engineering consultancy began to develop in the United States in the mid 1920's. These early consultants were known as industrial organization engineers or efficiency experts. In Spain, in the mid-1950s, with the end of the autarchy and the process of industrial modernization, consultants found fertile ground, especially in Barcelona (Ingeco-Gombert, 1952), Madrid (Técnicos Especialistas Asociados, 1952), and Bilbao (Sener, 1956, DOM, 1957). Cf. Historia Empresarial, Carmen Erro (ed.), Ariel, 2003; especially the chapter "Las consultoras y la organización de empresas en perspectiva histórica", by Matthias Kipping and Núria Puig.

[29] Cómo nace y se hace..., cit.

To achieve this ideal of independence, he had to pay the price of financial uncertainty, because the freelance engineer "does not receive regular remuneration from anyone, but rather income comes from work that they are regularly asked to do." But Rafael was willing to take the risk, underpinned by the great self-confidence acquired in the EOSA years.

> The primary confidence is that of a man in himself: without this quality, there is no professional work worthy of the name.[30]

5.1 Technical and Personal Leadership

The physical and budgetary dimension of the Bandas project was much larger than what Rafael was used to. It required knowledge that he did not have or that he had forgotten (industrial building structures, processes, materials, etc.) and he had a team, the Basconia team, that might not easily accept the authority of someone from outside.

Escolá spent many nights researching and studying, taking a crash course in industrial engineering. But he had an infallible weapon: his passion.

> Forgetting (the knowledge) is not very important if you keep your fondness for it, and thanks to this, I was able to pick it up again when I needed it, because I liked applied science.[31]

On the other hand, there was the problem of the inherited team. There were people younger than Rafael and also some older; all of them knew the company well, and each had their own way of understanding engineering. Would they accept his leadership?

To everyone's surprise, Rafa's authority was immediately accepted, not only because of his technical skills, but also because of his ability to connect with people: "he won over his audience within five minutes of meeting them".[32] They all welcomed with relief the idea that Escolá would lead them, taking responsibility for coordinating, setting schedules, unifying technical criteria, dealing with suppliers, etc.

One of his assistants on that project remembers him as a simple and approachable boss, whose authority flowed naturally from his own human depth and the trust he placed in his collaborators, no matter how young and inexperienced they might have been.[33]

Rafael Escolá completed the project for Bandas in 1959 to the full satisfaction of the client, as demonstrated by the fact that every time in the following years La Basconia wanted to undertake a new expansion of the production line, it entrusted it to Escolá.

---

[30] Deontologfa para ingenieros, cit., p. 197.

[31] Cómo nace y se hace…, cit.

[32] Conversation between Luis Olaortúa and Ana Cardenal. Year 2001. Rafael Escolá Foundation.

[33] Testimony of Luis Olaortúa. Rafael Escolá Foundation.

So, the independent consultant experiment had proved positive. For two years he had succeeded in realizing his professional ideal. But would he be able to maintain this ideal situation in the following years?

Businessman

To sign-off the Bandas project, Rafael had registered a company at the Patent and Trademark Office on January 1, 1959, with a constitutional value of 100,000 pesetas, with Escolá as its director and sole owner.[34] He did not complicate his choice of name too much, as he simply described the services provided: Dirección de Obras y Montajes (DOM).

Although he registered the trademark in 1959, the truth is that on that date he simply "baptized" the professional practice started in 1957, the year he always considered as the founding year of IDOM (he later added the "I" for Ingeniería *Engineering* to the acronym DOM).[35] This is how he remembered it at the end of his life:

> I had the "craving" for independence and in 1957 I started in Bilbao as an engineering "consultant", founding the firm IDOM (now 35 years old), of which I have been President for 22 years and in which I still continue my professional activity, always as an engineer, which is what I love.[36]

The registration of a trademark, moreover, should not be interpreted as a desire to create a typical company, with employees, organization chart, growth objectives, departments, etc. What's more, Rafael was clear that he wanted to continue working as an individual consultant, handling the work that came in. But was this possible?

Of course, he would not be short of work because, before finishing at Bandas, he had received news of other possible projects. So, there was a market for the type of service he wanted to offer: close to the client, comprehensive, of the highest technical level and guaranteed by his independence from the commercial and financial interests of third parties (Fig. 5).

But this perception contrasted with the opinion of the "experts". Escolá had consulted his idea of "free practice" with several business friends and they had all told him that the service he was offering was already covered by the companies' own technical offices. They had advised him against starting out as an independent consultant.

Nevertheless, when he finished his work at Bandas, he said goodbye to EOSA for good and continued the adventure of working "without bosses".[37] In 1959, he accepted two new projects: the construction management of a new aluminum plant for the company EKL Earle, and the integrated management of a hot rolling mill for Altos Hornos de Vizcaya.

---

[34] Pascual Montañés, La participación por competencia en el poder de la empresa, Doctoral Thesis, Pamplona, 1986, p. 233.

[35] From 1962 onwards, Rafael began to sign-off projects under the name of IDOM, since the "product" he offered was broader than the initial one: it included engineering projects and other studies.

[36] Cómo nace y se hace..., cit., p.3.

[37] Cómo nace y se hace..., cit., pp. 8–9.

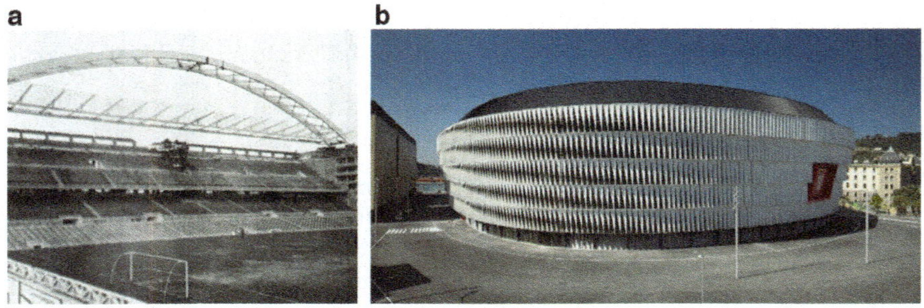

Fig. 5 **a** In 1961, Rafael was in charge of the lighting project for the old San Mamés stadium. **b** In 2015, IDOM carried out the architecture and engineering of the new San Mamés stadium, a facility that has received the "Elite" category, the highest awarded by UEFA (upper image). The company founded by Escolá is ranked 69th in the world ranking of architectural firms

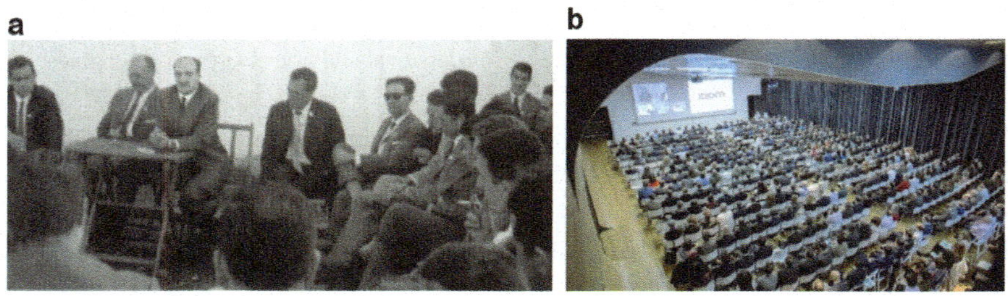

Fig. 6 **a** In 1965, IDOM had 165 people. **b** Today, it has more than 4,000, in 45 countries, ranking among the top 50 largest engineering firms in the world, according to Engineering News Record magazine (ENR, 2018)

The owners of Earle established two conditions that would determine a new twist in Rafael's career: for the aspects related to the aluminum process, he could count on the collaboration of an Earle technical engineer; but for the civil work, if he needed assistants, Rafael would have to hire them at his own risk.

This forced him to create an "entity", a term he used to avoid the word "company". But he understood that he would have to say goodbye to the fleeting dream of the individual consultant: he would have to return to the dynamics of hiring employees, paying salaries, assuming overhead costs, etc. In short, he would have to devote himself to subjects that would necessarily distract him from the practice of engineering.

> I was alone for a short time, as I soon surrounded myself with other people. Since then, as there were several of us, the professional work took shape within an entity that, although elementary, was already a whole, something different from the person who initiated.[38]

[38] Rafael Escolá, Perspectives of the former first president, March 1991. Rafael Escolá Foundation.

6 Minimal Structure

He was to create a company, but it would not be a normal company. It would be an "entity with a minimal structure", so that all its members could practice as if they were free professionals.

> I liked this free practice so much that I thought it would be good if everyone else could be as close to it as possible.[39]

What characteristics should the new entity have? Rafael put his analytical mind to identifying the obstacles that prevent the consultant from working with a sense of freedom in "normal" companies. And it seemed to him that the first obstacle was the pyramidal structure.

In ordinary companies, only the head of the company is in contact with the client. No matter how intelligent or creative they may be, "employees" cannot jump the ladder. If they have innovative ideas, they must "sell" them to their immediate superior. Only eventually, if the professional is accepted by his or her bosses, is he or she promoted to a position from which to realize his or her original aspirations.

Rafael felt that this process, in addition to being slow and tedious, required a certain calculating attitude, since diplomacy and tact are the fundamental qualities for getting along well with bosses. But he wanted to avoid a working environment where actions were driven by calculation and self-interest. If he had to surround himself with people, he would put them in direct contact with the client so that they would assume their responsibility and develop their creativity without depending on the opinion, the way of being, or the politics of their superiors.

These were the ideas that were buzzing around in his head when, at the end of 1959, Luis Olaortúa told him that he had been called up to do his military service in the University Naval Militia and Rafael had to hire a substitute, José Mª Ruiz Iturregui (Chema), a classmate of Luis, who joined the company on January 1, 1960.

At that time, DOM was already staffed by a senior engineer, an assistant on leave of absence and a substitute for the assistant. And if he accepted Earle's project, the team would grow. I could see that the formation of the "entity" was inexorably approaching.

In 1961, Rafael was in charge of the lighting project for the old San Mamés stadium (lower image). In 2015, IDOM carried out the architecture and engineering of the new San Mamés stadium, a facility that has received the "Elite" category, the highest awarded by UEFA (upper image). The company founded by Escolá is ranked 69th in the world ranking of architectural firms.

[39] Testimony of Luis Olaortúa, 04–30-97. Rafael Escolá Foundation.

7 Work "Without Bosses"

To develop the Earle project, in 1960, he had to hire two draftsmen and, shortly thereafter, a fourth engineer. He retained his status as an independent consultant, but the five people he had hired had a "boss". The problem intensified the following year when, with a growing order book, he hired more draughtsman, surveyors and clerks. In 1961 he had 17 people working for him. The contrast between reality and his ideal of professional freedom was becoming increasingly clear.

In this second period, the team concept was conceived. They were all my employees. I collected the fees in my *name and* paid out of my own pocket.[40]

He was determined to see a "team" where everyone else saw "employees". He wanted to organize things in such a way that everyone felt free. Initially, he brought this idea to fruition in a "radical"[41] way and instilled in the team a "boss-free" work style.

> The idea that I had in my head in the founding days was to look for a way of earning a living that would allow people greater freedom of movement; I came to think of everyone working in the way he or she wanted to work.[42]

To begin with, he tried to get everyone to banish the term "employee" from their vocabulary. He also avoided creating "positions" within the new team.

> We do not see the need to strongly emphasize hierarchical relationships or personal categories.[43]

Each person would achieve within the team the position that his or her own effort, personal worth and peer recognition would give him or her. Even those who had received formal appointments could not expect Rafa and the others to treat them differently:

> (...) with more status than that which is, in fact, recognized by their proven technical competence in projects and other engineering activities.[44]

Over the years, when the size of the firm made it inevitable to classify into job categories, Rafael still maintained this idea, and recommended to graduates that:

> They dispense with their status and form a team with those of middle or non-graduate level, and even invite the Manager and the President to form a team with others for a work, if it is deemed necessary.[45]

---

[40] Pascual Montañés, cit., p. 241.

[41] Some people described Rafa's egalitarian ideal as "radical" (Cf. Rafael Escolá, handwritten letter, dated Bilbao 11/III/1995).

[42] Perspectives of the former first president, cit.

[43] Rafael Escolá, Principles of IDOM, DB-0 n. 12 (1976). Rafael Escolá Foundation.

[44] The principles of IDOM, cit.

[45] The principles of IDOM, cit., n. 13.

And of course, he personally ensured that the difference in status did not create any distance with those around him. According to Maite Osta, his secretary since 1964, in a letter addressed to Escolá.

> You never made distinctions in treatment, nor did you look at class, nor qualifications. From the highest to the lowest (if there ever was one), you looked for how we could be useful, where we could be most needed, and what we were good for.[46]

7.1 The "Semi-Free" Practice

But, after colliding several times with the harsh reality, Rafael had to abandon the idea of "free practice.".[47]

In the early years, he realized that total freedom for the team was incompatible with service quality. The lack of punctuality, the informality of schedules, and in general the disorder in the work of many of his collaborators forced him to rectify his ideal of "work without bosses".

> Long before I left the presidency, I recognized that I had wanted to take the ideal of freedom too far.[48]

In short, on the one hand he wanted to avoid those business aspects that he described as the "harsh environment of commercial companies" (control, distance between managers and employees, Taylorist organization of work, radical separation between personal and professional relationships, etc.), but on the other hand he understood that his "entity" had to evolve towards a company more in line with the norm: with a minimum of formal authority, schedules and rules that would guarantee a working agreement.

Driven by his (Greek) idealism, Rafael had assumed that everyone would share his enthusiasm for engineering. He soon realized that this was not the case.

> The engineer who works in a company must be subject to the regulations and to everything indicated by those who can command them in their superior hierarchy.[49]

This is the stage in which he had to recognize some of the virtues of working "with bosses": first of all, it was convenient for every engineer to learn to obey before giving orders, since he or she would inevitably end up managing other people's work.

Moreover, working "with bosses" avoided the evils of professional loneliness: bosses can correct flaws in the work and discover errors in the projects before delivering the final work, when any rectification has a high cost, in all senses of the word.

But organizing a company "with bosses" made him a businessman, a status that he found annoying and distracted him from his beloved technical activity. At this

---

[46] Maite Osta, Letter to Rafael Escolá, 1997, p.3. Rafael Escolá Foundation.

[47] Cómo nace y se hace…, cit., p. 7.

[48] Perspectives of the former first president, cit.

[49] Rafael Escolá, Deontología para Ingenieros, 1987, p. 139.

point, he made another "radical" decision: to entrust the development of the typically entrepreneurial dimensions to his closest collaborators:

> I would not have developed an engineering firm on my own: with my mentality, I would have been more of an individual consultant.[50]

And he decided to retain for himself, as President, the development of his ideal of work, which he would henceforth call "semi-free". This would lead him to the definition of a basic structure, which would inevitably have to have leaders, but which would allow the profession to work with a sense of freedom. The internal documents that he drew up in the following years, "Philosophy" and "Style of Action", perfectly describe this ideal.

Just One of Many

In 1963, Escolá took another unusual decision in the business world: to distribute the Firm's net worth among the people who worked with him. This decision was not the result of a sudden decision, but rather a progressive maturing of his "company philosophy".[51]

We can distinguish at least four phases in this maturing process: the "moral" sharing of the company (1957–1962), the "material" sharing of ownership among graduates (1963), the extension of ownership to all persons without distinction of employment category (1965) and the formulation of these ideas in a document called the Associative Commitment (1965).[52]

8 Recognition of Work

Rafael describes the loneliness of the first stage of his journey (1957–1962) as follows:

> In those early days there was no company as we understand it today. I was alone as a consultant and, therefore, a group of people was not yet visible to the outside world.[53]

The truth is that from the very first day of Bandas he had had to rely on third parties. And even if his role in the project was decisive, he had avoided any personalism. Proof of this is that the DOM brand, unlike so many other brands of consulting firms, did not refer, directly or indirectly, to the surname or the person of the founder. Avoiding the limelight, in a way, Rafael had already begun to practice a certain "moral sharing" of the success achieved among the whole team.

In 1961 Escolá was in charge of 18 "employees", had a turnover of 1.25 million pesetas and personally paid his taxes to the Treasury through the College of Engineers.

[50] Letter from the President at the end of his term of office, para. 5, Rafael Escolá Foundation.

[51] Cómo nace y se hace..., cit.

[52] Associative Commitment. Internal IDOM document, DB-101. Rafael Escolá Foundation.

[53] Pascual Montañés, La participación por competencia..., cit., p. 241.

That embryonic "company" belonged entirely to him. He had had the idea, had given up a salary, had looked for assistants, had attracted clients and was in charge of the training of his collaborators, most of them recent graduates. But it was the "team" that carried out the work, each according to his or her possibilities. This had to be recognized (Fig. 6).

9 Distribution of Ownership

He soon found that "moral" recognition was insufficient. IDOM provided a service based on knowledge, something very different from the production of physical goods. In the production of this type of goods, capital (means of production) is the sine qua non of the final product. If there is no capital, there is no work and no product. In consulting, on the other hand, the pre-eminence is of labor, of "intellectual capital". Without intellectual capital, there is no final product.

> It was not like in the production of tangible goods; in this case [of IDOM], what the engineer designs is everything that is offered to the client.[54]

And the formula Escolá used in 1963 to recognize this preeminence of labor over capital in consulting firms was the division of ownership among the engineers.

How did he do it? As always: by looking for an ingenious solution that would adapt to the dynamics of people moving in and out of the company. First, he devised a method for determining the increase in the value of the company in a financial year, say "year n". This would allow him to distribute the increase in value among those who had worked during "year n". And the increase in value produced during "year n + 1" would be distributed among those who had worked during "year n + 1" (many of them starting in "year n" and others joining in "year n + 1").

Starting from year zero (when only Rafael was a graduate), he distributed the increases in value for each year. Fifty percent of the increase in value for the year was distributed among the graduates who had worked the previous year, in proportion to the percentage of ownership they already had. The remaining 50 percent was to be distributed among the new graduates, in proportion to their salary. It also established that in order to qualify for the distribution of the increase in value, four years' seniority was required.[55] (seniority which, in 1963, as the first collaborator, only applied to Luis Olaortúa).

No known legal figure expressed such an original idea of ownership distribution. Thus, in legal terms, Escolá would remain the sole owner of the brand. The distribution of ownership would be decided by private agreement. He thought that this forced duality was actually positive, because as long as he was the sole owner of the commercial enterprise, it was guaranteed that the ownership would be distributed

[54] Cómo nace y se hace..., cit., p. 10.

[55] Felipe Prósper, Testimony, 14/X/2003.

among all the members of the company, and that the ownership would be distributed among all the members of the company.

10 "Persons" Not "Employees"

He had already divided the ownership among the engineers. But what about the non-engineers? In 1961, two non-engineers had joined the team: a secretary and an assistant. And soon there were six people working alongside Rafael who had no technical qualifications but who, except for projects, "did everything": writing reports and letters, errands, etc.

In 1965, two years after the allocation to graduates, IDOM had 135 employees[56] and, seeing that they all contributed to the final product, Rafael's "radically logical" spirit came to a new conclusion: the fair thing to do was to distribute the property among all the people, in relation to their contribution.

But the ownership was already distributed and including the non-engineers, with their respective seniorities, implied "redoing accounts". This could open deep wounds. Rafael summoned the engineers who had been with him the longest and informed them of his idea of redoing the allocations, starting from zero.[57] When this was reported to the rest of the owners, the vast majority accepted the decision, although this meant a decrease in the share that corresponded to each one, since the same amount was going to be distributed among a greater number of people (Fig. 7).

11 The Associative Commitment

Rafael understood that the receipt of a "share in the value of IDOM" had to be matched by each partner with a "style of performance" that aligned with the "values of IDOM". These values had to be put in black and white. The senior partners helped him to draw up the document that substantiated the values, which would later be called the "Partnership Commitment". In 1965 they drew up a private contract.

> During two weekends we held the so-called "Muñatones convention", where the ownership sharing system was put into operation.[58]

One of the points included in the document stated that all IDOM employees, including himself, should sell their shares upon leaving the Firm.[59] In this way, there would

---

[56] IDOM, Pasado, presente y futuro, cit., p. 14.

[57] IDOM, Pasado, presente y futuro, cit., p. 18.

[58] Muñatones Castle is a building currently declared a national historic monument, the only medieval castle in Bizkaia. It was later used as a hospital and hostel, currently houses a restoration workshop. It is located in the municipality of Muzkiz, in the grounds of the Petronor refinery. Pascual Montañés, cit., p. 242.

[59] Partnership Commitment (Partnership Contract), cit.

Fig. 7 **a** At the University of Piura, Peru. Rafael voluntarily relinquished the presidency of IDOM at the age of 60 to devote himself to what he loved: engineering and the training of young engineers. He was able to pass on to his collaborators his "passion" for technology. **b** Multimegawatt wind turbine test bench for the Fraunhofer Institute (Germany). Today, IDOM is a leading referent in technology

never be external owners of IDOM, and above all, there would never be two parties: "owners" and "employees".

> Between us, in fact, there is no employment relationship. We are an Association of professionals: here there is no company, no two parties, one of which hires and pays the other one.[60]

The document also included Rafael's ideas about the best customer service, the importance of effectively delivering the service (as opposed to the work itself), professional competence, development, and training opportunities for all, and, in short, Rafael's entire "philosophy" on work and personal relationships.

What is truly unique about IDOM's associative partnership is that it did not come about, as is usually the case, by "merger", but by "elevation". Rafael did not combine consultants with a reputation equivalent to his own, but rather shared what he had among the young people he was training, regardless of their job category. The formula was successful. In 1993, thirty years after he had begun the process of sharing out the property, Escolá could write with visible satisfaction:

> At present, the Firm provides for 500 families, and 300 people have signed the Associative Commitment.[61]

12 "Just Like One of the Others"

On April 8, 1979, Rafael Esscolá turned 60 years old and, with the intention of "allowing younger people to take over management roles,"[62] decided to hand over the presidency of the company to Luis Olaortúa, while he would continue working in the company as "just one more". This he did for almost twenty years, until the time of his death.

Rafael handed over "the leadership" without flinching because, basically, it was something he had been doing from the beginning, when he left the running of the business side of IDOM in the hands of his collaborators.

> I was the President and carried out few of these functions, in order to devote myself almost entirely to engineering work and the training of young engineers.[63]

Escolá was unique and "free" even in the way he retired. He did it according to what he believed to be correct, regardless of what the laws or working methods dictated.

With the handover of the presidency, the founding stage of a company which has been studied as "The IDOM Case" in Universities and Business Schools came to an end.[64] Many have praised, and some have criticized, the unique structure of this

---

[60] Rafael Escolá, Testament of the President (1979). Rafael Escolá Foundation.

[61] Cómo nace y se hace…, cit., p. 49.

[62] Letter from the President at the end of his term of office, cit.

[63] Cómo nace y se hace…, cit., p. 17.

[64] Cf. the document "La filosofía de IDOM", 0–391-010, presented as a "case study" at IESE, prepared by Prof. Domènech Melé. Cf. also the comparative study by Pascual Montañés, cit.

Fig. 8 The world's largest solar telescope, the DKIST, located on the tip of Mount Haleakala (Hawaii), designed and manufactured (EPC) by IDOM. Rafael fostered professional ambition among young engineers by pushing them to reach the highest levels of technical development

company. But no one has imitated it. Why? Raphael himself, in a letter of 1995, the year of his death, offers us a reason.

> Everyone likes the idea that there should be no owners other than those who do the work, but it seems too radical to them that the person who initiated the business activities that led to the creation of the company should be "just like everyone else".[65]

How many company founders and presidents would be willing to give away for free what is theirs, the property, and to hand over "the control" while they are still in full use of their faculties? Many will share Rafael's reasoning: if IDOM's "products" come entirely from the engineers' heads,[66] the market value of these "products" also belongs entirely to them. But there is a long way to go from words to deeds. And perhaps that is why IDOM remains a unique case (Fig. 8).

Teacher

The "Escolá" training model, of course, was also original: it taught young people by engaging them in the work, thus opening up a wide variety of job opportunities. The creation of a company allowed him to develop this model to its fullest potential. When a young person came to him, he would put them to work alongside him, explaining that beyond a possible job or salary, the best that they could expect from him was training. At the beginning of their internship at IDOM, he used to say to the newcomer:

> Take advantage of the time you are here to learn, because that is what you will take with you when you leave.[67]

65 Rafael Escolá, Handwritten letter, dated Bilbao 11/III/1995. Rafael Escolá Foundation.

66 Cómo nace y se hace…, cit., p. 47.

67 Conversation between Luis Olaortúa and Ana Cardenal. 2001. Rafael Escolá Foundation.

12.1 Work and Learn

Luis Olaortúa was the first beneficiary of this training concept. And from then on, so were all those who entered Rafael's magnetic field. An engineering student at the School of Madrid who did his internship with Escolá in the summer of 1961 offers the following testimony:

> I remember meeting him on Easter Monday 1961. Without wasting a minute, he was curious about my interests, what I was doing and what I wanted to do and, although I was only an engineering student, he gave me the immediate opportunity to start working, to take advantage of my time a little more, to learn; that summer I was already at IDOM doing an "internship".

Consistent with his own life path, Rafael believed that project engineers should work and learn at the same time, taking on responsibilities beyond even what caution recommends. And the best time in life to do this is when they finish their studies, "and even a little earlier". That is why, as President of IDOM, he began to dedicate part of his time to students who had recently graduated or were in their final year at the Bilbao School of Engineering, where he taught the subject of Complementary Factory Installations.

> Some students came to our offices to do their Final Project and stayed for their internships until they found a job. The end of internships has always been celebrated with champagne.[68]

13 Human Values

The work with postgraduates culminated in a department that, from then on, was also the source that fed the flow of personnel to work at IDOM.

> In 35 years, more than 300 postgraduates have done internships in this Department (without belonging to the engineering firm): more than 100 of them have gone on to work at IDOM.[69]

He never carried out a selection process to recruit people to join the Firm, but rather put young people to work with him:

> (...) the best way to get to know a person is to see them work (...) a few months of working with a person shows much more than an interview or a " series of tests".

This selection method, although slower, seemed to him much better than hiring senior engineers from another company. Because the human values he wanted to transmit could not be learned outside IDOM. A former partner of the company describes the working atmosphere he experienced with Escolá in this way:

---

[68] Cómo nace y se hace..., cit., chapter "La formación de ingenieros jóvenes".

[69] Cómo nace y se hace..., cit.

(...) An environment in which people were cared for. An atmosphere of professionalism that could be felt in the approach to the works (always under the technical–economic aspect); in the search for proximity to the projects (teams on site); in the care for the professional progress of its components (study aids, training...); in the use of simple organizational methods to provide good customer service; in the invoicing by prices and fees based on criteria of technical or material difficulty, with total transparency. An atmosphere of good personal, professional, and economic relationships. An atmosphere of doing the daily work well, which was demonstrated in a great willingness to make extra efforts. An environment (...) in which there was a great transparency in information, in the participation of all in the building of the company.[70]

For Rafael, caring for people should not be confused with "paternalistic" protectionism. Everyone had to forge his or her own path:

When the first engineers were still very young, it was not difficult for me to leave them alone to deal with problems as they arose: they were trained in the ability to make decisions responsibly.

Association of Engineering Companies

For Rafael, professional independence was a necessary prerequisite for greater perfection in the service provided. And since "free practice" had the status of "ideal", that is, of universal worth, Escolá dedicated a large part of his life to promoting professional independence, not only among the people who worked with him, but also in other engineering firms.

In the 1960s, when he was already a well-known person in the Basque industrial sector, he initiated the "Montenegro Meetings" to meet on a regular basis with other Basque engineering companies, seeking to unify professional positions and criteria. The gathering began informally, in an old restaurant "txakoli Montenegro" located in Deusto, but after many years of perseverance, Rafael's ideas came to fruition in 1993 with the founding of the Basque Association of Engineering and Consulting (AVIC).

Prior to that, in 1975, Rafael had founded, together with Mario Romero, the Spanish Association of Engineering Consultants, created to bring together all independent engineering companies, of which he was president for almost five years. Two years later, that association was expanded to include the largest companies providing services based on the knowledge of applied sciences and technology, giving rise to ASINCE (Asociación de Ingenieros Consultores de España). According to another colleague of Escolá, Luis Fernández Tenllado, ASINCE:

It was the result of Escolá's vision of the profession, of what professional services should be in our sector.[71]

Among other things, the presidency of ASINCE offered Rafa the opportunity to deepen his knowledge of the Latin American market. In April 1977 he represented Spain at the Assembly of the Federation of Consulting Engineers of Latin American

[70] Cómo nace y se hace..., cit.

[71] Luis Fernández Tenllado, Testimony, 14/X/2003. Rafael Escolá Foundation.

countries, which was held in Brasilia. The integration of ASINCE in FIDIC (International Federation of Engineering and Consulting Engineers), at the Ottawa Congress in 1977, was also the work of Escolá.[72]

Raphael gave priority to the unifying dimension of human relationships and therefore tended to make everyone who approached him a "partner". He first made his collaborators partners. Then he encouraged associative structures in the engineering sector. And his breadth of vision was such that he saw Contractors, Suppliers and Financial Entities - written with a capital letter—as "cooperators in the execution of the work"; and Competitors as "components and drivers of the market, with whom we maintain professional relationships of coexistence, in loyal competition".[73]

International Activity

During the 1960s, Spain had been one of the five fastest growing countries in the world, with a sustained annual average of 7 percent. But the "world oil crisis" (1973) brought the country, which had no reserves of black gold, to a screeching halt. Consulting services companies, whose business is closely linked to investment, began to anticipate the uncertainty of the international panorama.

Rafael knew some Spanish engineers who had moved to live in South America, and without hesitation, he took charge of IDOM's internationalization, carrying out tireless sales work across the Atlantic. After his first trips, offices were set up in Venezuela and Ecuador, and he also won contracts in Argentina, Peru and Colombia.

In 1994, Rafael recounted his trips to the Southern Hemisphere: in total, he had lived almost three years on that continent. He had a fantastic recollection of the treatment he received, to the point of affirming that the development of a country should not be measured by economic parameters, such as GDP, but by human parameters, such as "the level of friendliness".

Escolá was able to impress on the team his concern for the internationalization of the company. IDOM currently has 45 offices around the world and operates in 125 countries.

Epilogue

Rafael Escolá deserves a place in the pantheon of illustrious engineers not only for what he did during his lifetime, but also for his contribution through IDOM to the international prestige of Spanish engineering. The innovative projects carried out by this firm all over the world, from the dome of the DKIST advanced technology solar telescope in Hawaii, to advanced engineering in the ITER international fusion energy experiment in France, to the new Riyadh metro in Saudi Arabia, to name but a few, are a source of national engineering pride. To these reasons we must also add that, thanks to his initiative, thousands of engineers and architects have found a channel for their professional development.

He created a collective project and put it ahead of his fame and personal gain. He was a visionary of the dignity of the person at work. He was in love with engineering.

---

[72] IDOM, Pasado, Presente y Futuro, cit., p. 13.

[73] Philosophy of IDOM, 1995, n.6. Rafael Escolá Foundation.

List of Main Works

Technical Publications

1. **Optimización de magnitudes en Ingeniería.** Rafael Escolá Gil. Ed. Cedel, 1982. ISBN 84–352-0553–3
2. **Seguridad en los proyectos de Ingeniería.** Rafael Escolá Gil. Ed. Bernardo Martín Hernández, 2010. ISBN 84–604-7489–5
3. **Vigas de grandes luces.** Rafael Escolá Gil. Ed. Técnicos Asociados SA., 1979. ISBN 84–714-6197–1
4. **Construcciones con defectos, fallos o peligros.** Rafael Escolá Gil. Ed. Bernardo Martín Hernández, 1993. ISBN 84–604-7488–7
5. **Cálculo de vigas de hormigón.** Rafael Escolá Gil, Angel Ayarza Ayarza, José Miguel Ramos Elezcano. Ed. Agrupación de Fabricantes de Cemento de España, 1975. ISBN 84–400-8343–2
6. **Cálculo de cimentaciones superficiales.** Rafael Escolá Gil, Alberto Oroviogoicoechea, Antonio Mª Mora, Amando Castroviejo. Ed: Agrupación de Fabricantes de Cemento de España, 1974.
7. **Depósitos de agua elevados.** Bernardo Martín Hernández, Ángel Ayarza Ayarza, Rafael Escolá Gil. La Zarza de Pumareda (Salamanca): B. Martín, 2003. ISBN 84–398-4344–5
8. **Resistencia al fuego y fiabilidad de las estructuras.** Walter Schaeidt, Rafael Escolá Gil. Barcelona: Editores Técnicos Asociados, 1982. ISBN 84–7146-229-X

Publications of a Humanistic Nature

9. **Deontología para ingenieros.** Rafael Escolá Gil. Ediciones Universidad de Navarra, 1987. ISBN 84–313-0983–0
10. **Cómo nace y se hace una empresa de ingeniería.** Rafael Escolá Gil. Ed. ES de Ingenieros Industriales y de Ingenieros de Telecomunicación. Bilbao, 1993. ISBN 84–600-8452–3
11. **El éxito en la profesión de ingeniero.** Rafael Escolá Gil. Ed. Bernardo Martín Hernández, 1993. ISBN 84–604-6279-X
12. **La personalidad.** Rafael Escolá Gil. Ed. Palabra, 1995. ISBN 84–8239-022–8
13. **Ética para ingenieros.** Rafael Escolá Gil, José Ignacio Murillo Gómez. Ed. Universidad de Navarra. EUNSA, 2000. ISBN 84–313-1744–2

Gilda Sara Fernández Levy (1944–1994)

J. R. Marty-Delgado and P. P. Hidalgo-Reina

Abstract There are few studies in Latin America and particularly in Cuba, on the presence and contribution of women to the history of mechanical engineering. The School of Mechanical Engineering in Cuba began its activities at the University of Oriente in 1949. Gilda Fernández Levy (1/09/1944–21/03/1994) was the first woman to receive the diploma of Mechanical Engineering graduate in Cuba; she soon stood out for her organizational skills, her work capacity, her human values and her sharp intelligence. She was a full-time professor of the Faculty of Mechanical Engineering at the Central University "*Marta Abreu*" of Las Villas, where she held various responsibilities. In recognition of her contributions and exemplary career, this article summarizes the life and work of this singular figure of mechanical engineering in Cuba, as a sample of the achievements and results obtained by the participation of women in engineering, an area of knowledge historically considered masculine, but which in recent years has shown a favorable evolution in the number of women enrolling in these studies.

Keywords History · Mechanical engineering

1 Biographical Notes

In spite of the indisputable achievements of Cuban women in the scientific and academic fields, this was not always the case. Women have progressively gained access to university studies and have maintained consistently high graduation rates in higher education, surpassing the levels achieved by men in various fields of study. Gilda Sara Fernández Levy is an example of this (Fig. 1).

J. R. Marty-Delgado (✉) · P. P. Hidalgo-Reina
Central University "Marta Abreu" of Las Villas, Santa Clara, Cuba
e-mail: jmarty@uclv.edu.cu

P. P. Hidalgo-Reina
e-mail: pedroph@uclv.edu.cu

R. López-García and M. Ceccarelli (eds.), *Distinguished Figures in Mechanical Engineering in Spain and Ibero-America*, History of Mechanism and Machine Science 43, https://doi.org/10.1007/978-3-031-31075-1_3

Fig. 1 Gilda Sara Fernández Levy. (Courtesy of Gilda Sara Fernández Levy Honorary Professorship)

The issue of gender has recently become an integral part of the scale of indicators of academic activity and research. The participation of women in the field of engineering, in scientific and technological development, is one of the main challenges of educational and professional equity (Leyva and García Hernández 1962). In Cuba, the incorporation of women in science has gone through a long way, particularly engineering that, in our opinion, reached its full maturity after 1959 with the full incorporation of women into the economic, political, social, cultural and scientific life in the country.

Mechanical engineering studies in Cuba began their activities at the University of *Oriente* in 1949. According to the academic structure of the time, in November 1959, mechanical engineering studies began at the school of the same name at the Central University of Las Villas, by agreement of the ordinary University Council of November 3, 1959, presided over by the acting Rector Dr. *José M. Ruiz Millar*.

The agreement also mentions the creation of the Schools of Electrical Engineering, Social Assistance, Psychology and Aviation. The Mechanical Engineering program began its teaching activities in the 1959–1960 academic year, starting classes in November 1959, with an enrollment of 84 students. Gilda Sara, at that time, had just turned 15 years old in the city of Santa Clara.

The years after 1959, meant a great transformation for Cuba and for the nascent School of Mechanical Engineering, in the technical, socioeconomic and political dimensions. In 1962, Cuban higher education underwent a university reform as a transforming scientific-humanistic element, adequate to the needs and historical context of the country. At that time, there was a latent need for a higher education that could respond to the new challenges of the country's industrialization. In this scenario of profound changes, the insertion of young people in the process of social transformation that the country was undergoing was an essential part of the preparation of future professionals.

Gilda's life and work are part of this transforming process, constituting a vivid expression of the potential of women, contextualized to the historical moment in which she lived, as well as her convictions and values formed in her family and in the heat of her work as a mother, pedagogue, leader and exceptional human being, who bequeathed to her peers a permanent example of etiquette in her personal and professional life.

There are few studies in Latin America, and particularly in Cuba, on the presence and contribution of women to the history of mechanical engineering. Gilda Sara Fernández Levy was the first woman to receive her diploma as a graduate of Mechanical Engineering in Cuba and very soon stood out for her organizational skills, her work capacity, human values and her sharp intelligence. Gilda was a full-time professor in the Faculty of Mechanical Engineering at the Central University of Las Villas, where she held various academic and political responsibilities.

In recognition of its contributions and exemplary career, this article summarizes the life and work of this unique figure of mechanical engineering in Cuba, as a sample of the achievements and results obtained by the participation of women in engineering, an area of knowledge historically considered male, but which in recent years has shown a favorable evolution in the number of women enrolled in this career. In which her sister *Guiselda Fernández Levy*, another mechanical engineer who deserves to be highlighted in the country, also stood out.

Enclosed in this background, this tribute of the authors to Gilda Sara Fernández Levy, of her co-workers and former students, intends to provide elements of her life that contribute to increase the culture on scientific development in Cuba, particularly of mechanical engineering that in 2019, the School of Mechanical Engineering of the Central University "*Marta Abreu*" of Las Villas, is celebrating its 60th anniversary.

2 Your Beloved City of Santa Clara

Santa Clara, capital of the former province of Las Villas[1] and head of the municipality of the same name, is located in the south-central portion of the province of Villa Clara, in the vicinity of the geographic center of Cuba. An essential cultural center and significant industrial and scientific enclave of the country, it is crossed by the *Bélico* and *Cubanicay* rivers, which belong to the basin of the *Sagua la Grande* river.

[1] Since 1878, by Royal Decree of the Spanish colony, Santa Clara was granted the status of capital of the province of Las Villas. The territory of the former province of Las Villas, since 1976, by the political-administrative division adopted in the country, was divided into the current provinces of Villa Clara, Cienfuegos and Sancti Spíritus.

According to legend and history, Santa Clara was founded on July 15, 1689 by the settlers of the town of *San Juan de los Remedios* who moved to the center of the island, near the banks of the Sabana River, today known as the *Bélico* River. With time, the people of Santa Clara raised in that place the hermitage of the *Señora del Carmen*, where a commemorative monument in the form of a descending spiral rises since 1951, around a tamarind tree, see Fig. 2, heir of the one that saw in its shade the foundational mass of the settlement.

Although during the colonial and republican period, the city of Santa Clara was characterized by an essentially commercial and agricultural development, after 1959 it became a significant industrial center that represents, approximately, a third of its mercantile production, see Fig. 3. Between 1960 and 1990, important national and provincial industries emerged in the areas of iron and steel, light industry, furniture production and, more recently, in the biotechnological area. The former province of Las Villas, now Villa Clara, has a significant weight in the national sugar tradition.

Today the city of Santa Clara is a fundamental node in the communications of the central region and an obligatory transit between the west and the east of the island.

Fig. 2 Recent image of the foundational site of the city of Santa Clara and its symbolic tree. Photo from: My Santa Clara Citizen Portal at https://misantaclara.gob.cu/monumentos

Fig. 3 Government palace, city of Santa Clara, mid-twentieth century. Photo from: My Santa Clara Citizen Portal at https://misantaclara.gob.cu/monumentos

It is a welcoming city and an obligatory reference in contemporary history in the scientific, cultural, economic and social fields of the country. It is also one of the most industrialized cities in Cuba.

The Central University "*Marta Abreu*" of Las Villas, the most multidisciplinary of the country, is a symbol of higher education in Cuba and of the province of Villa Clara, continuator of the best academic and scientific traditions in the central region, was the university where Gilda was formed and developed her fruitful academic life, linked to the teaching of mechanical engineering.

3 Early Studies

Gilda Sara Fernández Levy was born in Santa Clara, former province of Las Villas, on September 1, 1944. Although her parents and brother lived most of the time in the United States, Gilda Sara, like the rest of her family, always felt a very special love for her beloved city of Santa Clara.

According to the testimony of *Delvis Fernández Levy*,[2] Gilda's brother, *Sara*, the mother, was born on September 25, 1918, in the small town of Cascajal, in the former province of Las Villas and died on January 5, 2014 in San Luis Obispo, California. She was the eldest daughter of *Felicia Rodríguez Marzall* (native of La Palma, in Pinar del Río) and *Salvador Levy Levy* (born in Kirklisse, Turkey). She had three children: *Delvis Alejandro*, *Gilda Sara* and *Guiselda*.

In order to understand Gilda's avant-garde character and attitude during her adult life, it is necessary to understand the social political environment in which she attended high school and preparatory studies, as well as the family and social environment of that time.

During the republican period, a system of public, free and compulsory primary education was established in Cuba, although with some deficiencies. According to data from that time (Cuba 1961), in the 1950s, half of the school-age population did not attend school. In the cities, one out of every five people did not know how to read or write. Industrial education was provided throughout the island in only one center with a level equivalent to the medium technical level. The highest level of development was seen in the schools of commerce with the studies of economy and administration.

Gilda's first studies were in public schools in Santa Clara where she reinforced the ideals of social justice, law and dignity learned in the family, from the readings of *Félix Varela y Morales*, *José de la Luz y Caballero* and *José Martí Pérez*.

Secondary education was shared between the private school *Antolín González del Valle* and the *Instituto de Segunda Enseñanza de Santa Clara*, Figs. 4a, b and 5.

At the *Instituto de Segunda Enseñanza*, the most prestigious academic and cultural figures of the province collaborated as teachers or lecturers. Here Gilda excelled in science, especially mathematics.

---

[2] See Delvis Fernandez Levy's facebook profile.

Fig. 4 Institute of Secondary Education of Santa Clara. **a** Photo from the middle of the twentieth century. **b** at present, "Osvaldo Herrera" Pre-University Institute

Fig. 5 Exterior view of the newly constructed Technology Building (August 1962) where Gilda studied mechanical engineering and where she later worked as a professor. This building continues to be the headquarters of the mechanical engineering career at the Central University "Marta Abreu" of Las Villas. The University Hall that bears her name is located there (Courtesy of the UCLV historical archive)

She left the private school together with a small group of students because of disagreements with the school's management, about the little attention paid in the school to the social problems of the time.

From the 1930s until 1959, the students of the *Instituto de Segunda Enseñanza de Santa Clara* and part of the faculty took an active part in the struggle against the different governments of the time. It should be remembered that the Secondary School was, as it is today, the prelude to higher education.

This is a period of fervent revolutionary spirit in all the people of Cuba characterized, since 1957, by the presence in the east of the country, of Fidel Castro's guerrilla army. Meanwhile, in the main cities of the country, the clandestine struggle against *Fulgencio Batista* took on important heroic overtones. In the Aula Magna of this Secondary School, meetings, evenings and revolutionary activities were held. Some of its students took the path of the guerrilla struggle in the *Sierra Maestra*, among them *Osvaldo Herrera, Rodolfo de las Casas* and other young people. It is not difficult to imagine the impact that this revolutionary environment had on the formation of the character and will of the young Gilda Sara.

In 1960, the Contingent of Volunteer Teachers was formed as part of the National Literacy Campaign, and a plan was carried out to bring education to every corner of the country. Gilda, at the age of 16, with a bachelor's degree in science, was one of those 3,000 people (mostly young people) who left to educate peasants living in remote places throughout the country.

4 The Young University Student

The Central University of Las Villas, symbolically inaugurated on October 10, 1948, was the third university to be founded, before the University of Havana and the University of *Oriente*. Two months after Gilda turned 8 years old, on November 30, 1952, the first academic year began at the Central University.

We will never know for sure what motivated Gilda to choose the university career of Mechanical Engineering. The family environment and the social context that demanded the formation of engineers to support the industrial development that was projected in the country were surely decisive.

In 1961, among the many possibilities available to her, Gilda enrolled in the recently created School of Mechanical Engineering. She decided to carry out her university studies in her beloved city of Santa Clara, in a difficult and eminently masculine career for the canons of the time, a challenge that she faced with absolute naturalness. We imagine that while walking through the long corridors of the technology building, that beautiful young woman of more than 1.70 m tall, slow speech, open smile and penetrating gaze, was undoubtedly the admiration of the young people of the Chemical, Electrical, Industrial and Mechanical Engineering Schools that also studied there.

In the context of higher education, in that period, there was a great need to put an end to an education system inherited in part from the neo-colony, unbalanced, outdated, elitist and, above all, misaligned to the new situation of the country. Then, on January 10, 1962, while Gilda was in her second year of Engineering, the Reform of Higher Education in Cuba came into force.

The university reform of Cordoba can be summarized in the objectives of: opening the university to broader sectors of students regardless of their origin and social position, free attendance to facilitate the access of workers, welcoming all competent intellectuals and professionals, regardless of their ideologies and backgrounds, democratizing university government and the achievement of autonomy, and linking the university with the people and the life of the nation, which they called the university's social mission (Tunnermann 2008). The publication of the "*Manifiesto de Córdoba*" with these assumptions soon had great repercussions throughout the region, since Latin American universities and societies were facing the same problems. The debate that ensued brought the university and its place in national life to the forefront. The reform movement spread to countries such as Chile, Uruguay, Colombia, Mexico and Cuba, where it had the greatest impact.

Regarding the university reform in Cuba, it is necessary to ratify the Latin American perspective of the process, which has its roots in the university reform of Cordoba in 1918, where -to a certain extent- the critical path of higher education in the region was established. According to Rodríguez Rodríguez (2012), it is not necessary to point out that the Reform does not consist only of the approved bases. It is a complex and long process of which only the first stage was being lived in those years. Not everything adopted will remain and withstand the practical test. Not everything that is right and correct will bear, from the beginning, its optimal fruits.

In September 1962, the Central University had 296 hectares of land—8 more than a decade earlier, at the time of its foundation—and had 3316 students enrolled. It also had two other faculties, corresponding to Technology and Agricultural Sciences. It also incorporated pre-university training (preparatory and worker-peasant), of a pilot nature, with the purpose of adding students to increase their scientific-practical knowledge, in order to promote the mastery of instruments and machines in the future factories that the country would install. At that time, the University had access to a budget of 1,617,000 pesos for the assembly of constructive investments. Under these conditions, the University where Gilda studied was a hotbed of ideas and transformations that left none of those young people indifferent.

In 1962, while Gilda was studying, the School of Mechanical Engineering began to create Teaching Departments that grouped related subjects, being the first to be created the Department of Drawing, under the initial direction of the *Architect Justo Pérez Díaz*. Subsequently, the Design Department was created, under the direction of Engineer *José Francisco Regueiro*, who was also the director of the School. This was followed chronologically by the Department of Energy, under the leadership of Mr. *Rómulo Madrigal*, and finally the Department of Mechanical Technology, whose responsibility was assigned to Mr. *Domingo Artze*. The Drawing and Design Departments were the genesis of the Department of Applied Mechanics and Drawing in which Gilda developed all her academic work (Boada Carrazana 2006).

From that moment on, the ideas of the transformation of teaching and the interweaving of the University in the social development, penetrated in the thinking of the Cuban university student body and especially in Gilda who, from then on, would dedicate her professional work to the organization of Higher Education in Cuba, especially to the teaching of Mechanical Engineering, focused on the immediate future of the industrialization of the country.

The first graduation of the School of Mechanical Engineering was in 1964, where only six students graduated, out of an initial enrollment of 84 (*Héctor Artze, José Grave de Peralta, Juan Antonio Faget, Néstor Labrada, Francisco Martínez* and *Jesús Suárez Arias*). The second graduation took place in November 1965, graduating among those students were *Ortelio Boada Carrazana, Reinaldo Martínez Martínez, Juan Pozo Armas* and *Armando Morales Ruiz*, who were part of the faculty of the mechanics career and were Gilda's work colleagues. Each one of them deserves fair recognition for their contributions in the formation of hundreds of engineers who are today scattered throughout the island. *Reinaldo Martinez* is still active in teaching at the Faculty and is an inspiration for the new generations of mechanical engineers.

The years in which Gilda carried out her university studies were also marked by important social and political events in the country and in the province of Las Villas, in some of these events, she herself was a protagonist. Without abandoning her studies, Gilda taught at the *Facultad Obrera* and as an undergraduate instructor in the Design Department. In this way, together with other engineering students, she took on the preparation of technicians, workers and students of the province, with enthusiasm, responsibility and commitment.

When she finished her studies in 1966 with excellent results, she was selected by the direction of the School of Mechanical Engineering to be part of its faculty as a full time professor. That third graduation of the School of Engineering of the Central University of Las Villas, as a gesture of reaffirmation, she climbed the Turquino Peak, the highest elevation of the island, located in the former province of Oriente.

In 1966, Gilda married *Arturo E. López-Calleja Hiort-Lorenzen*, also a recent graduate of the School of Mechanical Engineering. From this marriage four sons were born: *Arturo*, *Jorge* and the twins *Ernesto* and *Enrique*.

From very early on in the University, Gilda was linked in her research work to the sugar industry, which for many years was a priority for the country.

Other investigations clarify that while Gilda was in her first year of university studies, another woman was in her third year; for unknown reasons, she dropped out without finishing her studies. The name of this precursor has been lost in time. Gilda Sara Fernández Levy graduated in 1966 as the first woman mechanical engineer in Cuba.

5 The Mechanical Engineer

Cuban higher education, following the precepts of the first reform, focused special attention on establishing a new organizational restructuring in 1976; in that year the Ministry of Higher Education of Cuba was created. In addition, a new career structure was developed and a significant expansion of the network of Higher Education Institutions (HEI) in the country was carried out. All this led to the expansion and creation of university capacities; and by the 1980s, the sustained increase in enrollment levels.

At this time, the School of Mechanical Engineering of the Faculty of Technology became the Faculty of Mechanical Engineering and a stage of consolidation of the career and its faculty at national and international level began.

In this new context, as a professor of the Faculty, Gilda took several postgraduate courses with Cuban specialists and from the former socialist camp (Czech, Soviet and German). She participated in many research and production-related works, with emphasis on those related to the sugar industry.

His teaching and scientific work was closely related to the specialty of Strength of Materials, for which he prepared auxiliary materials, study programs and methodological and scientific works. Most of these works have not been digitized. In this sense it is necessary to emphasize the book Resistance of Materials, a textbook in two volumes, which was for many years, the reference book for the study of that discipline in Cuba and other universities in Latin America in the careers of engineering and architecture. This is undoubtedly Gilda's most enduring written work. It is the work that identifies her in Cuba and Latin America, as an engineer and as a professor.

The textbook "*Resistencia de Materiales*" was first published in 1979, by the Cuban publishing house *Pueblo y Educación*, with several subsequent reeditions. In the prologue, Gilda states that it was written with reference to her classes at the Faculty of Mechanical Engineering and the recommendations and help of *Alina Expósito*,[3] her sister *Guiselda* and her husband *Arturo*.

In the first volume of the book, the general principles of strength of materials, simple stresses, as well as the theories of limit stress states are studied. The second volume begins with complex strength and contains specific topics such as: general determination of displacement, hyperstatic systems, calculation of vaults and plates, thick-walled cylinders, longitudinal bending, fatigue, dynamic action of loads and contact stresses.

Both volumes are illustrated with solved problems, chosen in such a way as to highlight the application of theory to the solution of practical problems. At the end of each chapter, problems to be solved with their corresponding answers are proposed and a series of questions grouped by headings are formulated as a support to the study of the topics.

An idea of the magnitude of the work written by Gilda is summarized below in the contents of each of the chapters of the book (Table 1):

Gilda developed an extensive contribution in the methodological teaching work. She prepared the official programs of the subject Resistance of Materials of the first study plans of the mechanical engineering career, after the creation of the Ministry of Higher Education, she also presented a great number of works in the methodological conferences of the University. During all the stage of her work at the Central University of Las Villas she was the main professor of the subject Resistance of Materials, directing the methodological work of the same. She developed methodological consultancy to the University of Camagüey, the Higher Institute of Holguín, the Higher Agricultural Institute of Ciego de Avila, the University Center of Moa and the University Center of Matanzas. For several years he was a member of the Methodological Commission of the University.

As part of his scientific work, he participated in more than 7 concluded research tasks, directly linked to the sugar industry and other mechanical industries in the central territory of the country. He participated as speaker in a group of national and

[3] Alina Expósito Claro, mechanical engineer and glory of the Cuban sports movement. She was Gilda Sara's teacher and co-worker for many years.

Table 1 Contents of book "Strength of Materials" written by Gilda Fernández

Chapter number	Chapter title	Chapter contents
Chapter 1	General information	General Overview of the history of strength of materials. Real system and analysis scheme. Deformations
Chapter 2	External and Internal Forces	Method of sections. Graph of normal forces and torsional moments. Types of supports used in beams. Bending moment and shear force in a beam section. Graph of bending moments and shear forces. Graphs of internal forces in plane and special frames. Stresses. Strength conditions
Chapter 3	Mechanical testing	Mechanical properties of materials Deformations and stresses in bars subjected to tension and compression. Tensile and compression tests. Deformation mechanism. Allowable stress. Factor of safety
Chapter 4	Axial tension and compression	Different types of problems that can be solved in bars subjected to axial tension and compression. Influence of self-weight. Statically indeterminate constructions. Systems carrying loads above the yield stress
Chapter 5	Theory of stress and deformation state	Linear stress state. Plane stress state. Principal stresses. Principal areas. Maximum and minimum principal stresses. Circular diagram of the stress state Generalized Hooke's law Potential strain energy
Chapter 6	Resistance theories	First, second, third, fourth resistance theory. Morh's theory. Mechanical theory
Chapter 7	Shear stresses	Pure shear stress state. Hooke's law. Deformations. Potential strain energy. Rivets and welded joints
Chapter 8	Torsion Stresses and strains in circular members subjected to torsion	Calculation of circular members subjected to torsion. Torsion in non-circular members. Torsion in thin-walled members. Torsion in systems working above the yield stress
Chapter 9	Bending Normal stresses in members subjected to bending	Shear stresses in members subjected to bending. Strength testing of a beam. Reinforced beams. Behavior of thin-walled members. Stress concentration in bending Deformations in bending. Variable section beams
Chapter 10	Composite Strength Oblique bending	Flexure with tension or compression. Flexure with torsion. General case of stresses. Plane bending in a curved beam
Chapter 11	Displacements in members produced by an arbitrary system of loads Work of external loads	Potential energy of deformation. Theorem of reciprocity of work. Theorem of reciprocity of displacements. Morh's method for the calculation of displacements. Vereshiaguin's method. Determination of displacements and stresses in helical springs

(continued)

Table 1 (continued)

Chapter number	Chapter title	Chapter contents
Chapter 12	Hyperstatic systems	Base system and equivalent system. Canonical equations. Method of forces. Continuous beams. Equation of the three moments. Hyperstatic beams working above the yield point. Rational location of beam supports
Chapter 13	Thick-walled cylinders and rotating disks	Fundamental equations for axis-symmetric bodies. Determination of stresses and displacements in a thick-walled cylinder. Calculation of stresses in composite cylinders. Disks of constant thickness rotating at high velocities
Chapter 14	Vaults and Plates	Fundamental properties of vaults and plates. Calculation of symmetric vaults. Bending of circular, rectangular and elliptical plates subjected to symmetric loads. Flexure of cylindrical vaults under symmetric loads
Chapter 15	Longitudinal bending	Euler's problem. Limit of applicability of Euler's formula. Yasinsky's formula. Choice of the most rational type of profile and material for columns. Simultaneous longitudinal and transverse bending
Chapter 16	Fatigue	Fatigue, properties. Characteristics of variable stress cycles. Wohler curves. Fatigue limit. Restricted fatigue limit. Fatigue limit diagram. Factors affecting fatigue limit. Ways to improve fatigue strength of machine parts. Fatigue strength safety coefficient and its determination. Calculation of variable stress resistance under unstable conditions
Chapter 17	Dynamic action of loads	Calculation of stresses during accelerated motion. Fundamental concepts of vibration theory. Free vibrations of a system with one degree of freedom. Forced vibrations of a system with one degree of freedom. Resonance
Chapter18	Contact stresses General	Calculation of contact stresses. Surface fatigue. Surface fatigue limit
Chapter19	Experimental methods to determine stresses and strains	Strain gage method. Mechanical or lever extensometers. Optical extensometers. Electrical extensometers. Optical method for determining stresses

provincial scientific events and published more than 13 scientific articles in technical magazines.

He taught different postgraduate courses related to his specialty, applied for a group of patents and was granted 4 application reports. An important step in her scientific career was in 1984 when she obtained the scientific degree of Doctor in Technical Sciences, under the guidance of Professor *Ortelio Boada Carrazana*. The research work was related to the origin and development of the failure of mill shafts in sugar mills. In fact, Gilda is also the first female mechanical engineer to successfully defend the degree of Doctor in Technical Sciences.

The unforeseen breakage of sugar mill trees has been an event of great interest to specialists in the sugar industry, both economically and technically. The economic aspect is justified by the high cost of this part, the time required to replace it and the limitations that are created in the extraction work by putting the machine to work in a limited regime of loads to prevent the breakage of the tree. The technical interest is determined by the fact that a machine element, designed according to known theories and proven in practice, fails in an operating time very different from the expected in some cases and in others it works uninterruptedly, until the wear in the supports or in the hammer limit its operation.

The first reported scientific study aimed at solving this problem was carried out by the Cuban *Edmundo Herrera* in 1956. Concluding that these trees fail by fatigue, an aspect that he proved theoretically and analyzing the form of breakage of a considerable number of pebbles that corresponds to the typical form of fatigue fracture, he addresses the issue of stress concentrators, improvement of the surfaces in the breakage zone and offers some recommendations for exploitation that would increase the resistance of the pebbles to fatigue.

In 1982 Gilda Fernández Levy in her doctoral thesis (Fernández Levy 1984) deals specifically with this subject, experimentally classifying the material of the pebbles, and developing a more complete analysis scheme than the previous ones, in addition to carrying out resistance and fatigue calculations of the pebbles. He characterizes the material for the manufacture of these trees and gives a series of recommendations for increasing the fatigue resistance of the pebbles. In addition, it recommends the improvement of the loading scheme to make the strength and fatigue calculations more accurate. This result constituted a scientific and economic contribution of high impact in the country.

Gilda was a member for several years of the Scientific Council of the University and of the National Scientific Grade Tribunal. She also held several administrative responsibilities such as: Professor Group Leader, Head of Department (1968–1969), Deputy Director of the School of Mechanics (1969–1970) and Vice-Dean of the Faculty of Technology (1970–1971). She performed each of these academic responsibilities with passion and infinite dedication. Her students spoke of an excellent teacher, and of a demanding and tender woman.

As it has been recognized in other bibliographic sources, the year 1984 proved to be the right moment to officialize a Higher Education Institution dedicated to industrial design in Cuba, due to the development reached by Cuban higher education, which had been restructured in 1976 and already had an important experience, which was reinforced by gathering for its moment, the best of the international trends that could be analyzed by a National Commission for Technical Sciences, which visited several of the best universities in the world in 1985. The Higher Institute of Industrial Design (ISDI) opened its doors to its first academic year in October 1984, with 50 students distributed in equal number, for the careers of Industrial Design and Information Technology, who worked in a provisional headquarters in the Havana municipality of Playa, until October 1985, when the activities officially began in the current headquarters. Its founding Rector, in the 84–85 academic year, was Gilda Fernández Levy (Cuendias Cobreros 2014).

Between 1985 and 1986 Gilda worked as a technical teaching advisor to the Ministry of Higher Education of Cuba and in 1989 she was appointed director of the Budgeted Unit of the National School of Cadres, of the then Ministry of the Sidero-Mechanical Industry (SIME) and finally, she was promoted to director of Science and Technique of this organization, a task she performed with good results until her death in 1999.

She joined the Communist Party of Cuba (PCC) in 1970, occupying from the beginning responsibilities as secretary of organization, member of the Bureau of the PCC of the UCLV and later, of the leadership of the same. She was a member of the university committee of the party since 1971, and belonged to the Provincial Committee of the Party from 1977 to 1985. She was elected as a delegate to the first and second congresses of the PCC.

6 Testimonials

Ortelio Boada Carrazana, Mechanical Engineer in the second graduation of the School of Mechanical Engineering. Doctor of Science, Gilda's co-worker for many years:

> As part of the rich history of our Faculty, it is important to highlight even if only in the form of a review, the role played by the unforgettable Gilda Fernández Levy, one of the most emblematic professors of our Faculty and who deserves to be highlighted for her human value, for her capacity, intelligence and revolutionary spirit. Her passage through our Faculty left an indelible mark, esp ecially to those who had the privilege of knowing her and sharing with her the daily and daily work in the foundational work of our center. She was the first Mechanical Engineering graduate of this Faculty and very soon stood out for her organizational skills, her work capacity and her sharp intelligence (Boada Carrazana 2006)

Luis E. Rabassa López-Calleja, Mechanical Engineer. Gilda's nephew:

> Aunt Gilda was an incredible woman, she will always be in my memory, thanks to her, following her example I studied Mechanical Engineering, whenever an obstacle appears on the road I think of how she would overcome it and I adapt it to my circumstances to overcome it, she is always in my thoughts, her values and her strength. What can I say about Chicha, she was everyone's grandmother and always energetic and right in everything (Comment taken from Arturo López Levy's Facebook profile).

Eusebio Pérez Castellanos. Mechanical Engineer. Doctor in Science. Gilda's former student and co-worker:

> It would be very difficult to condense in one paragraph all the things I can say about Gilda. She was my professor of the subject Resistance of Materials I and II in the course for workers. In that group we were all student-workers with deficiencies in general education since practically none of us came from a high school. His pedagogical mastery, his power of communication and above all his human qualities made us understand one of the most

difficult subjects of the course. He got along with the group in an incredible way. Later I started working as a teacher and I was lucky to have her close to me. She taught me things that I still remember perfectly about the interrelation with the groups of students and the pedagogy of teaching (September 2019).

Ángel Rubio González. Mechanical Engineer. PhD in Science. Gilda's co-worker:

I met Sara Gilda Fernández Levy recently graduated, year 1973, I was never her student, but I was her co-worker for several years and friend until her physical departure. Gilda was a special person, excellent in everything she did. I knew her in her political functions fighting for the poor of the earth, because she was a devoted Martiana and an insatiable reader; in her facet as a teacher -as I was her dean- she was exceptional in this, an evaluation that I verified many times with her students; In her mission to write the textbook on Strength of Materials for Cuban higher education, to which she devoted herself with all her intelligence and energy, and in her research for the sugar industry, which led her to successfully defend her doctoral thesis, research with a high scientific level and relevance. I also knew her in her family environment, an extraordinary woman and mother. I will never forget when, after having two sons and looking for the female she longed for, she gave birth to male jimaguas. She assumed, with extraordinary fortitude, the upbringing of four children and never abandoned her work at the university. I remember Gilda as one of those people who with simplicity, enthusiasm and joy, assume any task or function and do it well, engaging everyone around her and always being an extraordinary example of professionalism and guidance at work. I will always remember her with her eternal smile and her slow speech (September 2019).

Reinaldo Martínez Martínez. Mechanical Engineer. Gilda's co-worker:

Gilda was a very sociable woman, ahead of her time. She had a very good relationship with her classmates and later with the teachers when she was already a worker. She was dedicated to study and to the quality of her classes. In the positions she held she was responsible and intelligent in dealing with her subordinates. An exceptional woman (September 2019)

7 Gilda Sara Fernández Honorary Professorship

The Honorary Chairs at the Central University "Marta Abreu" of Las Villas, promote the work of University Extension from the perspective of an important activity in the university or community, as well as based on an outstanding figure from the academic point of view, research or otherwise both in the university activity, the territory or the country. The Honorary Chair "Gilda Sara Fernández Levy" was created in July 2004 by Rector's Resolution.

The fundamental objectives that this Chair will be in charge of will be the following:

1 To Rescue, Investigate and Divulge the Life and Work of Such an Outstanding Professor, as Well as the History of the School of Mechanical Engineering.
2. To promote teaching, scientific and cultural activities, in its broadest sense, in order to disseminate the activities, in all aspects, carried out by the Faculty at the Central University "*Marta Abreu*" of Las Villas and its surroundings.
3. To convene extension courses, postgraduate courses, conferences, lectures by Cuban and foreign specialists on topics that contribute to the integral formation in its broadest sense.

Fig. 6 Current exterior appearance of the University Hall Dr. Gilda S. Fernández Levy. Technology Building, Central University "Marta Abreu" of Las Villas

4. To promote the exchange with other chairs of our university or of other centers and countries that have the same purposes exposed in the previous objective, emphasizing the university extensionism and the promotion of our national culture.

To provide that this Chair will be located in a designated classroom of the School of Mechanics, which will be enabled for these purposes, Fig. 6.

This Resolution was signed by the Rector of the University on that date, Dr. José *Ramón Saborido Loidi.*

Gilda is a model university professor. True to her ethical principles, flexible and deeply humane. She never negotiated neither rights nor dignity. Despite the authority that emanated from her personality and demeanor, she did not exercise any form of authoritarianism. Committed to the great problems of the teaching of mechanical engineering in Cuba, she was an exponent of the struggle for women's rights and their role in society. Those of us who were her students or co-workers remember her as a teacher who left no room for injustice or fraudulent attitudes towards life.

She could be found in the corridors, on the sports fields or in meetings expressing without hesitation her criteria on the most diverse topics. The Rebel Youth Association first, the University Student Federation and the Communist Party later, were scenarios for her to discuss any kind of problem concerning the agitated national period in which she lived or the most relevant international issues. She took on numerous responsibilities, developed curricula for different subjects, always prepared for each class as if it was the first time she was teaching it. He educated his children in the constant search for truth and in love.

As *Graciela Pogolotti* pointed out, it is worth remembering that the teacher must be moved by a vocation of service that goes far beyond the mere transmission of knowledge. He is a trainer of conscience founded on unwavering ethical principles, an active interlocutor of young people who emerge to life, in whom he needs to encourage the need to understand the world, to encourage the defense of national sovereignty, the will to continue building a country oriented to social justice and solidarity among human beings, equipped with the necessary tools for the exercise of criticism in the face of wrongdoing, a sower of spiritual wealth, thirst for knowledge and fibers of sensitivity.

References

A. L. Leyva and A. García Hernández, "La reforma universitaria de 1962: un hito para la educación superior cubana," Revista Cubana de Educación Superior, vol. 1, pp. 64–74 (2018)

República-de Cuba, Gaceta Oficial de la República de Cuba. Primera Sección, vol. XI (1961)

C. Tunnermann, "90 años de la Reforma universitaria de Córdoba (1918–2008)," Consejo Latinoamericano de Ciencias Sociales, Buenos Aires, Argentina (2008)

C. R. Rodríguez Rodríguez, "La Reforma Universitaria," Economía y Desarro-llo. Universidad de La Habana. La Habana, Cuba, vol. 148, pp. 273–293 (2012)

O. Boada Carrazana, "Apuntes sobre la fundación de la Escuela y Facultad de Ingeniería Mecánica de la Universidad Central 'Marta Abreu' de Las Villas." (2006)

G. S. Fernández Levy, "Investigación sobre el origen y desarrollo de la falla en árboles de molinos de centrales azucareros," Tesis en opción al grado cientí-fico de Doctor en Ciencias Técnicas, Universidad Central "Marta Abreu" de Las Villas, Santa Clara, Villa Clara, Cuba (1984)

J. Cuendias Cobreros, "La formación de diseñadores. Lo particular en 25 años de la experiencia cubana," A3manos. Revista de la Universidad Cubana de Diseño, vol. 1, pp. 4–24 (2014)

Carlo Filangieri (1784–1867)

Marco Ceccarelli

Abstract This chapter presents Carlo Filangieri (1784–1867), who with King Ferdinando II of Bourbons was promoter and founder of the Royal Bourbon Machinery Factory in Naples at the beginning of the nineteenth century. Although he was not an engineer, but being a political-military leader, his vision and support were decisive for the development of the Royal Bourbon Machinery Factory as the first Italian plant in the Industrial Revolution. The cultural-social-technical heritage of the Royal Bourbon Machinery Factory is recognized of historical importance both in industrial production and in technical training schools and today is preserved as the Italian Railway Museum. The legacy of Carlo Filangieri can be considered in his view and supporting activities for development of industrial frames as means for society growth and welfare improvements.

Keywords History of MMS · History of machine design · History of machinery factories · History of locomotives · Carlo Filangieri

1 Introduction

Relations between Spain and Italy date back to before the Roman Empire and were vigorously revived during the Renaissance with scientific exchanges between inventors, designers and machine operators. Those relations were further strengthened in southern Italy with Spanish kings, both Aragonese and Bourbons with a communion of attitudes and developments.

Peculiar was the time of the beginning of the Industrial Revolution in the Kingdom of the Two Sicilies that was promoted by political plans of Bourbon kings, as mainly of Ferdinand II, through the first modern industrial and technical developments, supported and organized by visionary leaders. One of them and perhaps the most influential one was Carlo Filangieri, whose activity was particularly successful in

M. Ceccarelli (✉)
Tor Vergata Rome University, 10133 Rome, Italy
e-mail: marco.ceccarelli@uniroma2.it

R. López-García and M. Ceccarelli (eds.), *Distinguished Figures in Mechanical Engineering in Spain and Ibero-America*, History of Mechanism and Machine Science 43, https://doi.org/10.1007/978-3-031-31075-1_4

fostering the first modern Italian industrial factory in Italian machine design and production, by including also infrastructures for the formation of proper technicians.

This chapter is an attempt not really to discover his personality and his activities (although much must be rediscovered from the forgotten past), but rather to give visibility to that historical development that is little known in Italy as well (for various reasons that is not the case to discuss here). Carlo Filangieri is well known in Italian history mainly for his military and political careers with his activities and roles within which he addressed attention also to technological developments with personal entrepreneur attempts and governmental plans. Although those actions in technological developments are usually evaluated marginal within his profile, Carlo Filangieri can be considered a figure that well deserves attention as distinguished figure in the history of mechanical engineering since he well represents the integration of political leadership and technological vision that has enabled industrial development in Italy as in other countries for the welfare of mankind.

2 Biographical Notes

Carlo Filangieri, Fig. 1, was a military officer from an aristocratic family, who experienced political leadership in the kingdom of the Two Sicilies during the first half of the nineteenth century. His biography has been subject of interest and reviewed in detail since his autobiography (Calà Ulloa et al. 2015; Lorenzo 1997), although still today his figure is rather unknown to the large public, and in foreigner technical communities. There is quite a considerable literature on him, mainly in Italian, related to his activity and achievement as military officer and then as politician and leader for governing military corps. In particular, due to his technical-military training and career, he was also a promoter and influential politician in the technological development of the kingdom, an activity that is usually considered of less importance and effects in those biography works on him.

Carlo was born in Cava de Tirreni (Salerno), on May 10, 1784. His childhood training ended with interest towards a military career that he began in 1797. He tried to arrive in Spain as following the links of the two kingdoms, but due to the events of the French Revolution he managed to study with success at the French Military Academy which was later replaced by the Ecole Polytechnique. Thus, he began his military career in the Napoleonic army, which he served in numerous battles with success, achieving high command ranks. Later, he also served in the Kingdom of the Two Sicilies, as a State officer.

After the Restoration, the Bourbon kings kept him with political positions, also giving him the opportunity to act in his first entrepreneurial activities with his family properites. At that time, he married Agada Moncada, daughter of the Prince of Paternò, in 1820 in Palermo. For reasons of his different political vision when the king established relations with Austria, he lost political positions and between 1821 and 1837 he dedicated himself to the founding and development of his first industries within his properties. In Calabria, where a first significant technological pole was

(a) (b)

Fig. 1 Carlo Filangieri (1784–1867): **a** as general of military corps; **b** in his maturity as political leader

started in those years, (Ceccarelli 2014; Rossi et al. 2016), he settled up a metallurgical plant, a soap factory, a steam-powered floor mill, a textile factory with the intervention of technicians and materials also from other Italian states.

With the new king Ferdinand II since 1831 he recovered his military and political leadership position, and from 1831 to 1849 he was active dealing in particular with the artillery corps and engineering corps. In this period of political-technical office he encouraged new developments such as the topographical institute, the establishment of the arsenal, the military foundry, and the construction of several military buildings. In particular, he supported King Ferdinand II's plan for the foundation and development of the Royal Bourbon Machinery Factory with a theoretical-practical school for the training of technicians.

In 1849 he returned to military command to suppress an insurrection in Sicily, where he later remained governor until 1855 when he had to resign due to his divergent political positions and suspicions of fraud. He returned to Naples in 1859 from his retirement in Ischia when the new King Francisco II appointed him President of the Council of Ministers and Minister of War. But his different political views with the court led him to resign in 1860 before Garibaldi's expedition in Sicily. In the same year he had to leave the kingdom for Firenze where he remained until 1862 when he returned to Naples to participate in the funeral of his wife in Naples.

His merits and capabilities were also recognized in the new state of the Kingdom of Italy that was established with the absorption of the Kingdom of the Two Sicilies. In fact, in 1865 he was reinstated in military positions to reform the Italian army. He died on October 10, 1867, in San Giorgio a Cremano, a town near Naples.

Carlo Filangieri addressed attention to the Royal Factory continuously over time, also promoting training courses within the Royal Factory itself. This great commitment was also translated into continuous attention to the level of technical training that could promote the technological development of the kingdom, mainly with activities at the level of state structures. The school of the Royal Factory was a first example of this modern vision, which is still little known and indeed not well investigated for the cultural and social effects, beside the start of teaching a technical culture in state frames.

A plaque in Naples remembers the establishment of a technical school for the formation of machinists with the following text (Calà Ulloa et al. 2015):

> In order to make foreign labor not necessary in building steam engines, training of young Neapolitans to recover Italian inventions was organized by founding this school for machinist students by Ferdinand II, in the XI year of his Kingdom, with the governor of learned arms Carlo Filangieri, Prince of Satriano.

3 The Royal Bourbon Machinery Factory in Naples

The Royal Bourbon Machinery Factory, named and better known in Italy as "Officine di Pietrarsa" (Ferrovie dello Stato Italiano 2010; Middione 2014), began in 1830 by order of King Ferdinand II with Carlo Filangieri as advisor, as a first industrial activity for production of mechanical systems for military needs. The Royal Bourbon Machinery Factory was established by order of King Ferdinand II in Pietrarsa, Naples, in the year 1842, Fig. 2, with more wide purposes and indeed for development of the transportation infrastructures of the kingdom. But the factory began already in 1830 as a small plant serving military needs in Torre Annunzia, just outside of Naples, and then in 1837 it was located in larger plants within the same King House in Naples. Initially it was a factory only for weapons material and rather for the construction of incendiary balls that were invented by the captain Luigi Corsi. With a vision towards modernity, the king promoted the development of industry in the kingdom and one of the most advanced aspects were the plans for the development of railways transportation with a more appropriate location of a factory for production in Pietrarsa, Fig. 3, near sea water and in a place that was and is also reachable by road and easily available for railways network.

Figure 2 shows representations of the development of the industrial plant with buildings and with the first arrangements for Machinery production with an industrial settlement already planned for further development. Relevant in Fig. 2b on the right of the building is the representation of a long axis with several connections to pulleys that operate machines of various types, a large crane in the background and a steam engine at the entrance on the right.

Figure 3 shows the still existing buildings of the Royal Factory that today host the Italian National Raliways Mueum (Ferrovie dello Stato Italiano 2010; Middione 2014). From Fig. 3a with a view from the sea, the considerable extension of the factory can be appreicated with a location very close to the sea with a considerable

(a)

(b)

Fig. 2 Paintings of the buildings of the Royal Bourbon Machinery Factory in Pietrarsa, Naples at the time of its foundation in 1840s'

(a)

(b)

(c)

Fig. 3 The Royal Bourbon Machinery Factory in Pietrarsa, Naples today: **a** view of the set of buildings seen from the sea; **b** buildings aside the main factroy interna road; **c** buildings around the internal square in front of the sea

number of buildings representing the maximum expansion of the factory. Figure 3b shows the center of the factory with the two main buildings separated by a wide road which was also used for moving the locomotives and carriages under construction or outgoing for delivery. Figure 3c shows the large space that it was used to rotate built wagons and locomotives with a moving rail system that still exists today. The entrance to the land-side plant near the railway line still exists today with a specific train station called the Pietrarsa plant.

On October 3, 1839, the first railway line in Italy between Naples and Portici was inaugurated with a distance of 7.5 km as the first part of the line to Salerno, Fig. 4. Figure 4a shows the inaugural trip that was rung using the Bayard locomotive entirely built at the Royal Factory in Pietrarsa. The King and his entire family participated in the inaugural voyage both to confirm the safety of the new means of transport (differently from what occurred in other European countries!) and to support the corresponding technological development that was accepted with great enthusiasm from the population, as noted in the pictorial representation of Fig. 4a where the locomotive Bayard, Fig. 5, has been painted in detail. Just after, up to December 31, the line had already served 131,116 passengers.

The first locomotives, called Bayard, Fig. 5, and Vesuvio, Fig. 6, used on this first line, were built at the Royal Factory following the project of Armand Bayard de la Vingtrie, who was expressly contracted to design and build that first realiway line in Naples using a replica of George Stephenson's locomotive. Figure 6 shows the Vesuvio locomotive as advertised in a newspaper of the time.

In Fig. 5a the Bayard locomotive is illustrated in a photo of the time, which can be well recognized in Fig. 4a, as a construction of the project of Fig. 5b in its original drawing. Figure 7 shows a reconstruction of the locomotive that is exhibited today in the Italian National Rawilwasy Museum of Pietrarsa with all its wagons.

The success of the Bourbon project for the development of railway technology suggested immediately an expansion and upgrading of the factyory for the construction steam engines for more industrial application, and for finalized versions both for the navy and its ships and for the incipient industry. In 1842 the Royal Factory already had a large building, Figs. 2 and 3, fully operational with more than 200 employees under the direction of artillery captain Luigi Corsi (famous for the invention of incendiary balls as a maritime weapon of more 3 km range).

As early as 1843, it was proposed that projects and constructions of new locomotives with their own designs would be developed at the Royal Factory in Pietrarsa. In addition, a school was started internally to train machine technicians for the Kingdom's Naval Force and then specifically for the Royal Factory production. At the end of 1843, with the production in Pietrarsa and in collaboration with the Mongiana mining industry within the Calabria technological pole, the railway line between Naples and Caserta was inaugurated. In 1845 the first seven locomotives fully developed in Pietrarsa were finished and were named Pietrarsa, Corsi, Robertson, Vesuvio, Maria Teresa, Etna and Partenope with a well-established production, (Fang et al. 2022). These locomotive construction activities were fully manufactured in Pietrarsa in the central building that was equipped with all kinds of machinery, including

(a)

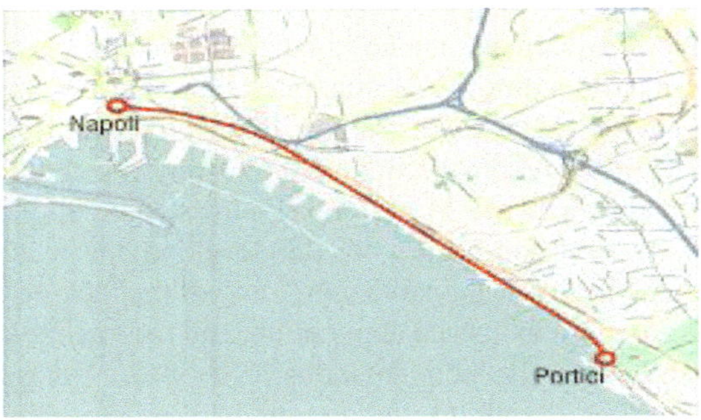

(b)

Fig. 4 First railway line between Naples and Portici designed and built by the Royal Bourbon Factory: **a** a zoomed view of the painting of the inaugural voyage on October 3, 1839; **b** current plan of the route

considerable mechanical transmission lines, large cranes, and areas for component assembly and repair, Fig. 3.

Figure 8 shows an engine from that time still exhibited in the museum as an example of the technological capacity of machinery development in the Royal Factory in Pietrarsa not only for its own necessity.

The success of the production of the Royal Factory in Pietrarsa as its own organization was also an inspiration for other countries. For example, Tsar Nicola I of Russia, after a visit to Pietrarsa in 1845, commissioned a similar plant in Russia.

(a)

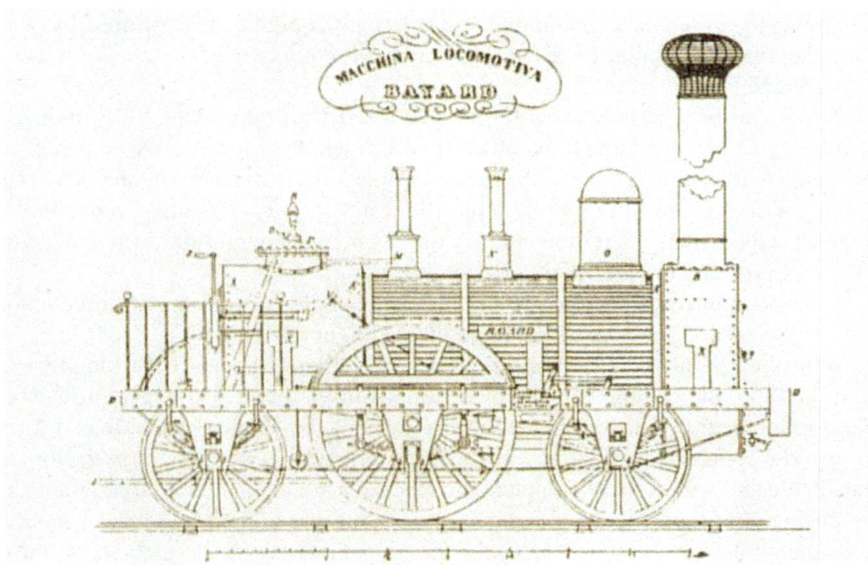

(b)

Fig. 5 The first locomotive named Bayard built at the Royal Bourbon Machinery Factory in Pietrarsa; **a** photo of the locomotive at the time; **b** original drawing

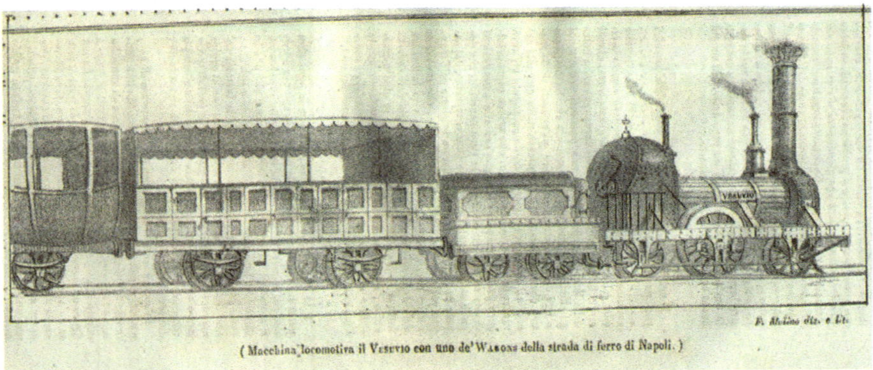

Fig. 6 The locomotive called Vesuvio built in the Royal Bourbon Machinery Factory in Pietrarsa for the line between Naples and Portici according to a drawing in a newspaper of the time

The good condition of the labor organization determined a tribute to the king with a statue made in Pietrarsa by foundry on May 18, 1852, Fig. 9. The statue was made of iron and not bronze as usual by the will of the king himself in order to celebrate the quality of the kingdom's technology (considering the high quality of metallurgy in Mongiana) and particularly the excellent production not only of material of railway in the Royal Factory of Pietrarsa.

The Royal Factory plant was further expanded with more than 700 workers in steam engine production and maintenance activities, with locomotive construction being the main activity. In 1853 the Royal Factory reaches the current 36,000 square meters of surface, resulting in the largest industry in Italy with also production of engines and components for the industry of the kingdom and for the Arny, even with export of products to other European countries.

Two locomotives are shown in Fig. 10 as examples of that quality production before 1860 that is the year of Italian reunification in one kingdom.

With the fall of the Bourbon kingdom and its annexation to the kingdom of Italy in 1860, the decline of the Royal Factory in Pietrarsa also began in favor of the development of new companies in the north of the country. Already in 1861 a ministerial commission of the new Italian kingdom indicated the Pietrarsa Factory as unsuitable and with few development prospects, decreeing a reduction program and proposing plans until its complete demolition. But the Pietrarsa Factory remained a technological development center for the Italian railway industry above all those constraints and political detrimental decisions until the beginning of the twentieth century, although its production capacities were slowly reduced, rather for state programs in favor of other factories in the north of Italy.

During twentieth century the Royal Bourbon Factory in Pietrarsa still played an important role in the development of the Italian railway system, that is still much to rediscover and revalue from its technical history, referring to construction and maintenance of lines and locomotives until its final closure in 1975.

Fig. 7 The first train of the Royal Factory in Pietrarsa as reconstructed in 1939 at the today Italian National Railways Museum during a visit by the author in 2022: **a** the Bayard locomotive; **b** the king vagon; **c** passenger wagons

(a)

(b)

(c)

Fig. 8 Steam engine of 6 HP for machinery built in the Pietrarsa office in 1846 in the Museum

(a) (b)

Fig. 9 King Ferdinado II: **a** portrait; **b** Fuso iron statue produced in Pietrarsa on May 18, 1852

(a)

(b)

Fig. 10 Locomotives built in Pietrarsa before 1860: **a** photo from the time of the 'Duca di Calabria' locomotive; **b** railway maintenance service locomotive in the current museum

Already in 1977 the Royal Bourbon Factory in Pietrarsa was considered the appropriate location for the Italian Railway Museum which was officially inaugurated to the public in 1989, [5.6], as tribute to its history with the achievemnts in the development and construction of locomotives and components for the railways in Italy.

Figures 11 and 12 show the current structure of the Italian National Railways Museum, (Ferrovie dello Stato Italiano 2010; Middione 2014), with indications also of the original functionality of the rooms in Fig. 11 and with the current arrangement of the museum material with the most significant locomotives in the Italian development. Explanatory displays include illustrative where unfortunately limited reference is made to the initial planning and development of activities of the Royal Bourbon Factory by leaders and technicians as well as to the important and pioneering activity of a specific technical school of an industrial type.

Since its foundation, the Royal Bourbon Machinery Factory in Pietrarsa has been an industrial center not only for production but also for technical training at various levels with a school in the same industrial plant as strongly planned and supported by Carlo Filangieri, that is still not well known and needs consideration due to the role it played in the success of the production of the Royal Bourbon Factory and more.

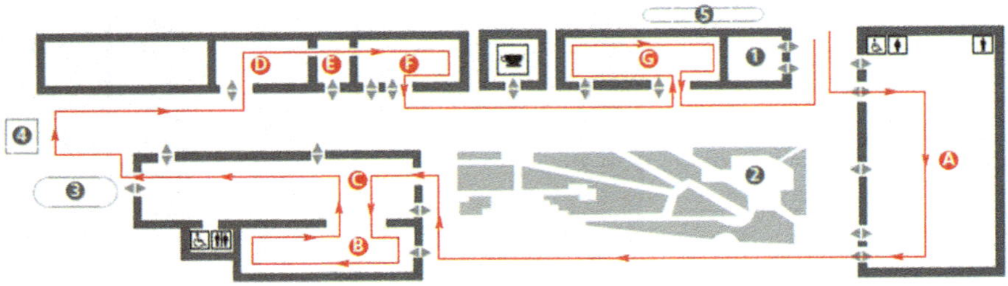

Fig. 11 Map of the Italian National Railway Museum today, in the Royal Bourbon Factory of Pietrarsa with indications of the functions of the buildings in their operation (A: assembly; B and C: furnaces and boilers; D, E, and F: area with forges, tools and production sectors; G: lathe area)

4 Legacy and Today Interpretation of Contributions

Carlo Filangieri is an emblematic figure of the first Italian modern developments, rather related to Spanish culture due to the Bourbon kingdom of the Two Sicilies. His personality blends the capacities of a political-military leader with modern visionary activities for technological promotion in the development of society and the well-being of the population. Carlo Filangieri can be considered the promoter and founder of the Royal Bourbon Machinery Factory in Pietrarsa, Naples, which has been a company of great influence in advances in the development of machine technology, not only in Italy.

His major contributions in the field of mechanical engineering and specifically in the development of machines can be recognized in the organizational aspects related to an autonomous and independent industrial design and production starting from a technological transfer from external sources and to the recognition and therefore implementation of the specific technical training for machine technology. These fundamental contributions in the starting period of the industrialization of the Kingdom of the Two Sicilies but also with a national and international vision have allowed not only an original development and prominence of the experience with the Royal Bourbon Factory in Pietrarsa but have created the conditions for more extensive developments also in other fields with a clear intention for the development of society and the improvement of the living conditions of the citizens of the Kingdom. These same contributions can be interpreted as due not only to the needs and political-strategic vision in his role as leader in governmental structures but also and above all due to his technical training received in the institution that later was the Ecole Polytechnique of Paris and to his youth experiences in the local area of the Kingdom of the Two Sicilies. Thanks to this training he was able to understand the importance and need of machine technology in the pursuit of objectives and promotions with incisive developments in much wider areas than the military ones.

Considering the above, the legacy of Carlo Filangieri can be recognized in his vision of considering machine technology and technological development in general, of primary importance for not only an improvement of society but also for the development of cultural, technical and social aspects as based precisely on the technology

(a)

(b)

Fig. 12 The Italian National Railway Museum today, in the Royal Bourbon Factory of Pietrarsa: **a** interior of the machinery building, with exhibition boxes; **b** main building with main locomotive collection form nineteenth century

of machines in the processes of the industrial revolution already started in the countries of Northern Europe with which the experience of the Royal Bourbon Factory in Pietrarsa was in a certain way confronted with results that can certainly be defined as emblematic and of success. Another aspect of the legacy left by Carlo Filangeri can be recognized in the fact, not always properly considered, that technical training, in particular those on machines, can result and was above all in the period of the industrial revolution of fundamental importance and efficiency in the development of the technique with significant repercussions on the economic and social frames of society. His technical training combined with his military career has allowed him an organizational capacity for the development of the technique with an entrepreneurial vision, with modern characteris that are still current today, experienced both at the state governament level and at a personal level as can be recognized from what is reported in the biographical notes.

References

P. Calà Ulloa, Carlo Filangieri, D'Amico Editore, Cava de' Tirreni, 2015. ISBN 9788890904936

R. De Lorenzo. Filangieri Carlo, Biograhic Dictionary of Italians, , Enciclopedia Treccani, Turin, Vol 47, 1997. (in Italian)

M. Ceccarelli. Early Machine Industry in the Kingdom of Two Sicilies in 19-th century, XX Congreso Nacional de Ingenieria Mecanica AEIM, Malaga, 2014, CD Proceedigns, paper 06-02.

C. Rossi and M. Ceccarelli, Science, Technology and Industry in Southern Italy Before the Unification, in: Essays on the History of Mechanical Engineering, History of Mechanism and Machine Science 31, Springer, Dordrecht, 2016. pp.159–180. https://doi.org/10.1007/978-3-319- 22680–4_10

Ferrovie dello Stato Italiano. Brochure: The National Museum of Railways in Pieterarsa, Roma, 2010 (in Italian)

R. Middione (editor), Guide: The National Museum of Railways in Pieterarsa, Prismi editirce. Napoli, 2014. (in Italian)

Y. Fang, M. Ceccarelli, Y. Chu, Earliest Locomotives in Italy and China from the Perspective of Technology Transfer. In: Niola, V., Gasparetto, A., Quaglia, G., Carbone, G. (eds) Advances in Italian Mechanism Science. IFToMM Italy 2022 pages 25–33. Mechanisms and Machine Science, vol 122. Springer, Cham (2022). https://doi.org/10.1007/978-3-031-10776-4_4

Eduardo Giró Barella (1940—Present)

D. Abellán-López and M. A. Oliva-Meyer

Abstract In the 60s and 70s of the twentieth century, the Spanish motorbike industry, which has almost disappeared today, was a world reference, first among the light two-stroke road motorbikes and later in the off-road motorbikes. In this context, the figure of the industrial engineer Eduardo Giró (1940) stands out, linked to the OSSA factory until the end of the 1970s, and who is known worldwide for having created a revolutionary motorbike: the 250 cc OSSA GP with the first magnesium monocoque chassis in a Grand Prix motorbike, which became the fastest single-cylinder motorbike of its time. This bike was able to challenge the world's major motorbike manufacturers and fight for the 250 cc world championship in the late 1960s, the bike was a compendium of innovations in the field of materials, chassis, suspension and engine. He created the OSSA trial champions and went on to develop the Cobas engines with which Crivillé won his first 125 cc world championship. His activity as an engineer was not limited to competition, but he was responsible for the technical management of OSSA and, finally, general manager of this brand in the years of greatest boom of this Spanish industry. This chapter pays a well-deserved tribute to the Spanish motorbike industry, which has now disappeared, through a figure of worldwide prestige.

Keywords History of MMS · History of IFToMM · Motorcycling · Engines · 2-stroke · Model airplanes · OSSA

D. Abellán-López (✉) · M. A. Oliva-Meyer
Miguel Hernandez University of Elche, Elx, Spain
e-mail: dabellan@umh.es

M. A. Oliva-Meyer
e-mail: ma.oliva@umh.es

R. López-García and M. Ceccarelli (eds.), *Distinguished Figures in Mechanical Engineering in Spain and Ibero-America*, History of Mechanism and Machine Science 43, https://doi.org/10.1007/978-3-031-31075-1_5

1 Biographical Notes

The figure of the industrial engineer Eduardo Giró is certainly the most relevant in the Spanish motorbike industry, which had its golden years between the 60s and 70s. From this period there remains today mainly the presence and dominance of Spanish riders in speed competitions (MotoGP), Trial and Raids, but the Spanish motorbike industry has practically disappeared. As head of the Technical Department of OSSA and with his collaboration in the JJ-Cobas team, he was responsible for numerous innovations and the development of mythical motorbikes for road, off-road and competition motorcycling.

Eduardo Giró always says that he was in the right place at the right time. Let's take a look at the circumstances that allowed the protagonist of this chapter to develop such a fruitful and creative career as an engineer from the very beginning of his professional activity.

1.1 Brief History of the Spanish Motorcycle Industry

After the Spanish Civil War, the need for private means of transport was pressing, in the 1940s the only vehicles available to the general public were bicycles. The Spanish car industry was developed at the end of the 1950s, making the car increasingly accessible to a wider public as the 1960s progressed. Thus, from the end of the 1940s and throughout the 1950s, the great window of opportunity opened for the motorbike industry in Spain. During these years more than a hundred manufacturers of different types of motorbikes flourished in Spain, some making their own engines and others assembling motorbikes with Hispano Villiers or A.M.C. engines. Although there were manufacturers all over the country, most of them were concentrated in

Fig. 1 Eduardo Giró Barella

the metropolitan area of Barcelona. The vast majority of Spanish motorbikes had 2-stroke engines, which were simpler and cheaper to manufacture than 4-strokes. The 2-strokes had found their development path in the patents of the German manufacturer DKW which, as war reparations, were now in the public domain.

Three of the most successful and world-renowned brands were born of the friendship between three Catalan industrialists: Manuel Giró, Pere Permanyer and Francisco Bultó. The first of these, who as manager of OSSA (whose main activity was the manufacture of film machines) had already manufactured outboard motors for Ricardo Soriano in the 1920s and 1930s, promoted the development of the first post-war Spanish motorbike prototype, a 125 cc 2-stroke single-cylinder which was presented at the Barcelona trade fair in 1942. For various reasons this prototype, the tooling they had for its manufacture and some components were ceded by Manuel Giró to Permanyer/Bultó and was the origin of the first Montesa in 1945. OSSA finally presented in 1947 the prototype of its 125 motorbike with an engine inspired by the DKW RT 125. At the end of the 50's the differences of criteria between Permanyer and Bultó about the withdrawal of the Montesa racing team resulted in the birth of Bultaco.

The 1960s saw the irruption of the automobile in a market eager for this means of transport and a critical situation for motorbike manufacturers. During these years, the great majority of Spanish manufacturers disappeared, leaving in the Barcelona area: OSSA, Montesa, Bultaco and Derbi (very focused on mopeds), all four with 2-stroke engines of their own design, Sanglas and Ducati Mototrans as manufacturers of motorbikes equipped with 4-stroke engines. In the rest of Spain, they were built under licence: Vespa (2 T) in Madrid, Lambretta (2 T) in Eibar and MV(4 T) and Puch(2 T) brands in Gijón by Avelló. Other manufacturers, such as Moto Guzzi Hispania in Seville, gradually focused on the manufacture of mopeds (Herreros and Aznar 1998; Merino 2019).

2 Eduardo Giró: His Origins

Eduardo was known in the motorbike world as "el Tècnic" (the technician in Catalan), perhaps because his academic background and his methodical way of working showed that he was much more interested in technique than in racing. He was highly trained as an engineer, unlike other motorcycling protagonists of that time, and had the ability to transfer this knowledge to the day-to-day running of racing motorbikes. At a time when the racing teams were far from what they are today, Eduardo Giró implemented a working system that made it possible to incorporate great innovations and systematize the work of other great engineers such as Antonio Cobas. No one who knows the history of motorcycling can deny that Eduardo Giró was one of the great figures of mechanical engineering of his time.

Eduardo Giro Barella was born in Barcelona in 1940 into a family of Catalan industrialists. Originally, the Giró family was linked to the textile industry until 1928, when Eduardo Giró's grandfather, Joan Giró Prat, founded Orpheo Sincronic

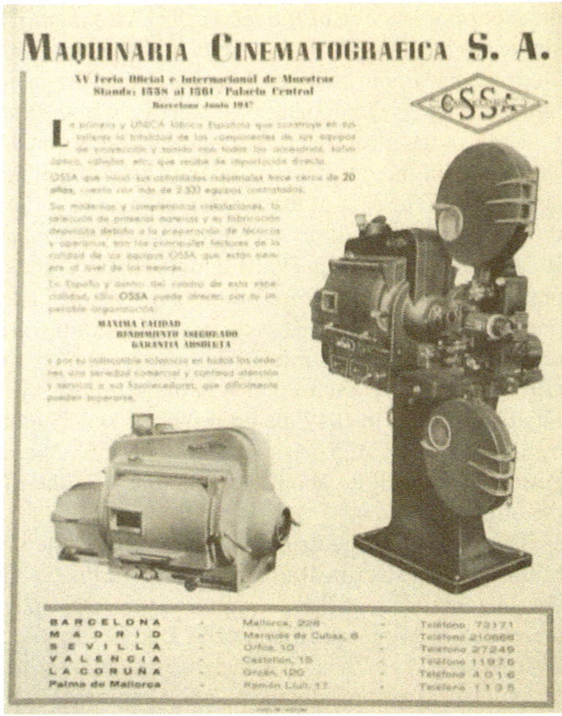

Fig. 2 Information leaflet of Maquinaria Cinematográfica, S.A., 1947

Sociedad Anónima. The company was initially dedicated to the manufacture, first of cinema projectors and, with the appearance of sound film, also of sound equipment.

After the Spanish Civil War, the company's name changed to Maquinaria Cinematográfica, S.A., although it kept the acronym OSSA as its trade name. It was not until the 1950s that Ossa also began to produce motorbikes. Manuel Giró Minguella, the son of the company's founder and father of Eduardo Giró, was responsible for the start of Ossa's career in the world of motorbikes. Eduardo's father was a merchant seaman and a great fan of motorcycling. In fact, he even competed in some motorbike and motorboat races. After getting married, he left his job as a merchant seaman to join the family business. The origin of Ossa's motorbike manufacturing was the agreement that Manuel Giró reached with the multifaceted businessman Ricardo Soriano in the 1930s to produce outboard motors for him at the Ossa factory. The two businessmen knew each other from the time when Manuel Giró took part in motorboat racing competitions. Although, as already mentioned, Manuel Giró created his first 125 cc motorbike prototype in 1940, various difficulties meant that Ossa did not start marketing motorbikes until 1951.

Eduardo Giró grew up in a family of industrialists dedicated to film machines and motorbikes. Like other companies of the time linked to motorbikes, OSSA was a family business, but unlike the Bultó family, the Giró family did not share a love of motorbikes.

> In my house there was little talk of motorbikes because my mother didn't like them. It wasn't like the Bultó family, a shared hobby. My father raced sidecars until he broke a bone at Montserrat. My mother said that racing was over.

Eduardo Giró was forbidden to ride a motorbike. His last "authorised" experience with a motorbike was as a teenager when he was on holiday in Mallorca. His father sent some 50 cc Motopedal produced by Ossa in the 1950s, popularly known as "Ossitas". Giró's mother found out from some neighbours about Eduardo's enthusiastic use of the Ossitas around the island and forbade him to ride a motorbike again. From then on, his few contacts with two-wheeled vehicles were in secret.

As he was not allowed to ride a motorbike, despite the activity of the family business, Eduardo started a hobby with aeromodelling engines. When he was about 12 years old, he began to build model aeroplanes at home. This hobby, self-taught by studying texts on aeronautics, engines, fluid mechanics, aerodynamics, materials,

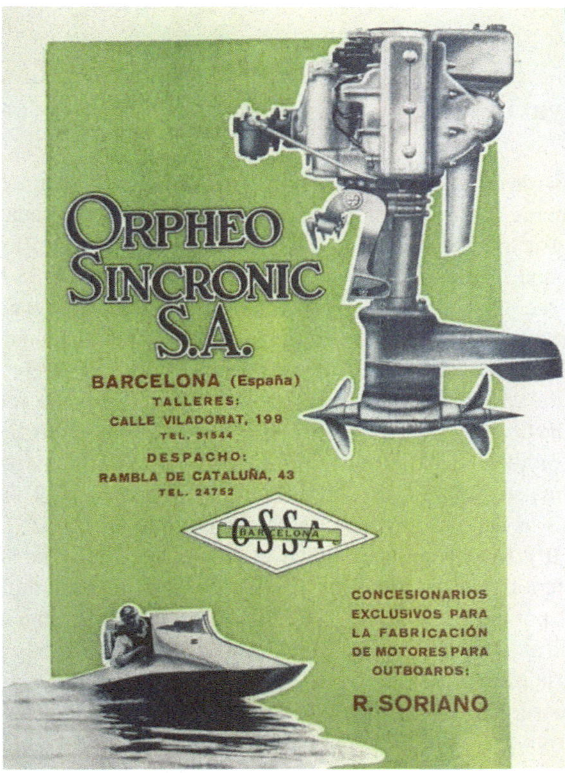

Fig. 3 Poster advertising the outboard motors produced by Ossa for R. Soriano. *Source* https://www.soriano-outboard.com/history/

etc., laid the foundations that allowed him to design his own aeromodelling engines when he was only 17 years old. In his room he had a lathe, a milling machine and a drill with which he was able to manufacture his engines, and when he needed some special process, such as heat treatments, he took the small parts to OSSA. He began to study Industrial Engineering at the School of Industrial Engineers in Barcelona, without abandoning his hobby of aeromodelling, even taking courses in gliding. Such was his enthusiasm for aeronautics that he intended to travel to the USA when he finished his degree to work in an aircraft factory. He even had some contacts at the Pratt & Whitney engine company, although he finally stayed in Spain. It is interesting to note that this hobby and knowledge of aeronautics is common to another of the great figures in the development of the 2-stroke engine, Walter Kaaden (MZ engineer) who came from the development of missiles in Peenemünde during the Second World War.

At the age of 24, while continuing to study industrial engineering, he joined the prototype department of the family company encouraged by his father, at OSSA he did not have access to aircraft engines, but he did have access to motorbike engines. His knowledge and love of the aircraft industry was later reflected in some of his motorbike designs.

3 His Arrival in Ossa

By the time Eduardo arrived in the prototype department of Ossa, the family company had been producing a utility motorbike, the OSSA 160, for a year, which was not selling as well as expected. It was a motorbike of modern conception whose design was commissioned to the Italian engineer Sandro Colombo. OSSA manufactured motorbikes of recognised quality and reliability, but the motorbike market was changing its orientation towards more performance models. Eduardo Giró undertook the task of creating a sportier motorbike that could compete in performance and sales with the Montesa Impala Sport and Bultaco's Metralla models. From the Colombo[1] engine of the 160 (10 HP) he created a sportier, larger displacement engine, redesigned the motorbike so that the chassis could support the increased power and look sportier, thus creating the Ossa 175 Sport (19 HP) in 1964. This motorbike had better sales and was followed by the Ossa 175 Sport Special (21 HP) and in 1965 the Ossa 230 (24 HP). In the engineer's own words, at that time (the 1960s) the usual method to create a larger displacement engine "was to increase the diameter of the piston, which is not entirely correct". He also stated that, given the need to keep the Colombo engine as a base (for economic reasons), he felt "his hands were tied" when he designed these motorbikes.

Initially, Eduardo Giró carried out his developments from the prototyping department and not from the engineering department. One of the advantages of working

[1] This is the name given to the engines developed by Sandro Colombo, an engineer who also worked for Gilera and Lambretta.

in the family company was that they allowed Giró a certain amount of freedom: he proposed certain changes to the designs which, if they worked, which was quite common, were put into production. It could be said that he had the good fortune of joining a family business whose activity was very much in line with what he liked to do, and moreover, at a time when the company needed a certain amount of young blood.

4 The Ossa Monocoque

The design for which Eduardo Giró is known worldwide is undoubtedly the Ossa 250 VR (rotary valve), commonly known as the "Ossa Monocoque". This motorbike, ahead of its time, was a technical feat that still produces admiration today in those who scrutinize its technical details. The Ossa 250 Grand Prix was used by the legendary rider Santiago Herrero to fight for the motorbike world championship in the 250 cc category, battling on equal terms with the great brands of the time (Herreros and Aznar 1998).

Although it was not the most powerful bike of the championship, with it Santi Herrero managed to beat the Yamaha (2-cylinder 2-stroke), Benelli (4-cylinder 4-stroke), MZ (2-cylinder 2-stroke), some of which had more power. The main strengths of this bike were its extraordinarily stiff and light magnesium alloy monocoque chassis and its single-cylinder engine with a wide range of use and very reliable.

In the design of the Ossa GP, Giró did not follow the usual lines of work, but following his convictions he wanted to develop the whole bike together, as a single homogeneous set. According to Eduardo Giró, this bike was not born to compete, but

Fig. 4 The OSSA monocoque and its rider Santiago Herrero in Opatija, Crocatia, 1970. *Source* https://chopperon.com/50-anos-sin-santiago-herrero/

was a "technical exercise". After the success developing the Ossa 175 and 230 cc, Giró was given carte blanche to develop an experimental racing engine, something that Eduardo had had in mind for some time. Those motorbikes, which were designed starting from the series models, obtained important sporting successes such as the Monjuic 24 h in 1967 where they obtained 1st place overall with the OSSA 230 (with 35 laps ahead of the 2nd classified: a Triumph 500 cc). For the conception of the Ossa GP, Eduardo Giró did not start with any kind of conditions, only with the economic limitations that a small factory like OSSA had.

> At that time, the chassis was designed on one side and the engine on the other. We decided that this could no longer be the case and we built a homogeneous unit. A lot of people from the factory took part and they were very enthusiastic because it was something really innovative. At that time, it was something so advanced that you could believe in it or not, because nobody had done anything like it.

> I was lucky enough to be in the right place at the right time. If I wanted to make a three-wheeler, I did it, I didn't even have to ask for permission.

4.1 The Engine

The story began with the engine. Originally the engine was not designed for Grand Prix racing. The engine was derived from a 400 cc engine that Eduardo Giró conceived with American desert racing in mind. That engine was not intended to be marketed, but was a "technical exercise" carried out on his own initiative and for the pleasure of designing an engine without the constraints of a previous engine. It was the first motorbike engine that he had designed completely from scratch, without any technical constraints. In 1970 the number of gears and cylinders of the GP 250 bikes was limited. It was then that Eduardo Giró considered the design for this category. In the initial version, the air-cooled cylinder had the exhaust to the rear (like the MZ designs) and the clutch in an oil bath. Given the heavy workload of the OSSA draughtsmen, this engine was completely drawn by Giró, who did all the exploded views. Normally he would draw a general plan of the engine or the assembly and the draughtsmen would take care of the complete exploded views. The engine materialized and the first time it was put on the dyno Eduardo Giró, who was not present, was called to see what had happened. He feared that something had broken, however, the reason for the call was that the engine gave a power output of 35 HP when the best 250 cc competition engines they had developed at OSSA were under 30 HP. At this point Giró understood the potential of this engine in racing.

This first version of the 250 cc engine was tested in a hill climb race in the Desierto de las Palmas (Castellón province) by the Spanish endurance champion rider Pedro Millet, better known as "Petrus", showing great performance. In the first tests, the new 250 cc engine was mounted on a 230 chassis. The nearly 40 HP of the new engine was too much for the rigidity of the 230 chassis, which was used to support less than 30 HP. This meant that the bike was not exactly easy to ride in a race. In fact, before Petrus the bike was tested by Ossa riders Luis Yglesias and Carlos Giró, who

Fig. 5 Detail of the 250 cc air-cooled engine of the monocoque OSSA (Polo 2002)

described it as unwieldy and even dangerous. Petrus, while criticizing the chassis, claimed that it was the fastest bike he had tested. This rider's opinion was important for El Tècnic as Petrus was a rider with a good technical background and had ridden the 250 GP Montesas. Giró had hit the nail on the head with the engine, a chassis for that engine was missing.

> We already knew that the chassis was no good, because it was very similar to the 230 cc chassis, and obviously it was too little chassis for the almost 40 HP that engine gave. But when Petrus got off the bike he told us: "this bike runs much, much, much better than any other 250 cc bike out there". We were all surprised because at that time everybody was talking about a lot of power. It must have been that other people's horsepower didn't run as fast as ours.

Thinking about the version with his own chassis, Eduardo Giró made the main modification to this engine, which was to orientate the exhaust towards the front. The exhaust was approximately 1.3 m long and occupied a considerable volume, so that it could be directed forward, allowing it to be tucked under the engine, leaving more freedom in the design of the chassis (the MZs with a unit displacement of 125 cc did not have this problem, their exhaust was about 90 cm long). Another important modification was to change the oil bath clutch to a dry clutch. The displacement of 248 cc was achieved by a piston diameter of 70 mm and a stroke of 64.5 mm (Polo 2002). The cylinder was made of light alloy with a cast iron liner. He wanted such a large exhaust port that in the end he opted to split it in two. It had two side transfers plus a secondary one at the rear which also helped to cool the connecting rod small end and piston. The single piston ring was designed by Giró and cast by the Tarabusi company. The piston pin clips and the cage for the connecting rod big end were specifically designed. In the case of the cage, there was none that could withstand the forces to which it had to be subjected at the 11,500 rpm at which the

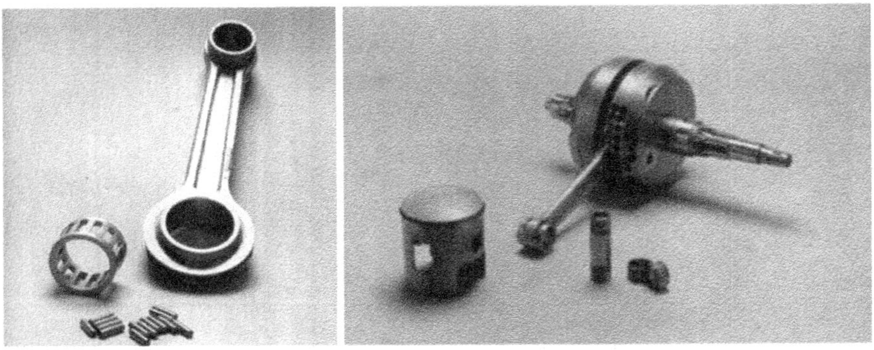

Fig. 6 Some of the engine components. Left: connecting rod together which are cage bearing designed specifically for the engine. Right: crankshaft with balancing holes (Burgaleta 1989)

engine spun. As explained by G. Jennings and expressed by Giró himself, the secret of these 2-stroke racing engines lies in the design of the connecting rod with its guide and in the piston crown and piston ring (Jennings 1973).

The crankshaft and connecting rod have an independent lubrication system made by means of bores in the crankshaft that create a circuit for the oil (Burgaleta 1989). This system allowed direct lubrication of the connecting rod big end bearing which was tightly enclosed in the crankshaft to achieve perfect guiding of the connecting rod. Initially, Giró tried to make the lubrication system work with changes in crankcase pressure, a technique he had already tried on aeromodelling engines for fuel supply. This system did not allow the oil flow to be adequately regulated and a commercial Mikuni pump was chosen. The most important feature of the Giró engine was that it was fitted with a rotary valve. This type of intake system for two-stroke engines was already used in the 1950s on German and Czech racing motorbikes. The advantage of the rotary valve over the traditional piston-skirt intake is that the opening and closing angles do not have to be symmetrical with respect to top dead center, but the ideal timing can be chosen.

The cylinder head was made of light alloy with a hemispherical combustion chamber. The compression ratio could be varied by changing the head gasket, but was usually around 13.8:1. The crankshaft was also designed specifically for this engine, with internal lubrication ducts and perfectly balanced, which was difficult to achieve considering that the rotary valve was attached to the crankshaft.

4.2 *The Chassis*

The monocoque chassis was made of magnesium. This was not a material frequently used in motorbikes, but it was used in aviation. It was not the first time that "el Tècnic" worked with this material: he had already used it in his youth in his model airplanes.

Some European brands had previously tried to make monocoque chassis without success. Eduardo Giró used his great talent to design a greatly effective monocoque chassis. The engine was hanging from the chassis, being very accessible. The fuel tank was integrated into the chassis and at the rear, under the saddle, was the oil tank. One of the main difficulties they encountered was that they had never previously welded the magnesium alloy which can catch fire extremely easily (Personal communication 2019). At OSSA they designed and manufactured a TIG machine and were able to solve the problem with in-house operators. The magnesium sheet was 1/4-inch-thick at the thickest parts and 1/8 inch thick at the thinnest.

The result was an extraordinarily rigid chassis weighing about 7.4 kg. Two chassis were produced: one with an 18-L tank and one with a 23-L tank. The large one would only be used in long races (such as the Isle of Man TT) to avoid refueling, as the rider Santiago Herrero preferred the small one. This was better suited to his 1.59 m height. Carlos Giró, Eduardo's cousin, in the Spanish 250 cc road championship, used the "big" chassis assiduously. Carlos won the Spanish championship in 1969, and he also took part in some World Championship races with Santiago Herrero. The final weight of the bike was around 90 kg in the air-cooled version and 100 kg in the water-cooled version.

Fig. 7 Above: the OSSA GP without fairing. Below: Eduardo Giró witnessing the construction of the monocoque chassis (Burgaleta 1989)

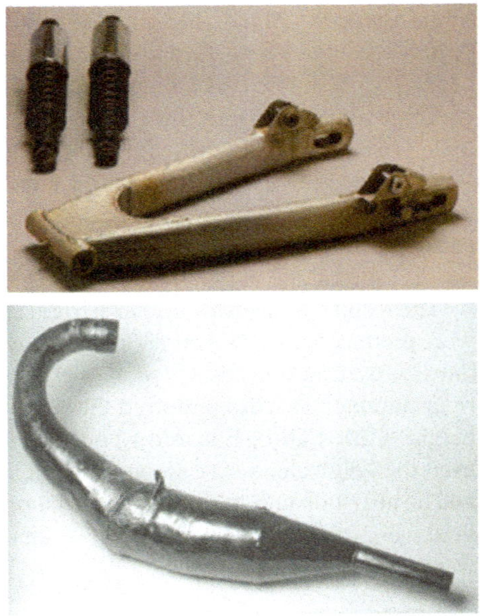

Fig. 8 Other monocoque components: Left: Girling rear shock absorbers and oval section swingarm. Right: exhaust pipe (Burgaleta 1989)

4.3 *Other Components*

The engine had generous cooling fins. Actually, for a single cylinder engine it was a wide engine. The crankcase was also profusely finned, from his experience at the 24 h of Montjuich Giró realized that keeping the crankcase cool was essential so that the power did not drop throughout the competition. On the 230 endurance race winning motorbike he had painted the inside of the crankcase with a resin and asbestos based insulating paint made by a paint manufacturer specifically for this purpose. The air-cooled engine proved to be reliable, with a blind hole in the cylinder head near the spark plug which allowed a temperature probe to be inserted to adjust the carburation. This solution was more accurate the closer the pits were to the end of a "flat-out" straight and the less the temperature readings dropped at the pit lane entry. A solution that was copied by Yamaha's technicians for tuning their bikes. In order to keep the engine performance constant, Eduardo Giró began to develop a water-cooled version of the engine. The crankcase of this engine was made of magnesium, instead of the original aluminum one, and it was even mounted in the "big" chassis, achieving good results in the hands of Carlos Giró, as mentioned above. It must be said that the water-cooled engine was not used by Santiago Herrero in the world championship races, but it would have been the evolution for the following season.

Also striking is the oval section swingarm designed by Giró for this motorcycle, extraordinarily rigid, made of chrome-molybdenum steel with a vertical internal reinforcement and a weight of 3.1 kg. Both this material and the aeronautical magnesium sheet metal were imported from the U.S. through the importer of the brand in this country: Yankee. As for the suspension, the bike was initially fitted with a Telesco front fork, a brand closely related to OSSA. Eduardo Giró, once again inspired by the aeronautical industry, conceived an oil-pneumatic rear suspension for the Ossa monocoque. With the help of Telesco, he developed springless shock absorbers. These had a membrane that separated the oil from the gas (nitrogen) and valves with which to independently tune the rebound and compression stiffness. This system when it worked provided a magnificent performance, but lacked development giving problems in the seals so that air entered and mixed with the oil producing an emulsion and losing efficiency. After several races, in the 68 Spanish Grand Prix when Santiago Herrero was leading the race, the rear suspension lost performance causing the driver to fall. These failures, the long time required for tuning, and the lack of interest of Telesco to continue the development, motivated the change to Girling shock absorbers, the leading brand of that time. In this respect Eduardo Giró introduced another innovation by fitting a hydraulic rotary type steering damper instead of the traditional friction type. This was a component of his own design that obtained good results. Finally, in order to achieve maximum competitiveness, the front fork was replaced by a Ceriani with a double front brake designed for competition.

In the fairing of the Ossa monocoque, Eduardo Giró was once again ahead of his time, introducing techniques from the aeronautical industry into motorcycling. Taking advantage of his knowledge in aerodynamics Giró optimized the fairing starting from a NACA airfoil. The design was inspired by the fairings of radial aircraft engines. He knew that using this type of geometry could improve aerodynamic drag. It was necessary to meet the requirements of the regulations that did not allow a fairing on the rear of the motorcycle and that forced the complete wheels, hands and legs of the rider to be seen. He moved the fairing as far forward as he could and made an effort to combine improved penetration with air channeling to cool the cylinder.

In order to study the improvements "el Tècnic" made several tests on circuit. In these tests the bike was put at 180–200 km/h accompanied by a car that allowed to observe the bike from up close. The fairing was covered with strands of wool glued

Fig. 9 The mechanic Esteban Oliveras, Santiago Herrero with the Ossa GP, Eduardo Giró, and the Seat 1500 pickup, Monza. *Source* http://btc-built-to-go.blogspot.com/2010/06/santiago-herrero-40-6-06-1970.html]

Fig. 10 Santiago Herrero's first test with the monocoque, March 1968. From left to right: Eduardo Giró, with his father Manuel Giró on his back, Esteban Oliveras crouched next to them and Santi Herrero with helmet (Walker 1986)

Fig. 11 The Ossa monocoques sleeping in the hotel room before the Grand Prix of Yugoslavia (Burgaleta 1989)

that allowed to observe the air flow. He even put this wool in the interior ducts to study the air flow that cooled the engine. He also applied his knowledge in this field to the optimization of the carburetor inlet nozzle that improved the flow compared to the commercial ones.

Fig. 12 OSSA Enduro AE-71 250 of 1971. It became famous for appearing in the movie " The Summertime Killer ", by Antonio Isasi. *Source* "Creative Commons" by Peprovira under license BY-SA 3.0

4.4 The OSSA Monocoque with S. Herrero in the World Championship

Santiago Herrero previously competed with the Spanish brand Lube until its closure due to financial problems in 1965, achieving a second place in the Spanish 125 cc championship. After a season racing as a privateer he joined Ossa. In the Spanish championship of 1967 the Ossa were unbeatable, but among his riders Santiago Herrero stood out and finally took the title in the 250 cc category. This made him the one chosen from among the riders of the house to ride the Ossa monocoque in international competitions. In addition, he had experience as a mechanic, as he worked as such before and during his time as a rider. This facet made him a great help in the development and tuning of the innovative bike.

> Santiago was a very good racer, but he also not only raced, he knew. He had previously been a test rider for Lube, and he helped set the bikes up, he knew the mechanics and he knew how to solve problems.

After testing the bike on some tracks near the Prat airport (Barcelona), the results could not have been better. The bike was fast and stable, the chassis went well, and the rider was able to get its full potential. They had the rider, the bike and the team, the only thing left to do was to go to the races.

The team was formed by Eduardo Giró, Esteban Oliveras as mechanic, Santiago Herrero as driver and a Seat 1500 modified as a pick-up that transported the bike, spare parts and team members. Nothing to do with the luxurious and comfortable motorhomes that can be seen today in the motorcycle world championship.

In the mid 60's the World Championship was dominated by the Japanese and Italian brands with the addition of the East German MZ. At that time, the dominance of 2-stroke engines was beginning (especially in the lower displacement classes). The bikes that were eligible to win a Grand Prix were official team bikes with multi-cylinder engines and more than 6 gears (an extreme example would be the

Fig. 13 Above: Mick Andrews riding an Ossa MAR in 1971. Below: Ossa Mick Andrews Replica 250, 1974 version [catalog photo]

Fig. 14 Eduardo Giro and Antonio Cobas with the Kobas MR1 Grand Prix

Suzuki 50 cc with 3 cylinders and 15 gears). In this context, the more modest brands had practically no opportunities to compete (Kaaden's MZ were an exception). The presence of Spanish brands was practically limited to Bultaco and its 2-stroke single-cylinder TSS (one of the first "racing-customer" bikes). The Bultacos in the hands of private riders "could be among the first 6 classified in a GP" in different cylinder capacities according to F.X. Bultó. This brand was the first Spanish to win a Grand Prix (Ulster GP in 250 with Ginger Molloy in August 1966) and the first to give victory to a Spaniard in GP's (Spanish GP in 125 with Salvador Cañellas in May 1968).

In this year 1968 the OSSA 250 GP debuts in the world championship, with Santiago Herrero as pilot, the season can be summarized as follows:

Debut of the bike in the winter GP of Alicante (not scoring for the world championship but that gathered the great international figures of the moment) with abandonment (brake problems and experimental rear suspension) but only surpassed in speed by the official Yamaha 4-cylinder. This year Santiago traveled alone to the Grand Prix, acting as mechanic and driver, in the 1st GP in Nürburgring (RFA) he finished 6th, in the following GP in Montjuich he fell because of suspension problems and Giró decided to abandon the hydropneumatic system. He traveled alone to the Tourist Trophy (TT) of the Isle of Man, where he had support from the OSSA importer and finished in 7th position. His bike and his riding caused a sensation, he was awarded the prize for best rookie of the year and the British press called him "Spanish Flyer". Santiago Herrero was 6th in Assen (Holland), 5th in Spa (Belgium) and 3rd in Monza (Italy) obtaining his first podium. It should be noted that, except for the suspension problems, the bike showed great reliability, even in ultra-fast circuits like Spa, a circuit where the bike went accelerating full throttle, taking advantage of the slipstreaming, a large part of the time.

For the 1969 season, OSSA forms the aforementioned structure with Giró and Esteban Oliveras. With the experience of the previous year, the withdrawal of the 4-cylinder Yamaha (they switch to 2-cylinder), and the knowledge of the circuits by Santiago Herrero, the prospects are good. In the winter race in Alicante they win without opposition. The first GP of the season at Jarama (debut of this circuit in the world championship) they win under the rain. The next race in the very fast Hockenheim (F. R. Germany) OSSA fought for the first positions with the twin-cylinder Yamaha and MZ, until the electronic ignition broke. At Le Mans (France) he won with 41" advantage over Gould's Yamaha. He was third at the Isle of Man ("Herrero the Hero!" the British press called him). Also third at Assen and first at Spa at an average of 190 km/h, proving that the OSSA monocoque was capable of beating more powerful bikes on ultra-fast circuits. In the following GP at Sachsenring (GDR) he qualified in 2nd position. With the results obtained Santiago Herrero was leading the championship. During the race in Brno (Czechoslovakia) they were supplied with low octane gasoline which produced the first engine failure. In Imatra (Finland) Santiago Herrero crashed fighting for the first position and restarted the race arriving 6th. In the Ulster GP he crashed in the rain breaking his hand and the season was definitely complicated. A victory in the two remaining GP's would have given him the world championship, but in Italy he raced with a badly injured left hand and could not do a good result. His last chance to win the World Championship was in Opatija (Yugoslavia) but he crashed when he stepped on a wet line due to a light drizzle. The 250 cc World Championship was won by Carruthers with the Benelli 4-cylinder 4-stroke (last victory of a 4-stroke engine in this category). Santiago Herrero was 3rd in the 250 world championship and OSSA also won 3rd place in the constructor´s category. This year was the first year that a Spanish motorcycle and a Spanish rider won the world championship: Derbi with Angel Nieto in 50 cc. For this victory, the help of Santi Herrero, who raced the third Derbi 50 cc of the team, was fundamental (Willoughby 1982).

In 1970 the new regulations came into force, the rivals would be the twin-cylinder Yamaha and MZ. The season reached the Isle of Man with three races: a victory, a second place and a breakdown. For Santiago Herrero winning the Tourist Trophy was almost as important as winning the World Championship. In the race, after an inconsequential fall, the OSSA rider started a comeback and suffered a second crash that caused his death. Santi was only 28 years old. The Tourist Trophy circuit claimed the lives of six riders that year.

Fig. 15 Cobas BMW K100R, known as "the shaver" because of its sponsor [Source: www.motosclasicas80.com]

Fig. 16 Official presentation of the Braun JJ-Cobas BMW team. The first two from the left are Eduardo Giró and Antonio Cobas. *Source* https://elgeniocobas.jimdofree.com/historia/nace-jjcobas/la-jjcobas-bmw/

Herrero's death was a hard blow for Ossa members, especially Eduardo Giró and mechanic Esteban Oliveras. The three of them had gone around the world fighting for the world championship, taking on the big brands with a bike that became a legend along with its rider. The project of competing in the world championship and, to a large extent, the development of the bike had revolved around Santiago Herrero. After the loss of the Madrid-born rider, Eduardo Giró decided to leave the World Championship.

> We raced for Santi (Santiago Herrrero) and we won't race again without him.

To see a Spanish rider climb to the top of the podium in the 250 cc category, we had to wait 14 years until Sito Pons did it, precisely with the JJ-Cobas team of which Eduardo Giró would be a member.

The OSSA 250 GP caused a sensation. Vic Willoughby (one of the great specialists in GP motorcycles) interviewed Giró in '69. He was particularly impressed by the size of the carburetor (42 mm) and the exhaust. The leading 2-stroke engines in the Grand Prix had unit displacements of 125 cc (MZ) or less, and the proportions were different. Giró told him that this was necessary to ensure equivalent volumetric performance (the areas of the ports increase with the square of the dimensions, but the volume increases with the cube of the dimensions). He also commented that he would have preferred a 45 mm carburetor but that it was not available. He emphasized the size of the cylinder fins, the wide range of use of the engine (from 6500 to 11,000 rpm), the 42 HP of power, the reliability, the original chassis, the 214.5 km/h timed at the TT (232 km/h would be the top speed on the flat obtained in Albí (France)) and a consumption of 9.4 l/100 km. Possibly in these data was the key to the success of this revolutionary bike, the engine was not the most powerful but could follow in the wake of the most powerful bikes in fast circuits, in twisty areas its elasticity was superior to the multi-cylinder and its chassis gave enough confidence to the rider not to slow down where others did. Another key was reliability (greater than the MZ and even the Yamaha), when 20 m/s average piston speed was considered a limit that should not be exceeded, Giró's OSSA reached 24 m/s and did not break!

In Belfast, during the Ulster GP, Professor Gordon P. Blair of the Queen's University of Belfast (possibly the most influential academic in engine design and simulation (Blair 1996)) contacted Eduardo Giró because of the spectacular results that the OSSA monocoque was obtaining in the GP's. Immediately an understanding and mutual interest in their respective work arose (at that time Blair was starting to develop computer applications on 2-stroke engine simulation). Professor Blair invited Giró to visit Queen's University where he was impressed by the facilities available compared to the Spanish technical schools of the time. The contact continued over the years in epistolary form and with meetings at OSSA and Queen's U. Giró turned to Blair and his calculation programs for the tuning of the OSSA engines. While Walter Kaaden when he was at the helm of MZ was very reserved, after his retirement and during Giró's tenure with the Cobas team they were able to exchange knowledge and find topics of common interest beyond rivalry and pure competition. Giró has always been more attracted by the technique and its advancement than by the competition, which depends on many factors external to the engineer's work. Thus, in '69 and '70 three of the most relevant engineers in the development of 2-stroke racing engines met.

The development of the Ossa monocoque was cut short with the death of its rider. Before Santiago Herrero's death, Giró was developing a 500 cc engine based on the Yankee. It will never be known how far the two could have gone together: Santiago Herrero and Eduardo Giró.

5 Leaving the Asphalt: The Success of Off-Road Motorcycles

To a certain extent, the evolution of the market made OSSA's strategy, with Eduardo Giró at the head of the technical department, turn towards the production of off-road motorcycles. At the end of the sixties in Spain most of the cinemas had OSSA machines (machines that if well maintained are practically eternal), and with the appearance of television the market suffered. The same happened with the market for utility motorcycles whose sales dropped with the appearance of the Seat/Fiat 600 (and other affordable vehicles for the middle class such as the Citroën 2CV and the Renault R4). The same shift towards off-road motorcycles was made earlier by the Bultaco and Montesa brands. Between the three brands, they "filled" the field with motorcycles. The pioneer in opening up this market was F.X. Bultó who, at the beginning of the 60's, took a look at the Trial specialty, which was practically only known and practiced in the United Kingdom. Bultó signed Sammy Miller with whom he developed the Bultaco Sherpa model. This model revolutionized the Trial specialty and made the British 4-stroke motorcycles obsolete. At the same time, the light Spanish single-cylinder sport bikes OSSA, Montesa and Bultaco opened a niche in the North American market. The versions for this market adapted to their legislation and tastes: higher handlebars, scrambler versions were a starting point for the off-road models of these Spanish brands.

The off-road bikes developed by Eduardo Giró were successful both in racing and in sales. Paradoxically, OSSA did not make motorcycles with racing in mind, even "el Tècnic" stated on one occasion that he did not care so much about racing, but about technique. What he liked was to develop engines. Nevertheless, OSSA motorcycles were winning races and that showed in sales.

In the mid-1960s, OSSA had an incipient range of off-road motorcycles (trial, off-road and motocross) with a common base. Among them it is worth mentioning the appearance of the OSSA Enduro, a series of models whose versatility was especially valued and which were one of the pillars of OSSA's sales in Spain, Europe and the U.S.A. An essential characteristic of these motorcycles was their design, the shapes (tank, seat/hill) and the daring choice of colors clearly differentiated from the competitors due to Giró's creativity. "el Tècnic" has repeatedly recounted the importance of the choice of dates on which, based on drawings made by him and his team, he began to make plaster models for the construction of the molds. Since the services of model makers were hired for this purpose, it should be taken into account that when the dates of the Fallas of Valencia approached, they would disappear to work on these monuments.

Another development for which Eduardo Giró has gone down in the history of motorcycling is the OSSA Mick Andrews Replica, known as the MAR or simply as Mick Andrews, of which more than 15,000 units were sold. Presented in 1971 and marketed the following year, it was the result of the development that Eduardo Giró carried out with the British rider Mick Andrews, signed by OSSA in 1967, to ride its off-road models (not only trial, but also enduro and cross). The design of this bike was a second revolution in trial after the appearance of the 2-stroke by Bultaco. The British rider settled in the Barcelona area and frequently went with Giró to Mount Tibidabo (Barcelona) to test the modifications that were introduced in the prototype. There Giró and Andrews could make a circuit that resembled those of British competitions. Finally, Andrews took the bike to the United Kingdom to finish the tuning for the prestigious Scottish Six Days competition. Mick Andrews won in 1970 and 1971, the same year he also won the European Trials Championship (the World Championship did not yet exist). These results were repeated in 1972 with a version equal to the commercial one (not in prototype phase). The OSSA MAR dominated the world trial. The bike, with the chassis interrupted under the engine with structural functions, was clearly shorter than its rivals (Bultaco Sherpa and Montesa Cota), forcing a more forward riding position, with the engine and footpegs further back in conjunction with an engine with more power at high revs that allowed a more "juggish" driving than its competitors. This behavior meant a change in the riding and in the design of trial motorcycles from then on. This bike also broke with the prevailing aesthetics of the trial bikes of the competition: independent tank and side covers (as opposed to the single set of the competition), wide and comfortable seat (and not the symbolic one of the Cota and Sherpa), white color with a green stripe (not red like the others) and is a clear example of Eduardo Giró's talent as an industrial designer. The MAR 250 cc was on the market for 3 years without significant modifications and models directly derived from it continued to be marketed until 1977 in 250 cc and 310 cc versions.

Eduardo Giró developed with Mick Andrews a revolutionary new prototype of trial bike that had a 7-speed gearbox (compared to the 5 of the MAR), demountable chrome-molybdenum steel frame. A revolutionary bike, very small and light, from which the OSSA TR 80 (known as "the Yellow") was developed. Although it did not inherit all the improvements devised by Giró, it was also a bike that modernized the trial.

In the 70 s, Giró and his team of designers developed the Super Pioneer (successor of the Enduro), the Desert, more focused on off-road competition, the Phantom cross, and the 250 road. The Phantom were the basis of the new off-road models with the virtues that Giró wanted to imprint on this type of motorcycle: lightness, power and handling. They were the first production bikes to incorporate magnesium parts, such as the crankcase covers, they also used 1.7 mm thick Cr-Mb steel frames and in some versions aluminum swingarms. "A dirt bike has to be under 100 kg, otherwise it's a piece of junk."

The off-road motorcycles, besides working well in the domestic market, were the key to expand in the foreign market, especially the U.S. market where they sold 60% of their production. In those last years of Ossa, Eduardo Giró created the largest displacement motorcycle produced by OSSA, the most powerful of the Spanish motorcycles, and one of the most remembered today, the Yankee 500. The engine was originally commissioned by the American Yankee Motor Company, with the intention of marketing a motorcycle in the U.S. with the chassis manufactured there and the engine produced in Barcelona. Yankee Motor Company was a company created by the businessmen who at that time were in charge of importing Ossa motorcycles to the United States. The project had been offered to Bultaco, Montesa and Ossa, but it was finally the latter that accepted the challenge. What the American company wanted was an engine for desert and dirt track[2] racing. In these tests, one of the main handicaps was the lack of traction. And at that moment was when the rigor and ingenuity of "el Tècnic" came to light again. Eduardo Giró traveled to the United States with a team of high-speed film cameras (Personal communication 2019). They were dedicated to filming the rear wheel with the aim of capturing the sliding of the wheel with different engine configurations: cylinders with simultaneous explosion, at 60°, 90°, 180° …, different engine configurations and power curves. The slow motion movies allowed to see when the wheel skidded (dust came out in the explosion phase and how the pilot could get the maximum motricity by controlling the gas when the explosions of the two cylinders were simultaneous. The premise studied by Eduardo Giró, that of improving the traction of the traction wheel by concentrating the explosions of the cylinders along the crankshaft rotation instead of distributing them evenly along the rotation, was integrated in Grand Prix motorcycles almost two decades later: the famous "Big Bang" of Mick Doohan's Honda.

> They made the chassis and we made the first Big Bang engine so we could have traction in the desert. We invented the Big Bang!

Ossa delivered the first units of the Yankee engines. The American company, with economic problems, stopped paying for the engines supplied by Ossa and its tooling. After years of negotiation, not without disagreements, both companies reached an agreement so that Ossa could use the 500 cc engine in its own version of the Yankee, oriented to asphalt. With this, OSSA was able to use the capital it had tied up in this project. The result was the fastest touring bike produced in Spain, equipped with a 6-speed gearbox (the single-cylinder OSSA contemporaries of the Yankee had 5), the crankshaft of the Spanish Yankees for asphalt was 180° (betting on cyclic regularity), 6-speed gearbox derived from the 250 GP and had separate lubrication with Mikuni pump (conventionally, not through the crankshaft as in the Monocoque). However, the consumption of a 2-stroke at the time of rising fuel prices after the oil crisis was a handicap and the market demanded 4-stroke engines.

> For me what had to be made in Spain was a very light, very powerful and very sporty sport bike. But we didn't make the Yankee because we wanted to, but because the Americans

[2] This is a type of racing originating in the United States in which vehicles compete on an oval circuit with clay, dirt or ash terrain. The circuit is usually flat or with very little camber.

> decided they didn't want any more mountain bikes because the Japanese were selling them cheaper.
>
> On the plane back to Barcelona I thought I had to make a bike. Actually, the Yankee was what the Japanese did later with the RD, Kawasaki three-cylinder, crazy engines, light bikes. We put separate lubrication, redesigned the engine to adapt it to road use. We made a chassis, bodywork, suspensions, disc brakes, rims, etc. The bike came out late, but we couldn't do any better because we had work to do on servicing other bikes. The Yankee could have been part of our salvation if it had come out sooner.

When the technical director at Ossa left the company, Eduardo Giró took charge of engineering. Giró not only took charge of the technical aspect, but, after his father, it was he who took over the management of the company. As CEO he also performed management tasks at the Ossa factory. Far from being bored by these tasks, he found them fun and interesting. Inseparable from the design process of a part is the production process. When designing a motorcycle component, one must think about manufacturing, type of machinery to be used, production capacity... Eduardo Giró therefore managed different departments from the technical office, which meant that his industrial rigor extended to the entire company (Medina 2015).

6 Downfall of the Motorcycle Industry in Spain and Exit of Ossa

At the end of the 1970s, Ossa, the company owned by his family and in which Eduardo Giró had developed his career as an engineer, went into receivership. The big Japanese manufacturers had broken into the American market in an extremely aggressive way, radically reducing the price of their off-road motorcycles, which at that time did not have the quality of Spanish motorcycles, and offering importers of Spanish brands the range of Japanese road motorcycles. Sales in the USA fell radically and these represented more than 60% of the total sales of OSSA, Montesa and Bultaco. These companies were left with oversized workforces (with no possibility of adjustment) at a time of high labor unrest, in the midst of political transition and found themselves doomed to closure or alliance with a foreign manufacturer (Japanese rather, the rest of the European manufacturers were not in much better conditions). There was no significant support from the State for these industries. They were offered the possibility of becoming Labor Corporations, the Giró family ceded the shares to the labor mass and a cooperative called Maquinaria Cinematográfica Sal was created. The objective was not to lay off the more than 300 people working in the factory. The Sociedad laboral formula worked well at that time in some companies in the Basque Country, but it did not in the case of Ossa. The rest of the Spanish manufacturers opted for other formulas to prevent the crisis affecting the sector from forcing the factories to close and destroying jobs. In Gijón, the Avelló family factory, where MV

Agusta and Puch were produced, was acquired by Suzuki in 1983.[3] Montesa reached an agreement with Honda thanks to which they kept the factory in Catalonia and the Montesa brand for trial motorcycles and some off-road motorcycles. Eduardo Giró had some contacts with Yamaha to try to get them to buy the factory, something that Bultaco and Sanglas also did. Finally, it was the company founded by the Sanglas brothers that was absorbed by the Japanese giant, which used it to penetrate the Spanish market, ceasing in 1982 to produce under the Sanglas brand.

At some point there were meetings between the Spanish companies: Montesa, Bultaco and Ossa, to discuss a possible union of the three. Such a union never took place, among other reasons due to a lack of real interest of the parties and the enmity that separated Bultó and Permayer for years.

Eduardo Giró did not stay in Ossa. As soon as the cooperative was created in 1979, he left the company. Partly because he did not share the new form of administration of the company and partly because he saw the end coming. At this point Giró had designed his last engine for OSSA, an engine for off-road motorcycles from 250 to 400 cc that was initially going to be manufactured in its version with piston skirt intake, but which was planned for a modern version with reed intake (which could be reliably simulated thanks to the work of G. Blair) and of which Giró was particularly proud: "I was in love with this engine", he declared in an interview. In 1982 Ossa closed its doors[4] (Herreros and Aznar 1998).

7 After Ossa

A member of SAE, Eduardo always remained linked to aeromodelling, developing two-stroke engines. He even developed a small gearbox (weighing 20 g) for the propellers of model airplanes that prevented the entry into sonic speeds of the tips of these. Eduardo Giró has several publications in aeromodelling magazines where he presented his advances. In the period after Ossa, the aeromodelling developments that meant Giró's return to the engines, he made them in the back room of an aeromodelling store located in the Rambla of Barcelona owned by a friend of his. There he set up a test bench to study the engines. He mounted pressure sensors in the combustion chamber and exhaust pipe, optimizing the performance of these small engines. Techniques that he would later apply to motorcycles.

Eduardo Giró returned to the world of motorcycles, specifically to competition, in the early 80 s. With Antonio Cobas and Manolo Burillo he joined the Tecomsa team (which built a 250 GP bike, the Kobas MR1, whose double beam chassis has marked the design of many of today's road and competition bikes). In this team and in his

[3] Suzuki completely acquired Avelló, S.A. in 1988 changing the name of the company to Suzuki Motor España. Unfortunately, in 2012 the multinational announced the closure of the Gijón plant.

[4] In 2009 some businessmen from Girona relaunched the OSSA brand by manufacturing off-road bikes. The experience only lasted until 2015 (Merino 2019), a year after its merger with Gas Gas and the definitive closure of both brands.

next job for JJ Cobas "el Técnic" he kept a low profile in front of the media. In fact, he had his professional dedication outside these teams to which he dedicated his nights: "he worked from 9 to 12 at night", "in the races, based on illusion, a lot is done". For the media, his involvement was focused on the preparation of the Rotax engine, but his experience as an industrialist was also vital for these teams: to achieve reliability, to ensure that the bike always works the same way, to rationalize the manufacturing methods. The rider of this team, Sito Pons, achieved with the Kobas bike in 1982 his first podium in Imatra (Finland) and the victory in the 250 cc category in the Spanish GP at Jarama, 14 years after the last victory of Santi Herrero in Yugoslavia. In this GP Antonio Cobas saw from the box of the JJ Cobas team (his new team) the first victory of a Kobas motorbike, and how his rider Carlos Cardús obtained the pole position in practice with the Cobas. The good results of Sito Pons with the Kobas consecrated him as a rider and helped him to win the 250 cc world championship in 88 and 89 with Honda.

In the mid 80's, Giró also moved from the Kobas team to JJ Cobas (team of Jacinto Moriana and Antonio Cobas) with a similar dedication and leaving the media protagonism to Cobas. This period, despite the scarcity of means, was very fruitful, culminating with Alex Crivillé's world championship in 125 cc with the JJ-Cobas TB6 in 1989. One of the keys to obtaining this championship was the improvement of the engine where Eduardo Giró found a way to improve the crankcase cooling of the Rotax engine (Cathcart 2019). Taking advantage of a double wall of the crankcase he managed to devise a system to cool the crankcase, something fundamental to improve the performance of a 2 T engine. He also applied his experience in aeromodelling to make a correct pressurization of the intake, his modifications to the engine made the difference. Before Alex Crivillé joined the JJ-Cobas team in 1989, Eduardo Giró was developing a 250 cc V-twin engine. Starting from a Rotax whose tandem architecture was not the most suitable for a GP bike, he made the crankcases of the V-engine and used all the elements of the Rotax engine that he could, so that, even without a dyno, the engine worked well at the first time (once again!). This engine aroused the interest of several top drivers in the world, Giró saw great potential in it. But the plans of Jacinto Moriana and Antonio Cobas went in the direction of getting 250 cc Honda engines, abandoning this V engine, a decision that proved to be wrong.

In the last days of this team, Cobas began to apply telemetry to the chassis, being pioneers in this technique in the GPs. Eduardo Giró also applied temperature probes to the exhausts to help in the tuning of the bikes. This system that forced the rider at the end of a straight to look at the value marked by the instrument (it was not telemetry) provided much more reliable information than the system of measuring cylinder head temperatures in the "pit lane" of the OSSA era.

One development of that time that is worth mentioning is the JJ-Cobas BMW 1000 cc. This motorcycle was based on a commercial motorcycle, the BMW K100. It was one of the fastest touring bikes on the scene at the time, but transforming it into a racing bike was no easy task. Taking advantage of the engine block and swingarm they designed a new tubular chassis of only 6 kg where the engine/transmission was a structural part of the chassis and redesigned a progressive suspension geometry to avoid the problems associated with the cardan shaft transmission. Eduardo Giró designed the exhaust system, mounted new camshafts and modified the intake and exhaust ducts. In a few months he managed to increase the engine power from 90 to 108 hp. BMW did not want to provide the codes that allowed the Bosch electronics to be configured, so the team had to "cheat" the electronic control unit to increase its performance.

BMW's lack of cooperation with the team was the cause of electronic problems in its participation in the 24 h of Montjuich. Subsequent development of the bike for speed competitions reached a power output of 122 hp and won all six races in the Formula 1 category of the "Motociclismo Series".

After a stint in motorcycling and a brief three-year stint at Motor Ibérica, now Nissan, Eduardo Giró worked for more than twenty years in a port company in Barcelona. There they developed and implemented container terminals in ports around the world. These terminals included the organizational elements in which they were pioneers. They were the first to use terminal activity modeling. From the management program developed in this port company derives the one used in the Barcelona airport after its expansion for the Barcelona Olympics. After his retirement, Eduardo Giró has continued for several years linked to aeromodelling and reaping success in this field.

In this chapter, we have tried to pay tribute to the figure of Eduardo Giró through, above all, his activity in the field of motorcycles for which he is known worldwide. He enjoyed unique opportunities when he was born into the Giró family, but it was his great talent for mechanical engineering that allowed him to take advantage of them. From the smallest of the three flagship factories of the 2-stroke motorcycle in Spain, he achieved goals that no other engineer has ever reached.

We also wanted to pay tribute to the motorcycle industry in Spain. The end of these industries is sad, today there is practically nothing left of the industrial power of the 70's, which in this sector was a leader in the world. Other technicians such as Leopoldo Milà at Montesa, Francisco Tombas at Derbi, and F.X. Bultó at Bultaco also have a place in the history of Spanish motorcycling, but Eduardo Giró "el Tècnic" (the technician) is undoubtedly the great figure of engineering in this sector in Spain.

References

P. Carlos, La máquina perfecta Ossa 250 monocasco. Moto Clásica. Número 63, agosto 2002, Páginas 30–37

B. Pepe, "La Ossa de Santi Herrero. La leyenda al descubierto". Solo Moto. Número 82, diciembre 1989, Paginas 38–61

M. Alex, "Eduardo Giró: «Nuestro rigor industrial era mucho más alto que el de Montesa o Bultaco»". Solo Moto. Edición digital, julio de 2015. URL: https://solomoto.es/eduardo-giro-nuestro-rigor-industrial-era-mucho-mas-alto-que-el-de-montesa-o-bultaco. [Consulta: 8/12/2019]

C. Alan, "Álex Crivillé: 30 años de su primer campeonato del Mundo". Fórmula Moto. Edición digital, junio de 2019. https://www.formulamoto.es/pruebas-alan-cathcart/2019/06/07/alex-criville-10-anos-primer/24478.html. [Consulta: 8/12/2019]

F. Herreros, J.L. Aznar, *Historia del motociclismo en España* (RACC, Barcelona, 1998)

M. Nicolas, "La Leyenda continua 2019: homenaje a la marca Ossa". Fórmula Moto. Edición digital, diciembre 2018. https://www.formulamoto.es/zona-clasicas/concentraciones-eventos/2018/12/30/leyenda-continua-homenaje-marca-ossa/23173.html [Consulta: 8/12/2019

W. Mick, "Spanish Postwar Road and Racing Motorcycles". Osprey Publishing. Junio de 1986. Páginas 126–152. ISBN 0–85045–705-X

G.P. Blair, *Design and simulation of two-stroke engines* (Society of Automotive Engineers, Warrendale, PA, 1996)

G. Jennings, "Two-stroke Tuner's Handbook". HP Books 1973, ISBN: 978–0912656410

V.H. Willoughby, *Das Motorrad-Rennsportbuch* (Motorbuch Verlag, Stuttgart, 1982)

Personal communication: Interview in Barcelona 28th, october 2019

Personal communication: RetroMotro Barcelona, La Farga de Hospitalet, 29–31 de marzo de 2019. http://www.amoticos.org/t15568-eduardo-giro-el-tecnic-rompe-su-silencio

Alejandro Goicoechea Omar (1895–1984)

M. Sánchez Lozano

Abstract The Talgo train has deservedly become a symbol and benchmark for Spanish engineering. Its name is usually linked to that of the engineer Alejandro Goicoechea, whose revolutionary ideas were the seed that served as the basis for its development. The inventive activity of this engineer was reflected in the publication of a significant number of patents between the 1920s and 1980s, mainly related to the railway, some of which have been key to the development of railway engineering in Spain. It is true, however, that it was the vision of the entrepreneur José Luis Oriol, and his successors, which allowed, from the seed sown by Goicoechea, the development of a company culture in which the promotion of creativity and technological innovation has played a fundamental role. So it is fair to mention, in the last part of the text, the teamwork and achievements of the engineers and technicians who came afterwards. As a result, in its 70-year history, Patentes Talgo has gathered more than 100 patents related to railways and mechanical engineering, many of which were truly disruptive innovations in the sector. This chapter briefly describes some of these revolutionary ideas and novel developments in railway engineering. The aim here is not to gloss Goicoechea's biography, nor that of the engineers and technicians who came after him, but only to highlight the importance of their contributions to Spanish mechanical engineering.

Keywords Light articulated train · Independent free wheel · Guidance mechanism · Elevated rolling gear · Vertebrate train · Track gauge variation · Natural tilting · TALGO

M. S. Lozano (✉)
University Miguel Hernández of Elche, Alicante, Spain
e-mail: msanchez@umh.es

R. López-García and M. Ceccarelli (eds.), *Distinguished Figures in Mechanical Engineering in Spain and Ibero-America*, History of Mechanism and Machine Science 43, https://doi.org/10.1007/978-3-031-31075-1_6

1 Biographical Notes

Alejandro Goicoechea Omar was born in Elorrio (Vizcaya) in 1895, and studied military engineering at the Army Engineering School in Guadalajara between 1912 and 1918. He later joined the Railway Regiment, where he worked on the construction of the railway branch from Cuatro Vientos to Leganés, in Madrid. In 1921 he left the army and was hired shortly afterwards by the La Robla Railway Company. There, he held the position of head of materials and traction, until the outbreak of the Spanish Civil War. It was in this position that he became aware of the possibilities for improvement offered by the railway technology of the time, and began to develop and propose the ideas that will be described later, related to the lightening and increased safety of trains, which were soon to be applied.

At this point, it is useful to put into context the economic and social situation that favoured the acceptance of these ideas. During the 1910s and 1920s, Spanish industrial production grew strongly, especially after the First World War, which had increased the demand for goods and services from the countries at war. This had a direct impact on the Spanish economy, which opened up to the export of industrial products. At the same time, imports were further reduced as a consequence of the war economy prevailing in Europe, with the productive potential of the country being used for military industry. The First World War therefore causes a flourishing of Spanish industry to supply both foreign and domestic demand.

At the same time, the need to improve export infrastructures favoured naval and railway construction. On the other hand, with the policy of national protectionism established with the dictatorship of Primo de Rivera, the Railway Statute of 1924 obliged companies in the sector to source steam locomotives from national industry. This process of economic and industrial expansion particularly benefited regions such as the Basque Country, where numerous metal processing, mechanical construction and capital goods companies set up (machinery, engines, shipbuilding and railway industry).

On the other hand, at the end of the 19th and the beginning of the twentieth century, a generation of scientists and technicians flourished in Spain, such as Santiago Ramón y Cajal, Leonardo Torres Quevedo, Isaac Peral and Juan de la Cierva, who spread the idea of technical and economic progress as a formula for social improvement. And it was in this context, favourable to science and technical innovation, which unfortunately did not last long, that Alejandro Goicoechea's ideas were favourably received. This would probably have been unthinkable in another time.

The search for funding to carry out the idea of a light articulated train, on which he had been working since 1936, led Alejandro Goicoechea to meet with José Luis Oriol in May 1942. The latter's receptiveness to his ideas, and the trust he placed in them, led to the foundation in October 1942 of the company Patentes Talgo.

As part of the new company, Goicoechea participated in the development of the idea and the optimisation of the first prototypes. Until 1944, when the need to find an industrial partner became clear. The industrial environment in post-war Spain was in a more than precarious situation, aggravated by international isolation, and framed in a continent that was also suffering the devastating effects of the Second World War. Goicoechea went to the United States at the end of 1944 to help in the search for an American manufacturer interested in being that industrial partner. The expedition bore the expected fruits.

Nevertheless, on his return, in May 1945, Alejandro Goicoechea decided to abandon his participation in Patentes Talgo. He thus put an end to his link with the light articulated train, that had been his dream for years. He was therefore not a direct participant in the subsequent development and industrialisation, which was finally able to make the application of his revolutionary ideas in passenger rail transport a reality.

After leaving Patentes Talgo, Goicoechea continued to develop his creative activity, which was reflected in the application for various patents. These included those related to another revolutionary idea, the "vertebrate train", which led him in the 1960s to set up a company, the Sociedad Anónima de Trenes Vertebrados, in order to develop his invention. After several attempts to put it into practice, which included the start-up of a first experimental line, as will be discussed later, he was finally forced to abandon his project at the end of the 1970s. But he continued with his inventive activity until the early 1980s, when his last patents were published.

Alejandro Goicoechea died in Madrid in January 1984 at the age of 88.

2 List of Main Works

Patents related to the construction of wagons by electric welding:

System for the construction of railway wagons by electric welding. Introductory Patent No. 103453. Spanish Patent and Trademark Office. 1927.

Patents related to the light articulation system, with triangular guidance and independent wheels:

Combined light articulation system for organisation of trains on guided routes. Patent No. ES151396A1. Spanish Patent and Trademark Office. 1941.

Improvements in combined light articulated railway systems. Patent nº ES159301A1. Spanish Patent and Trademark Office. 1942.

Articulated railway vehicles consisting of light, short, low elements on independent wheels, with triangular traction-guided rolling gear. Patent No. ES163239A1 (already registered in the name of Talgo). Spanish Patent and Trademark Office. 1943.

Patents related to elevated rolling gear ("vertebrate train"):

A new railway transport system with elevated rolling gear. Patent No. ES141056. Spanish Patent and Trademark Office. 1936.

Improved procedure for the construction of vertebrate trains. Patent No. ES339112A1, ES344763A2, ES345424A2. Spanish Patent and Trademark Office. 1967.

Procedure for spherical vertebration in railways. Patent nº ES350085A1. Spanish Patent and Trademark Office. 1968.

Procedure for organic circulation and accesses in vertebrate trains. Patent n° ES350648A1. Spanish Patent and Trademark Office. 1968.

Braking system for high rolling railway vehicles. Patent n° ES457322A1. Spanish Patent and Trademark Office. 1977.

Multiple elevated train. Patent n° ES8306653A1. Spanish Patent and Trademark Office. 1982.

3 Review of Main Contributions

3.1 Lightening of the Train: Welded Wagons

La Robla railway company was mainly engaged in the transport of minerals, which was carried out in heavy and slow wagons, with high operating costs. Traditionally, it had been considered that the railway needed large axle loads to ensure adhesion and stability. This statement could be justified in the traction equipment, to ensure the necessary adhesion for traction and braking, but not in the towed equipment, where at most it can be considered an influential factor in preventing derailment.

Goicoechea saw here an opportunity to significantly reduce operating costs, by reducing the curb weight of the wagons. On the other hand, he knew that Belgians and Germans were beginning to use electric welding in the construction of bridges and battleships, a technique not used until then in the railway world. After carrying out tests in the workshops of Valmaseda, the blast furnace of Vizcaya and the School of Industrial Engineers of Bilbao, in 1926 he promoted the construction of an entirely welded wagon, without rivets or screws, much lighter and stronger than the usual ones, which was presented at the Bilbao Work Competition in 1927. The invention of the "system for the construction of railway wagons by electric welding" was registered by Goicoechea as an "introductory patent" for the welding technology already applied in other sectors in Belgium and Germany (Goicoechea Omar 1927).

Figure 2 shows the sketch included in this patent, which depicts a frame structure based on welded and braced profiles. The design may seem very simple from today's point of view, but it was an absolutely disruptive proposal for the time.

The patent was immediately exploited in the workshops of La Robla, and was transferred a year later to several industrialists from Valmaseda, who would end up founding the company Uniones Metálicas a Soldaduras (UMAS, S.A.). In the following years, the construction of new lightened freight wagons, thanks to the replacement of the traditional bolted or screwed wagons with welded ones, made it possible to progressively reduce the weight of the wagons, which went from representing 65% of the maximum load to only 25%, resulting in significant savings in operating costs.

Goicoechea became interested in passenger coaches in the early 1930s after witnessing, according to his own account, a collision between two trains, in which most of the injured were injured by flying wood splinters after the collision. His

Fig. 1 Alejandro Goicoechea next to the Talgo I prototype

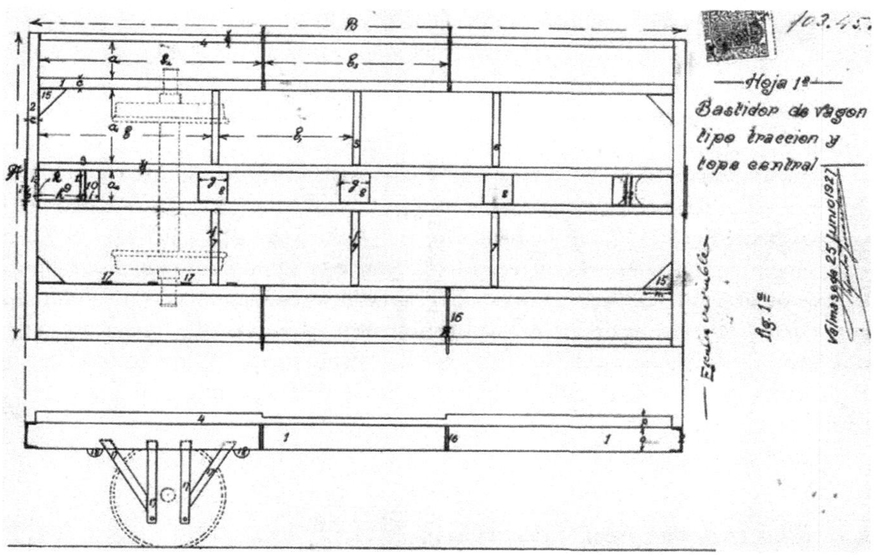

Fig. 2 Welded railway frame, proposed in the introductory patent nº 103,453 (Goicoechea Omar 1927)

idea was to apply the techniques of all-metal construction and welding to passenger railways, with the aim of making transport cheaper, safer and faster.

3.2 *The Articulated Train: The "Triangular Guidance"*

The lightening and increased speed of passenger trains could increase instability and the danger of derailment. To avoid these risks, Goicoechea devised a concept of a continuous articulated composition, replacing the traditional coaches with independent carbodies mounted on bogies. The elimination of the bogies would firstly allow a significant lowering of the centre of gravity, increasing stability in curves. On the other hand, the continuous connection between coaches would improve behaviour in the event of derailment, avoiding the overturning between coaches, or the possibility of derailment and overturning of each individual coach.

In the proposed articulated structure, each link of the composition is made up of a light and simple triangular structure, with a single axle at the rear, and whose front vertex rests as a trailer on an articulation point located in the centre of the axis of the immediately preceding triangle. According to the inventor himself, the idea for the guidance system was inspired by the behaviour of a row of children's tricycles hooked together, which he observed in the park of El Retiro in Madrid. The person in charge of collecting them rested the front wheel of each tricycle on the rear axle of the previous one. In this way, the direction set by the first tricycle was followed by the others, which followed the same trajectory without the need for any additional guidance.

The invention is covered by the patent "Combined light articulation system for the organisation of trains on guided routes" (Goicoechea Omar 1941), shown in Fig. 3, which was filed in 1941 after the hiatus forced by the Spanish Civil War.

By applying this simple configuration, guidance is therefore no longer produced by the contact with the rails of the conical profile wheels mounted on rigid axles, which in the classic conception of railway guidance are kept on the trajectory thanks to a known yaw movement. Goicoechea's proposal avoids transverse movements of the axles, which at high speeds can cause vibrations, discomfort and, ultimately, derailment.

3.3 *The Independent, Free Wheel*

With the concept developed, the inner and outer wheels have to turn at different speeds when curving, which prevents mounting on a rigid axle and forces the adoption of free and independent rolling of each wheel. This fact, far from being a disadvantage, should be seen as an advantage. The elimination of the physical axle, together with the elimination of the bogies or undercarriages, allows the floor of the coaches to be located at a very low level, and therefore a significant lowering of the centre of gravity, which increases stability and will increase curving speed. This low floor can furthermore be maintained in the space between the wheels, allowing both passenger access to the train and passage between cars at this level.

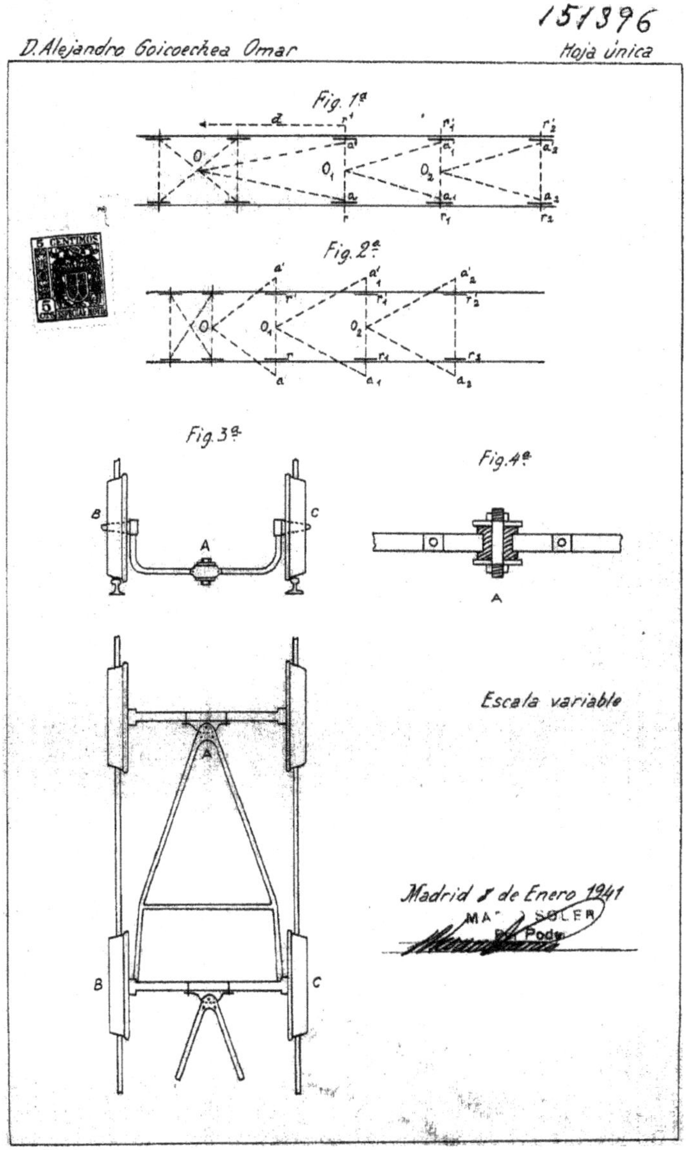

Fig. 3 "Triangular guidance" system, as proposed in the patent ES151396A1 (Goicoechea Omar 1941)

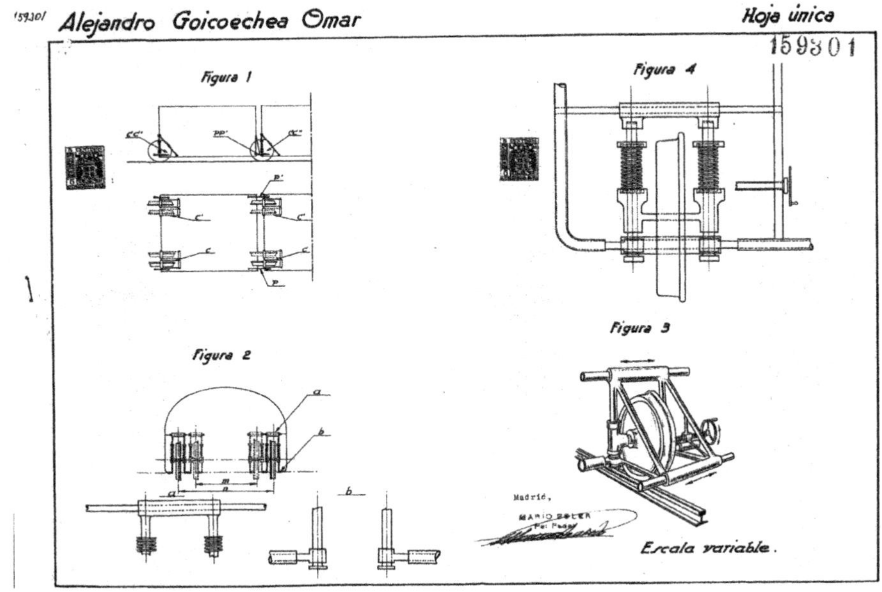

Fig. 4 Movable free wheel, as described in the patent ES159301A1 (Goicoechea Omar 1942)

The details associated with the adoption of the free wheel are set out in the following patent filed by Goicoechea in 1942, "Improvements in the railway systems with combined light articulation" (Goicoechea Omar 1942). This patent also includes the possibility of transversal displacement of the wheels, in order to adapt to different track gauges (see Fig. 4). As will be seen below, the development of this idea would represent another of the disruptive innovations implemented by Talgo in later decades.

3.4 Other Developments by Goicoechea: The Elevated Rolling Gear and the "Vertebrate Train"

After leaving Patentes Talgo, Goicoechea continued to develop his creative activity, which was reflected in the application for various patents. Among these inventions, the "elevated rolling gear for railways" is worth mentioning. As early as 1936, he had already published a first patent (Goicoechea Omar 1936), which describes an original idea of a light, articulated and non-derailable train, which would roll supported by pneumatic wheels on two elevated rails, between which it would be fitted (Fig. 5). This configuration would place the centre of gravity so low in relation to the rail that it would allow it to reach high speeds without risk.

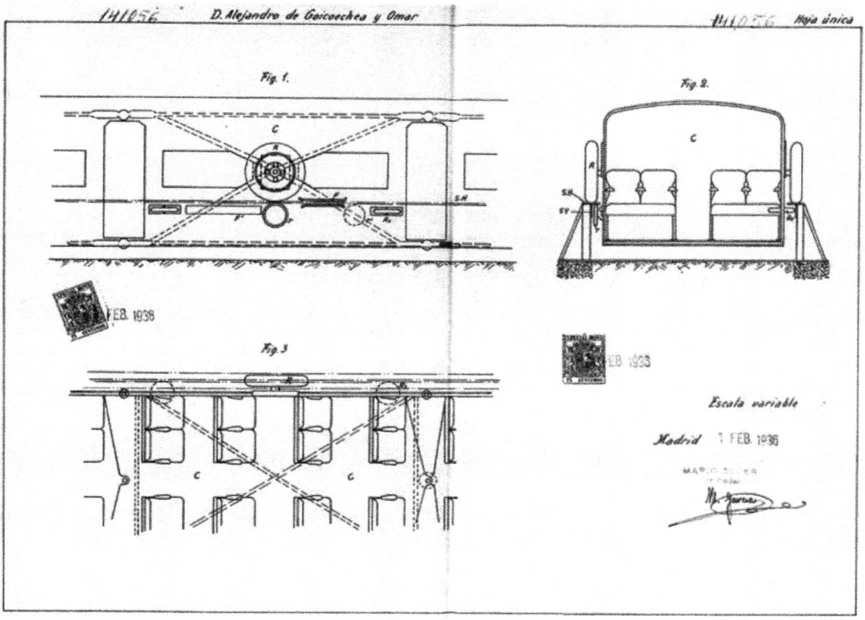

Fig. 5 Elevated rolling gear system, as proposed in the patent ES141056 (Goicoechea Omar 1936)

In fact, this idea predates that of the triangular guideway on which the Talgo train was based. Presumably, it was not decided at the time to develop this track as an option for guiding the light articulated train, due to the disadvantages it presented, mainly the incompatibility with the existing railway infrastructure, or possible problems associated with the maintenance and durability of the pneumatic wheels. However, he recovered this idea years later, and in the 1960s applied for several patents related to what he called "vertebrate trains" (Goicoechea Omar 1967a, b, c, 1968; Omar 1968). These patents dealt with the optimisation of the elevated rolling gear system, as well as solving problems related to traffic flow at intersections and train access. It should be taken into account that the rail was located halfway up the vehicle body, so it was necessary to have retractable sections at the height of the doors to allow passengers to enter and exit.

Goicoechea set up the company Sociedad Anónima de Trenes Vertebrados with the aim of developing his invention. He built a prototype with the collaboration of the French firm Brissonneau & Lotz, and managed to get FEVE (Spanish public company of narrow-gauge railways) to lend him the land of the former Basque-Navarre railway station in Santa Cruz de Campezo (Alava). There, in December 1969, he undertook a campaign to test the prototype on a section of track set up for this purpose.

Once the tests were completed, he tried to commercialise his product, undertaking unsuccessful negotiations for its implementation on various lines on the Iberian Peninsula. In 1971, the preliminary project and the application for the concession

Fig. 6 Experimental installation of the Vertebrate Train, in the Avenida Marítima de las Palmas de Gran Canaria (source: eldiario.es) (Otero Paz 2017)

of the Las Palmas de Gran Canaria-Gando-Maspalomas line were presented to the Ministry of Public Works. As a first step, an experimental installation of one and a half kilometres was carried out on Avenida Marítima in Las Palmas de Gran Canaria (Fig. 6), with the aim of carrying out a demonstration before the Ministry before it approved the system that would allow the definitive installation.

The project was criticised from the outset due to the visual impact of the elevated installation. This, together with the late start-up and the associated unresolved technical problems, finally brought the project to a halt. The experimental installation was finally dismantled in 1975 (Otero Paz 2017).

After several attempts at implementation, Goicoechea finally abandoned his vertebrate train project. However, he never abandoned his idea of the elevated track, and published patents on the evolution of this concept can still be found up to the 1980s (Goicoechea Omar 1977; Omar 1982).

4 On the Circulation and Implementation of the Contributions

The publications related to the introduction of electric welding in train manufacturing had a great impact and diffusion at the time. As has been said, the technology was implemented on an industrial level almost immediately, and its use became widespread in a short period of time.

The reception and success of the concepts of elevated running gear and vertebrate trains was much lower. As has already been mentioned, the project was not favourably received nor did it achieve large-scale dissemination, despite the fact that Goicoechea devoted himself to its development for decades.

But, undoubtedly, the works that were most widely disseminated, and to which Alejandro Goicoechea's name would be linked from then on, were those related to the light articulated train and the free-wheels rolling, which were the seed for the creation of the Talgo train.

4.1 The Birth of the Talgo Train

As early as 1936, Goicoechea had already submitted a project to the management of the La Robla railway to build a prototype of a light articulated composition, which did not receive a positive response. After presenting the idea at the 15th Congress of the Association for the Progress of Sciences held in Santander, in 1939 he requested help from the Army High Command to build and test a prototype, an attempt that was also unsuccessful.

Finally, in 1940, a first prototype of the idea was built in the Valmaseda workshops, using several front bridges from Russian lorries, joined to welded frames forming isosceles triangles. This rudimentary prototype, which can be seen in Fig. 7, was successfully tested for the first time on the Madrid-Leganés line at the end of the summer of 1941, exceeding a speed of 75 km/h.

Goicoechea then set himself the goal of developing the first prototype of a complete train based on this concept. He initially found the support of the recently created RENFE (Spanish National Railway Network) and the firm Hijos de Juan de Garay de Oñate, manufacturers of tubes and umbrellas, which allowed the construction of the prototype to begin in the workshops of RENFE itself.

The search for funding to go ahead with the project led him to meet José Luis Oriol in May 1942. Oriol's receptiveness to his ideas and the trust he placed in them led to the founding in October 1942 of the company Patentes Talgo. The name is an

Fig. 7 First triangular guidance prototype produced in 1940

Fig. 8 The prototype Talgo I

acronym for "Tren Articulado Ligero Goicoechea Oriol" (Goicoechea-Oriol Light Articulated Train).

Oriol's financial support allowed the development of Goicoechea's ideas from then on, and the first result was the completion of the prototype called Talgo I (Fig. 8), which made its first demonstration trip that same month on the Madrid-Guadalajara route.

From the creation of the company, patents related to the concept of the articulated vehicle were published by Talgo. In the first half of the 1940s, engineering efforts were devoted to the analysis of the triangular guidance system, the optimisation of the independent wheel suspension systems, and the joint and articulation systems between coaches. In this way, a number of improvements were introduced and tested on the first prototype, which was used for testing between 1942 and 1945.

When the company decided to manufacture the first light articulated train based on triangular guidance technology, and which could already be commercially exploited, the industry in post-war Spain was in a more than precarious situation, framed in a continent that was also suffering the devastating effects of World War II, and aggravated by international isolation. This led to the search for a North American manufacturer interested in the production.

Alejandro Goicoechea travelled to the United States at the end of 1944, together with the engineer Jaime McVeigh, to look for this industrial partner. The expedition bore fruit, and after months of study and negotiation, an agreement was finally reached with the company American Car and Foundry (ACF) for the development and construction of two trains and three Iberian gauge locomotives for service in Spain (which would later be called Talgo II), reserving the manufacturing rights for the United States.

On his return from his trip to the United States, in May 1945 Alejandro Goicoechea decided to abandon his participation in Patentes Talgo, thus putting an end to his involvement with the light articulated train that had been his dream for years. He was therefore not a direct participant in the subsequent development and industrialisation that was finally able to make the application of his revolutionary ideas in passenger rail transport a reality. In the second part of this article, we will talk about the technicians and engineers who succeeded him and made it possible.

5 The Legacy. Talgo's Evolution to the Present day

5.1 The Success of the Team

After Alejandro Goicoechea left Patentes Talgo, other engineers and technicians took over, and it was they who made possible the technical evolution of his ideas, the industrialisation and putting into service of the Talgo train. Without this subsequent evolution, Goicoechea's revolutionary ideas would probably have fallen into oblivion. The company culture developed by Patentes Talgo has been fundamental to this evolution. From the very beginning, it fostered and encouraged the continuity of this creative spirit and teamwork among its technicians. In the words of José Luis López Gómez, who inspired and collaborated in the documentation of this text, the technical achievements from that point onwards are the joint merit of a team of people, above and beyond individuality. There is no point in having a great idea if there is no team of people capable of developing it.

On the other hand, it should also be noted that the maintenance of the trains was carried out from the beginning by Patentes Talgo itself, according to an agreement with Renfe. The fact that the manufacturer of the railway material itself was in charge of maintenance was a novelty at the time at European level, and was undoubtedly beneficial for the evolution of technology, as it allowed for a thorough analysis of problems and breakdowns and served as a source of information for new technological developments. On the other hand, the responsibility for train maintenance opened up a line of development of control systems and specific workshop machines, which, as will be seen later, also proved to be fruitful.

The details of the evolution of the company and Talgo trains are beyond the scope of this article, and can be found in a multitude of publications (Española and de Patentes y Marcas. Galerías temáticas, grandes empresas: Patentes Talgo, S.A. 2020; Galán Eruste et al. 2018; López Pita 2010). The following are just a few of the people who played an important role in this evolution, from an engineering point of view, during the second half of the twentieth century. And, later on, only a brief outline of the main technological milestones achieved by this team a is given.

Ángel Torán Tomás, together with Francisco Martín Fernández de Heredia, were the two young industrial engineers who went to the United States after Goicoechea left to collaborate in the development and manufacture of the Talgo II. On their return, from 1950 onwards, they took charge of the management of the Talgo Patents technical office, which in the following years tackled the technological evolution that would be reflected in developments such as the reversible guidance, the gauge change system and the natural tilting suspension. Ángel Torán continued to be the technical director until the 1980s, although Heredia left the company some years earlier to take up important positions in other sectors.

José Luis López Gómez, a Technical Engineer from ICAI, began working at Talgo in 1967, initially occupying the positions of workshop manager and workshop engineering manager. In those early years, he became interested in the development of tooling and specific machinery for railway maintenance, opening up a line of development that has materialised in a large number of designs and components. From 1990 he held the position of Technical Director and, later, General Manager of Technology and R&D at Patentes Talgo until his retirement in 2004. Under his leadership, work continued on the optimisation of Talgo's technology, which led to the publication of a significant number of patents, and which is necessarily associated with its entry into the high-speed sector.

Finally, it is worth mentioning the contribution of José Isidro Nardiz Landa, who in the early 1950s started working as a draughtsman in Talgo's technical office. In the words of José Luis López, Nardiz stood out from the beginning for his vision and organisational capacity, which contributed decisively to the planning of the technological development of the Talgo train in the second half of the twentieth century, and later in the company's global strategy. It is for this reason that, despite the fact that he did not study engineering, his figure deserves to be mentioned here as an outstanding figure of Mechanical Engineering in Spain.

5.2 Key Technological Milestones

The Reversible Guidance System

The simple triangular guidance system proposed by Goicoechea presented two main problems. On the one hand, the rigid connection of the wheels at the rear of each car causes the virtual axis that joins them to be oriented in a direction that does not pass exactly through the centre of the trajectory. The wheels therefore do not remain tangent to the rails, but have a negative angle of attack towards the inside of the curve. This effect can be seen in the sketch included in the first patent applied for by Talgo for the guidance system in 1943 (Talgo 1943), shown in Fig. 9. In order to avoid reaching excessively high angles of attack, it was necessary to limit the length of the triangular elements, and therefore of the coaches, which a priori did not seem to represent an excessively serious problem as it was a continuous composition.

As a first consequence, this negative angle of attack in curves usually leads to the flange of the outer wheel resting on the rear part of the wheel-rail contact, which results in additional wear on the material. However, this contact at the rear causes a vertical downward frictional force to appear on the flange, which pushes the wheel against the rail, opposing derailment. This effect is the opposite of that experienced by

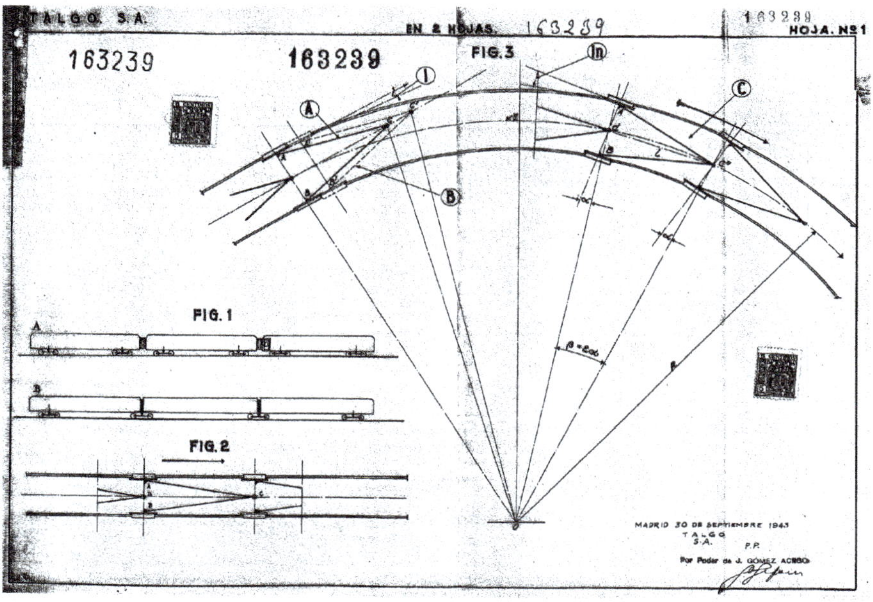

Fig. 9 Negative angle of attack obtained with the triangular guidance system, sketch included in the patent ES163239A1 (Talgo 1943)

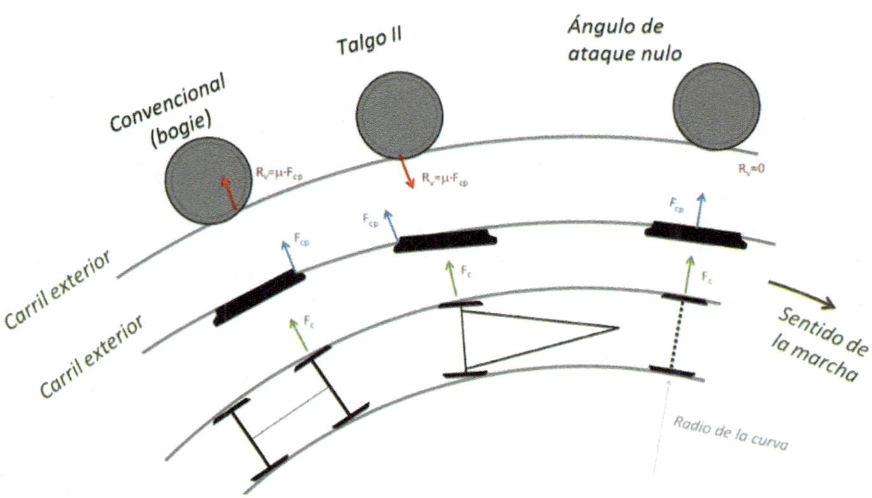

Fig. 10 Frictional stress on the flange as a function of the angle of attack on the outer rail. *Source* José Luis López Gómez

the front wheel of a traditional bogie, as shown in Fig. 11, and provides an additional improvement in derailment resistance.

But the big problem with the triangular guidance system was its irreversibility. In fact, the stable self-guiding of the articulated composition, and the anti-derailment effect described above, only occurs when it is pulled in a forward direction. The reverse driving direction causes the angle of attack of the curved wheels to be positive, towards the outside of the curve. In this way the flange of the outside wheel contacts the rail at the front, and the frictional force generated pushes the wheel upwards, so that it tends to go up the rail, and derailment can occur at moderate speeds. Reverse steering is only permissible for manoeuvring at low speeds. Reversing the train therefore requires the whole composition to be turned around, which can be a serious disadvantage from the point of view of future operation.

From the first tests with the Talgo I, the need to improve the guidance system began to be considered. As a result, as early as 1944, a patent was published on "Light rail vehicles with reversible triangular traction" (Talgo and S.A.: Vehículos ferroviarios ligeros de tracción triangular reversible. Patente nº ES166809A1. 1944). As can be seen in Fig. 11, the problem has not yet been solved in detail, but the patent lays the foundations for the requirements for a mechanism that would allow the wheels between the coaches to be kept permanently oriented in a direction tangent to the track, with a zero angle of attack. This would almost completely avoid the friction of the wheel flange with the rail, and the lateral sliding component in the wheel-rail contact, achieving what José Luis López Gómez likes to call "guidance on the track, instead of by the track".

After the commissioning of the Talgo II manufactured by ACF, the efforts of the team led by Torán and Heredia focused on optimising the suspension and the link between coches, and fundamentally on the development of a reversible guidance system. The company's management, after the difficulties encountered in selling the Talgo II in Switzerland, saw the need to make the train reversible. In the same sense, Nardiz correctly foresaw that the imminent disappearance of steam locomotives, also irreversible, would lead to the removal of the "triangles" (manoeuvring routes designed to be able to turn the trains around) from the stations.

As a result, in the early 1950s, several patents were published proposing mechanisms designed to allow reversible guidance (Patentes Talgo 1951; Talgo and S.A.: Rodadura guiada para vehículos ferroviarios. 1952). The first of these, "wheel pair guidance system", describes a simple system for axle guidance. As can be seen in Fig. 12, it has similarities with the Watt mechanism used in automobile suspensions, and achieves a near-zero angle of attack on the wheels, so that they are oriented tangentially to the rail.

This type of mechanism, which finally allowed reversible guidance, was implemented for the first time on the Talgo III, which began commercial operation in 1964. This was the first Talgo train in service manufactured in Spain (largely due to the demands of Franco's government, which imposed as a requirement for its circulation that at least 80% of the manufacture be national). But the system developed at that time, with the logical design optimisations, is the one on which the guiding mechanism of current Talgo trains is still based today.

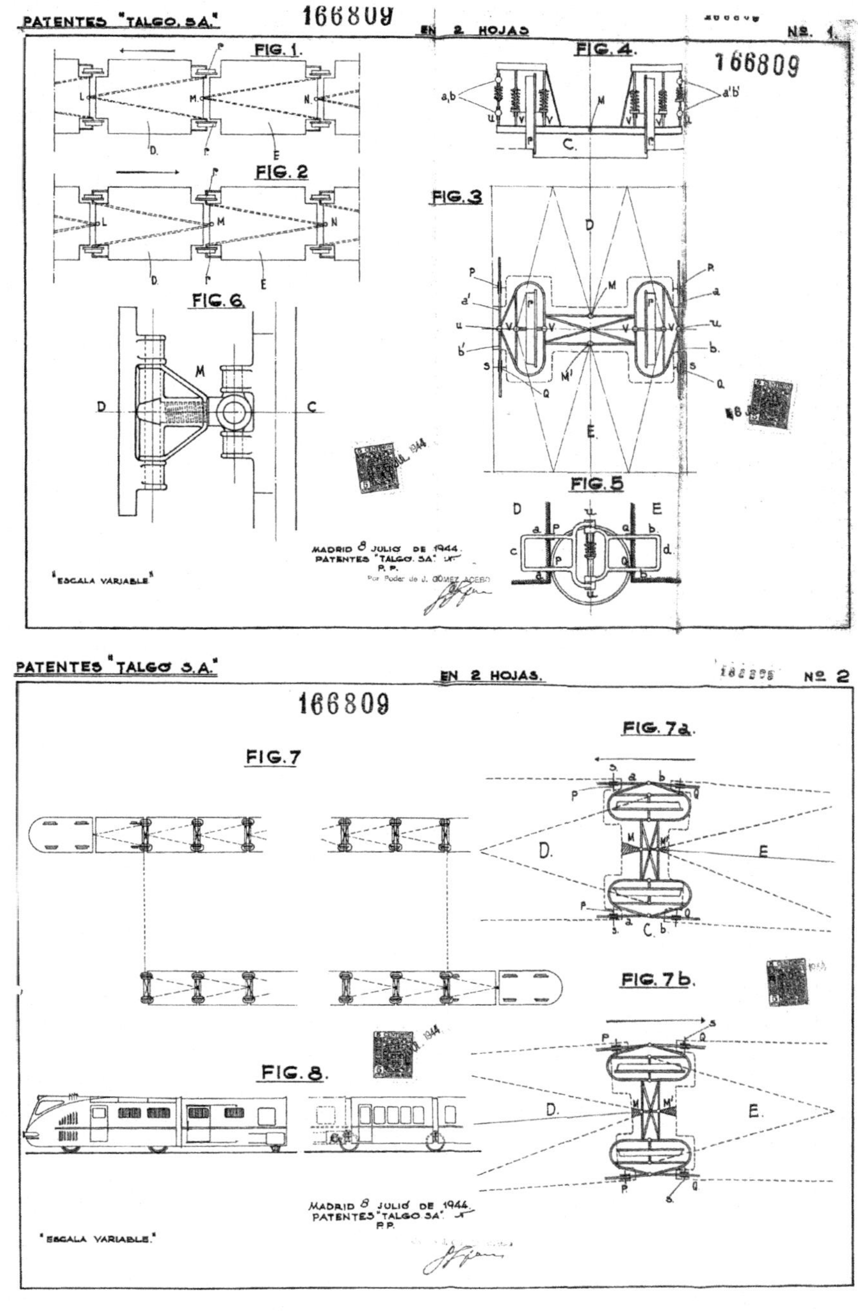

Fig. 11 First proposal for a "reversible triangular drive", in the patent ES166809A1 (Talgo and S.A.: Vehículos ferroviarios ligeros de tracción triangular reversible. Patente nº ES166809A1. 1944)

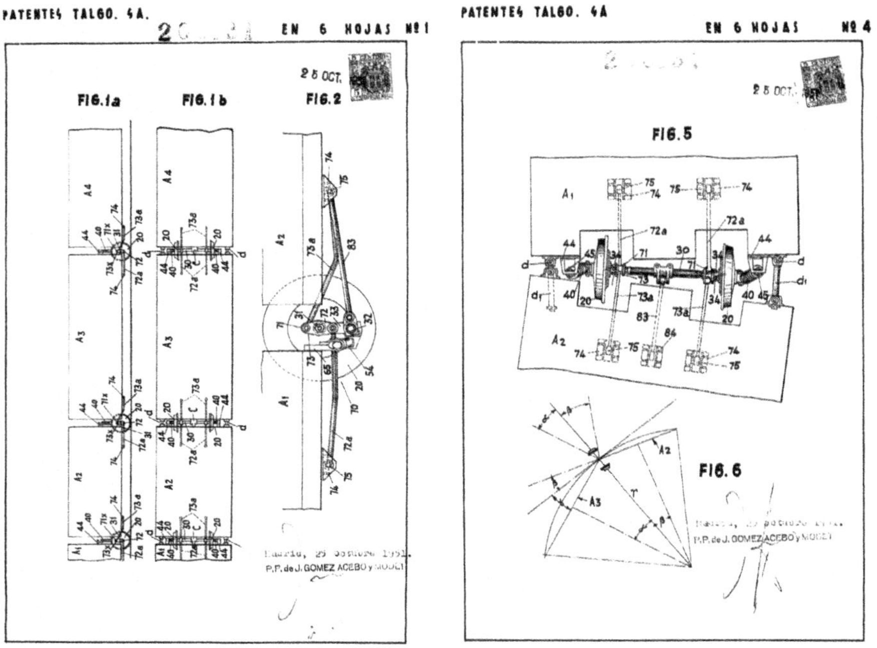

Fig. 12 Reversible guidance mechanism proposed in the patent ES200131A1 (Talgo and S.A.: Vehículos ferroviarios ligeros de tracción triangular reversible. Patente nº ES166809A1. 1944)

The Track Gauge Variation

The possibility of offering trans-Pyrenean rail routes, to connect the main Spanish and European cities, was mainly hampered by the difference between the Iberian gauge and the international gauge used in the rest of Europe. After the Talgo III was put into service, work was carried out on the development of systems that would allow the gauge to be changed quickly at the station itself, on the train in service, without the need to pass through a workshop.

As a result, in 1966 a series of patents were successively published proposing different systems for this change of gauge:

Patents ES323364A1, ES323365A1 and ES323366A1 (Patentes Talgo 1966a, 1966b; Patentes Talgo and Torán Tomás 1966a) propose a procedure for the rapid exchange of train axles for other axles with different track gauge. A multiple installation is also proposed that would allow the simultaneous exchange of the axles of a complete trainset. However, this technology was never implemented due to its complexity and high cost.

In patents ES332451A1 and ES332454A1 (Patentes Talgo 1966c, 1966d), based on the idea of the sliding wheel already pointed out by Goicoechea in 1942, a new system for gauge variation is proposed. With the train stationary, and while the wheels are kept raised and separated from the rail by means of a system of vertical actuators,

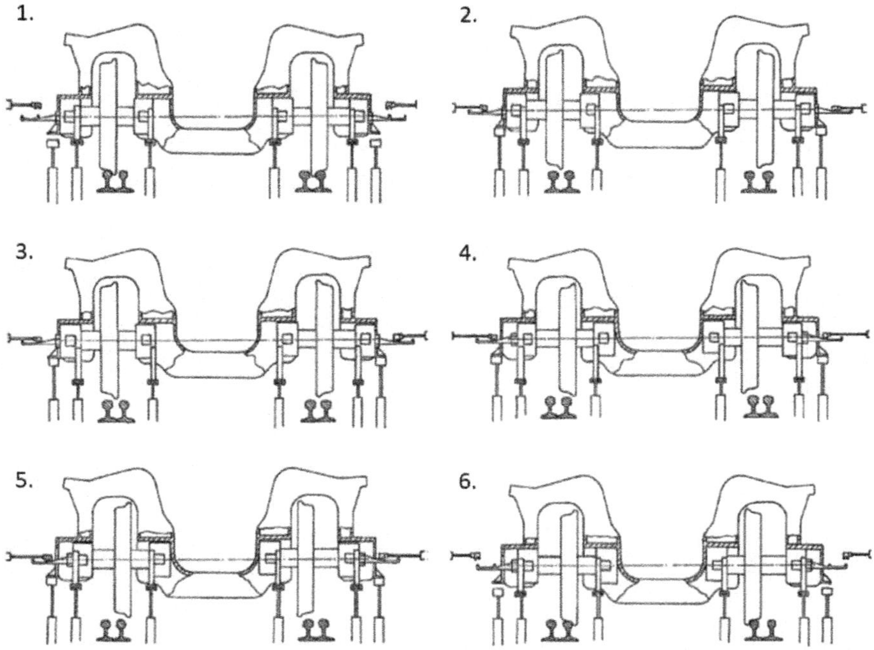

Fig. 13 Sequence of actions for track gauge variation, extracted from the patent ES332451A1 (Patentes Talgo 1966c)

a trigger system releases them from their transverse fixing. By means of a transverse actuator system, they are then brought to their new position, adapted to the new track gauge, and finally they are re-fixed by means of the triggers and brought into contact with the new rail. Figure 13 shows the proposed sequence of actions.

Patents ES332452A1 and ES332455A1 (Patentes Talgo and Torán Tomás 1966b, 1966c) finally propose a system based on the previous one, in which the vertical and horizontal actuators are replaced by a system of rails and slides of variable height and width as shown in Fig. 14. With this simple and purely mechanical system, the sequence of operations described above is carried out automatically while the train runs at low speed over the installation. In this way, the track gauge variation is carried out successively on each of the train's axles.

The system was implemented for the first time in an evolution of the Talgo III called RD (wheel displacement initials, *Rueda Desplazable* in Spanish), and it is basically the one that is still used in the track gauge change systems incorporated in current trains.

It is worth noting that an evolution of this technology, adapted for use at extremely low temperatures, enabled the Moscow-Berlin line to be put into operation in 2016, allowing the two capitals to be connected in just five hours. As a result, the Russian government awarded Talgo the Agustín de Betancourt medal in 2018.

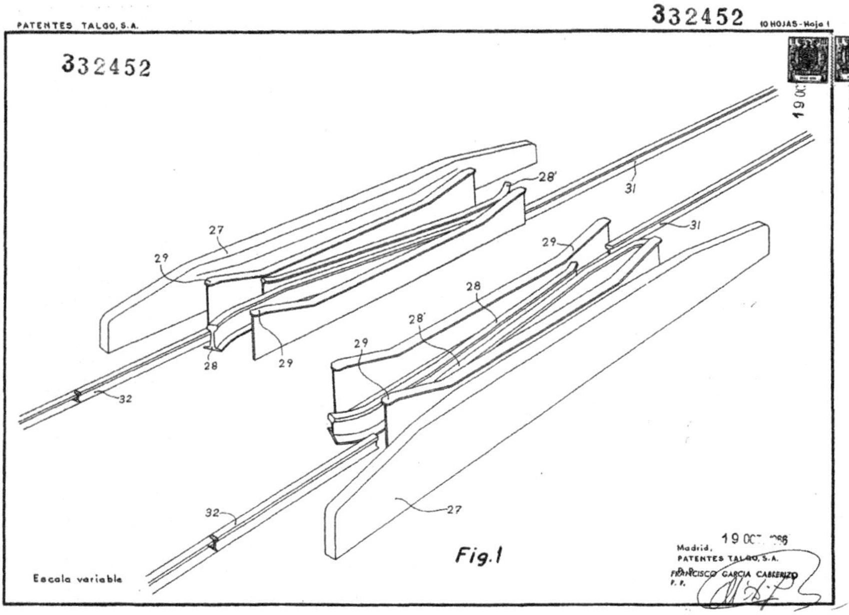

Fig. 14 Track and skate system for track gauge variation while running, described in the patent ES332452A1 (Patentes Talgo and Torán Tomás 1966b)

The Natural Tilting

The need to increase the speed of traffic in order to be able to compete with growing transport markets, such as road and air travel, led to the development of systems that would make it possible to increase the speed through curves without changing the railway layouts and without affecting passenger comfort.

As early as the 1950s, experimentation began in France with trains which, on entering curves, were inclined inwards, with an additional angle to the cant of the track. The aim was to reduce the centrifugal force component projected onto the passenger plane, and thus reduce the lateral acceleration experienced by passengers. This inclination was obtained by acting on the bodies by means of an active suspension system, while the pantographs were kept attached to the bogies in order to remain parallel to the running surface. Similar systems were also tested in England in the early 1970s, with unsatisfactory results. At the same time in Italy, Fiat developed a prototype that years later gave rise to the production of the well-known "Pendolino" trains which, in different versions and evolutions of the technology, have served and are still serving in different European countries. A similar tilting train prototype was also developed in Sweden at the same time.

The forced tilting system of the trains was expensive to maintain, and also required a sensor and control system that was relatively complex for the technology of the time, and presented problems of response delay. This meant that the trains developed on the basis of this technology in the different countries did not enter into real service until the second half of the 1980s.

In the meantime, Talgo developed the so-called natural tilting system, which simply consists of raising the plane of the suspension (and therefore its roll centre) far above the carbody's centre of gravity. In this way, the centrifugal force acting on entering a curve naturally causes an angle of inclination in the opposite direction to that which would be obtained with the suspension positioned below the centre of gravity. The principle of operation is shown in the sketches shown in Fig. 15, taken from the first patent for the natural tilting system (Patentes Talgo and Torán Tomás 1975) published in 1975.

In the first prototypes developed, two cylindrical columns were placed on the Talgo III's *rodals* ("*rodal*" is the name that Talgo uses for its rolling gears with independent wheels). Two pneumatic air springs were placed on the upper part of the columns, on top of which the roof of the body rested. The replacement of the coil springs with pneumatic suspension also made it possible to modulate the height of the cars and the rigidity of the pendular suspension, and to improve the comfort offered to passengers.

Moreover, the combination of the designed raised suspension with the carriage guiding system required the redesign of the articulation system between coaches, which was the subject of another patent published in 1976 (Patentes Talgo 1976).

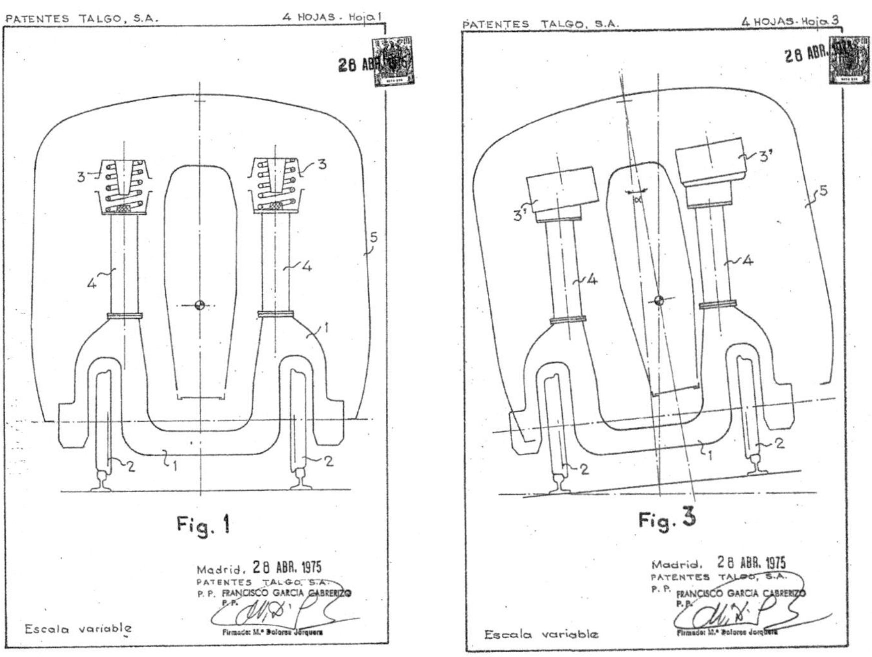

Fig. 15 Natural pendulum system described in the patent ES424615A1 (Patentes Talgo and Torán Tomás 1975)

A system is proposed as shown in Fig. 16, in which the front end of each carbody hangs from the suspension columns of the front coach by means of two vertical braces. The lateral restraint is completed by an extendable horizontal cylinder of adjustable stiffness, which joins the cars at the top and allows the cars to be tilted successively as they enter the curve.

The good results obtained with the first prototypes allowed the natural tilting suspension to be incorporated into service in 1980, in the so-called Talgo Pendular, which was the first train with tilting technology, being a few years ahead of the introduction of forced tilting trains developed in other countries.

The evolution of the natural tilting technology described here, with minimal changes to optimise its performance, is still used today in the high-speed trains manufactured by Talgo.

Pit Lathes and Other Workshop Machines

As mentioned above, the commissioning of train maintenance by the company itself led to the opening of a line of development of maintenance systems and workshop machines.

Among the developments carried out, the design of the pit lathes, such as the one shown in Fig. 17, which are located in a pit below the plane of the track (hence the name) and allow the train wheels to be reprofiled without being disassemble, undoubtedly stands out. The train is moved forward until the axle whose wheels are to be reprofiled is placed on the pit lathe. The rotation of each wheel is then triggered by friction cylinders, and the turning tool is operated automatically by a numerical control system.

The first talgo patents relating to pit lathes date back to the early 1970s. Their development was originally forced by the high level of demand from the French railway authorities, with very strict restrictions on the width measurement between internal wheel faces, as a condition for allowing Talgo trains to run in that country. The lack of confidence in the ability to maintain these measurement ranges using free wheels resulted in the early days in the obligation to carry out daily measurements of these dimensions on the trains in circulation. The development of the pit lathe made it possible to successfully overcome these requirements. And later, the evolution and optimisation of technology has allowed Talgo to open its own line of business as a supplier of this type of system.

The development of new methods for measuring diameters, camber, ovality and wheel profile parameters is also worth mentioning. As early as the 1980s, the measuring devices patented by Talgo introduced major conceptual changes in the way wheel parameters were measured, and in the precision obtained in the subsequent profiling of wheels on the pit lathe. The progressive evolution of these systems later allowed the integration of different technologies such as laser measurement, or the incorporation of other defect control methods such as ultrasonic analysis. As a result, Talgo's rolling maintenance systems are still considered to this day to be at the forefront of the sector.

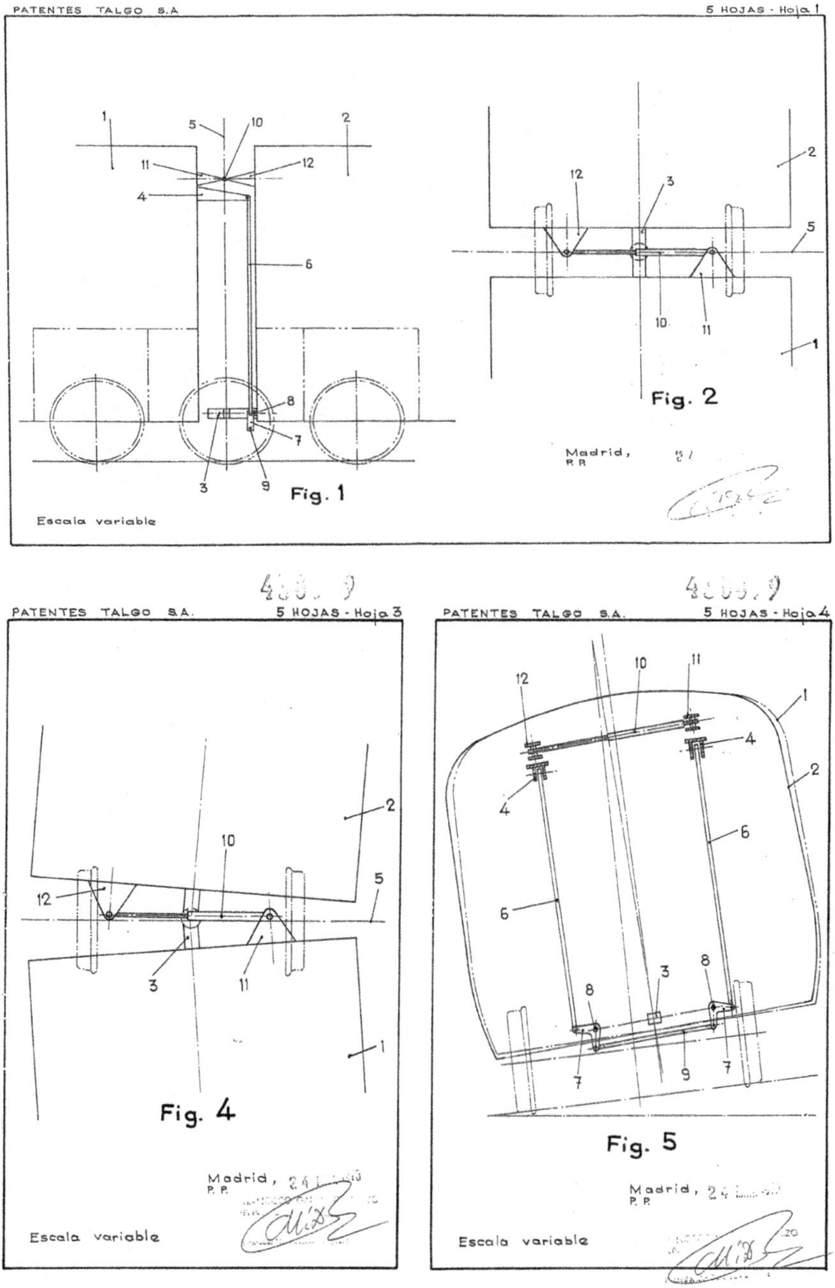

Fig. 16 System of articulation between the cars with natural tilting, as described in the patent ES430079A1 (Patentes Talgo and Torán Tomás 1975)

Fig. 17 A pit lathe

5.3 Leaping to High Speed and Beyond

From 1990 onwards, Talgo's research efforts focused on optimising the different systems for the development of a high-speed train. The coaches of the Talgo Pendular, with which the maximum speed of 500 km/h was achieved on the Munich test bench in 1990, were taken as the starting point. Then, a process of dynamic, structural, aerodynamic and vibro-acoustic optimisation was undertaken in order to meet the stringent requirements regarding active and passive safety, passenger comfort and environmental performance.

A detailed description of the developments in recent years is far beyond the scope of this text. It is important to note, however, that in its evolution to high speed, Talgo has maintained the basic mechanical principles described here, which have shaped its own distinct identity over the years. The evolution of Goicoechea's original disruptive idea, the "on the track" guidance mechanism, remains the fundamental technological difference of Talgo's high-speed trains,

In recent years, work has continued on optimising this guidance system by means of an active control system that acts on the guidance mechanism of each axle (Patentes Talgo, et al. 2001; Patentes Talgo 2006). Based on the comparison of the rotational speeds measured in each wheel, the system is able to detect small deviations in the trajectory, which are corrected in real time by acting on the length of certain bars of the mechanism. This ensures that the angle of attack of the wheels is zero at all times, minimising losses, friction and unwanted vibrations.

The evolution of other equally novel concepts developed in later years that have been mentioned here, such as the mechanism for changing the track gauge on the move and natural tilting suspension, are also incorporated into current trains. Optimisation of the suspension system first involved the incorporation of a primary suspension on the *rodals* (Talgo rolling gears), in addition to the pneumatic suspension at the top of the columns (Investigación y Aseguramiento Técnica S.A., Nardiz Landa, J.I., et al. 1994). Later, an active suspension system was incorporated, capable of optimising dynamic safety and ride comfort based on the information provided by sensors located along the entire length of the train.

The aim of this chapter has been to highlight, in the context of Talgo's engineering achievements, the importance of the disruptive ideas that served as the basis for its development. But also, and above all, we have sought to highlight the importance of teamwork in engineering in general, and in this one in particular, where, from the 1940s to the present day, it has allowed those ideas to be developed, to evolve and become a reality.

In this regard, it is worth mentioning in closing the prize for "European inventor of the year" awarded by the European Patent Office to José Luis López Gómez in 2013, for the optimisation of guidance systems mentioned above. This is undoubtedly a well-deserved recognition of the career of this engineer, as the penultimate representative of the saga of outstanding figures in Spanish mechanical engineering who, since Alejandro Goicoechea, have made a decisive contribution to placing Spanish railway engineering in the leading position it currently occupies worldwide.

Acknowledgements The collaboration of the aforementioned José Luis López Gómez has been essential for the writing of this text. We would like to thank him for the documentation and data provided, the knowledge shared and, above all, his illustrative and inspiring conversations about the railway.

References

M. Galán Eruste, M. Cano López-Luzzati, J.L. López Góme, Talgo, 75 años de espíritu innovador. Ed. Abomey Maquetren, S.L. (2018)

A. Goicoechea Omar, Sistema de construcción de vagones ferroviarios por soldadura eléctrica. Patente de introducción nº 103453. Oficina Española de Patentes y Marcas (1927)

A. Goicoechea Omar, Sistema de articulación ligera combinada para organización de trenes en rutas guiadas. Patente nº ES151396A1. Oficina Española de Patentes y Marcas (1941)

A. Goicoechea Omar, Mejoras en los sistemas ferroviarios de articulación ligera combinada. Patente nº ES159301A1. Oficina Española de Patentes y Marcas (1942)

A. Goicoechea Omar, Un nuevo sistema ferroviario de transporte por rodadura elevada. Patente nº ES141056. Oficina Española de Patentes y Marcas (1936)

A. Goicoechea Omar, Procedimiento perfeccionado para la construcción de trenes vertebrados. Patente nº ES339112A1. Oficina Española de Patentes y Marcas (1967a)

A. Goicoechea Omar, Perfeccionamientos a la patente principal nº 339112. Certificado de adición nº ES344763A2. Oficina Española de Patentes y Marcas (1967b).

A. Goicoechea Omar, Nuevos perfeccionamientos a la patente principal nº 339112. Certificado de adición nº ES345424A2. Oficina Española de Patentes y Marcas (1967c)

A. Goicoechea Omar, Procedimiento de vertebración esférica en ferrocarriles. Patente nº ES350085A1. Oficina Española de Patentes y Marcas (1968)

A. Goicoechea Omar, Procedimiento de circulación orgánica y accesos en trenes vertebrados. Patente nº ES350648A1. Oficina Española de Patentes y Marcas (1968)
A. Goicoechea Omar, Sistema de freno para vehículos ferroviarios de rodadura elevada. Patente nº ES457322A1. Oficina Española de Patentes y Marcas (1977)
A. Goicoechea Omar, Tren elevado múltiple. Patente nº ES8306653A1. Oficina Española de Patentes y Marcas (1982)
A. López Pita, Talgo y la alta velocidad, explotación técnica y comercial. ISBN: 978–84–89649–74–3. Ed. Patentes Talgo (2010)
I. Otero Paz, El tren vertebrado de Las Palmas de Gran Canaria, nacimiento y muerte en el barrio del Atlántico. http://www.eldiario.es (2017/01/06), last accessed 2020/01/15
Oficina Española de Patentes y Marcas. Galerías temáticas, grandes inventores: Goicoechea Omar, Alejandro. http://historico.oepm.es/museovirtual, last accessed 2020/01/15
Oficina Española de Patentes y Marcas. Galerías temáticas, grandes empresas: Patentes Talgo, S.A. http://historico.oepm.es/museovirtual, last accessed 2020/01/15
S. A. Talgo, Vehículos ferroviarios articulados formados de elementos ligeros, cortos, bajos, sobre ruedas independientes, con rodadura guiada por tracción triangular. Patente nº ES163239A1. Oficina Española de Patentes y Marcas (1943)
S.A. Patentes Talgo, Vehículos ferroviarios ligeros de tracción triangular reversible. Patente nº ES166809A1. Oficina Española de Patentes y Marcas (1944)
S.A. Patentes Talgo, Sistema guiado de pares de ruedas. Patente nº ES200131A1. Oficina Española de Patentes y Marcas (1951)
S.A. Patentes Talgo, Rodadura guiada para vehículos ferroviarios. Patente nº ES206t362A1. Oficina Española de Patentes y Marcas (1952)
S.A. Patentes Talgo, A. Torán Tomás, Procedimiento para la adaptación rápida en vehículos ferroviarios a diferentes anchos de vía. Patente nº ES323364A1. Oficina Española de Patentes y Marcas (1966a)
S.A. Patentes Talgo, A. Torán Tomás, Bogie ferroviario para la adaptación de vehículos ferroviarios a diferentes anchos de vía. Patente nº ES323365A1. Oficina Española de Patentes y Marcas (1966b)
S.A. Patentes Talgo, A. Torán Tomás, Mecanismo de vía para la adaptación de vehículos ferroviarios a diferentes anchos de vía. Patente nº ES323366A1. Oficina Española de Patentes y Marcas (1966c)
S.A. Patentes Talgo, A. Torán Tomás, Procedimiento para variar la distancia entre ruedas en vehículos ferroviarios. Patente nº ES332451A1. Oficina Española de Patentes y Marcas (1966d)
S.A. Patentes Talgo, A. Torán Tomás, Instalación fija múltiple para cambio de ancho de vía. Patente nº ES332454A1. Oficina Española de Patentes y Marcas (1966e)
S.A. Patentes Talgo, A. Torán Tomás, Procedimiento para variar la distancia entre ruedas en vehículos ferroviarios en marcha. Patente nº ES332452A1. Oficina Española de Patentes y Marcas (1966f)
S.A. Patentes Talgo, A. Torán Tomás, Instalación fija para cambio de ancho de vía de trenes en marcha. Patente nº ES332455A1. Oficina Española de Patentes y Marcas (1966g)
S.A. Patentes Talgo, A. Torán Tomás, Sistema de suspensión pendular. Patente nº ES424615A1. Oficina Española de Patentes y Marcas (1975)
S.A. Patentes Talgo, Perfeccionamientos en los acoplamientos entre vehículos de trenes articulados pendulares. Patente nº ES430079A1. Oficina Española de Patentes y Marcas (1976)
Investigación y Aseguramiento Técnica S.A., Nardiz Landa, J.I. et al., Suspensión de tipo primario para vehículos ferroviarios. Patente nº ES2061354A2. Oficina Española de Patentes y Marcas (1994)
S.A. Patentes Talgo, López Gómez, J.L. et al., Sistema para optimizar el guiado de ejes ferroviarios. Patente nº ES2195756A1. Oficina Española de Patentes y Marcas (2001)

S.A. Patentes Talgo, J.L. López Gómez, Método para optimizar el guiado de vehículos ferroviarios. Patente nº ES2316220A1. Oficina Española de Patentes y Marcas (2006)
A. Sáenz Sanz, J.P. Saiz González, Biografía de Alejandro Goicoechea Omar. Real Academia de la Historia. http://dbe.rah.es/biografias, last accesed 2020/01/15

Patricio A. A. Laura Casas (1935–2006)

Walter A. Montano and M. Gretchen Iorio

Abstract Patricio Laura is an outstanding figure in the Applied Mechanics world and deserves a space of recognition in the History of Science of Spanish-speaking leaders. His High School studies were carried out in a Jesuit institution, which left their ethics and scientific curiosity imprinted on him, because although he dedicated his life to abstract and applied science, he never left aside his Catholic thoughts. This led him to have epistolary discussions with Carl Sagan; among his more than 600 publications, he also wrote secular books. He graduated from the University of Buenos Aires (Argentina) in 1959, and shortly after, he went to the USA to study at Catholic University of America. Among the topics of his PhD, he published one of the first works on cable tension, determining the vibration induced by the drag of submarine radars. He returned permanently to Argentina in 1971. He was one of the founders of the American Academy of Mechanics and belonged to other Societies. He obtained the degree of Fellow of the Acoustical Society of America in 1976, among many other distinctions. He has inventions that have obtained international patents and developed devices for medicine. Patricio Laura passed away in 2006, and his colleagues and friends remember him as a character of great humor and intelligence. Inspired by the French Enlightenment, he was concerned about culture, religion, and education in addition to dedicating himself to science.

Keywords History of science · Acoustics · Applied mechanics · Numerical methods

---

W. A. Montano (✉)
Laboratory of Acoustical Research, ARQUICUST, Gualeguaychu, Argentina
e-mail: wmontano@arquicust.com

M. G. Iorio
Up-Wares, Inc. Waterbury, Waterbury, CT, USA

R. López-García and M. Ceccarelli (eds.), *Distinguished Figures in Mechanical Engineering in Spain and Ibero-America*, History of Mechanism and Machine Science 43, https://doi.org/10.1007/978-3-031-31075-1_7

1 Introduction

The author of this chapter, although a contemporary of Patricio A. A. Laura, did not have the fortunate opportunity to meet him personally. His biography and lifeline are constructed from conversations and exchanges of innumerable emails with his wife Yiyi, his sons and daughters, with his colleagues, and from the reading of his CV published in 1987.

At the beginning of the research, there was knowledge of the existence of the *Institute of Applied Mechanics* of the Universidad Nacional del Sur (National University of the South in Bahia Blanca City, Argentina), but there was no information about the professional and scientific life of Patricio Laura. After having the testimonies of his relatives and friends, it was found that he was a personality in the international scientific community for his pioneering contributions in applied mechanics methods, but the immensity of his humanistic thought and the legacy he left among his disciples and colleagues, is much greater than the scientific one.

The life of Patricio Laura also surprised many acousticians of Argentina, those who had used his books and papers written in English, to find out that he was Argentinean. Initially, the author assumed that Patricio Laura had developed his activity only in the Acoustics field. However, he was a researcher of many disciplines within Applied Mechanics; some of his work on mechanical vibrations and mathematical models is known as well as in acoustics.

Dr. Laura's doctoral thesis (sponsored and financed by NASA, available on the Internet) is considered in the scientific world as a pioneer in applied mechanics, rocket propulsion, among others. It is mentioned as a source of data in much other research, part of its content was later extended and gave answers to model the stress due to cable vibration. It helped to determine the mechanical energy exerted by the trailing cables of submarine sonars and improve their acoustic response. Fortunately, his family holds almost all the documentation; Patricio Laura's archive is vast, and they were asked to share pictures that they consider of greater emotion; some of them are published here.

2 Biographical Notes on Patricio Laura

Patricio Adolfo Antonio Laura Casas was born on June 13, 1935, in the city of Lincoln, a province of Buenos Aires, Argentina, which at that time had less than 8 thousand inhabitants and most of them were European immigrants, himself being of Italian descent.

Dr. Laura grew up in an educated family: His grandfather, Adolfo Laura, was an accountant, and his father, Lauro Olimpio Laura, was an outstanding Civil Engineer responsible for large infrastructure works in the Province of Buenos Aires. On his maternal side of the family line, his grandfather Absalón Casas was a great representative of the well-accepted politics of those times: Lawyer, jurist and Argentine politician, he became an interim governor in 1910 of the province of Córdoba, Argentina; from his grandmother Dominga Vázquez, he took many of the Catholic values that he would later defend to the utmost, as well as his great love and observation of nature; and one of his uncle was a well-know Jesuit priest.

Patricio's childhood was spent in a peaceful, religious, and calm social environment, mixed with his family's professional activity. Patricio attended high school between 1948 and 1952 at the Colegio del Salvador of the Jesuit Order, in the City of Buenos Aires, a post-war period with a small social welfare in Argentina–thanks to the country's participation in the Marshall Plan and in the years of Juan D. Perón's military government.

2.1 *Life as a College Student in Argentina*

After World War II, the race for development of nuclear technology began all over the world, and Argentina played a very important role in this. The Argentine Navy decided to create an institution with high intellectual and scientific content (similar to MIT of the US) to support the development of nuclear technology in the country. The Argentine government through Dr. Enrique Gaviola (eminent Argentinian astrophysicist) invited Nobel laureate Werner Heisenberg to head that future institute, but it was not possible, despite Heisenberg's acceptance, because the British occupation forces refused that he leaves Germany (Comastri 2013). Norbert Wiener was also unsuccessfully able to go to Argentina and direct that Institute (Puglisi 2011). Finally, in 1946, the Instituto Radiotécnico (Radiotechnical Institute) was created through an agreement between the Navy and the University of Buenos Aires, an educational center with a high academic and scientific level.

It is worth mentioning that the existence of the Radiotechnical Institute, which, although it had few years of existence, in this area the Navy officers studied engineering, and Patricio Laura shared his university studies with many of them; this was when scientific research began at the University of Buenos Aires.

Patricio Laura started to study engineering at University of Buenos Aires in 1953, (one year after the creation of the Engineering Faculty) which coincides with the opening and diversification of professional specialties. Patricio Laura obtained a Civil Engineer degree in 1959, but finally he dedicated his life to science and applied mechanics. During his university studies (according to the testimony of his wife Yiyi), Patricio Laura felt great admiration and interest in how people studied in the US and England; so, before finishing his degree, he established contacts and wrote to many universities in those countries to continue his postgraduate studies.

2.2 Life as a Postgraduate Student in the US

Patricio Laura, shortly after graduating as an engineer in Argentina, married Nélida Gómez Villafañe (known as "Yiyi," a Neonatology doctor) in August 1959. They both decided to travel to the USA with the financial help of Yiyi's father, because in those years it was not usual to emigrate outside Argentina to pursue postgraduate studies It was necessary to pay a deposit to obtain a US "Green Card" to stay in that country. Patricio immediately got a job with a construction company, Arthur Venneri Co., as a project engineer (he participated in the construction of a large library in Maryland and an atomic reactor, among other important projects).

Dr. Laura began his postgraduate studies in 1960 at the Catholic University of America in Washington, D.C. (let's not forget that he was educated by the Jesuits), specifically in the School of Engineering with the specialty of Solid Mechanics, in the subsidiary fields of Applied Mathematics and Fluid Dynamics.

He obtained his Ph.D. on May 4, 1964, presenting the thesis, "Conformal Mapping of a Class of Doubly Connected Regions" (Laura 1965), in the specialty of Fluid Dynamics, certified with a score of 100/100 by Dr. Chieh-ch'ien Chang Chang, Ph.D. (Laura 1965). Chieh-ch'ien Chang (an outstanding Chinese scientist in that specialty), mentioning that his thesis was financed by NASA and supervised by John H. Baltrukonis (a renowned scientist in solid fuel rocket dynamics), since his work was related to that subject.

2.3 *His Work in Scientific Research and Teaching in the US*

In 1961 Patricio Laura began his scientific research activity at the Catholic University of America (Laura 1987; p. 5) with the following:

- 1961–1962: Laboratory Assistant to Prof. A. J. Durelli, known worldwide in experimental determination of deformations and stresses. Work performed: development of methods for measuring deformations in solid fuel rockets, investigation of properties of viscoelastic materials, etc.
- 1962–1965: Research Associate with Prof. J. H. Baltrukonis. Work performed: developments of numerical methods in vibrations of continuous media with application to solid fuel rockets, determination of characteristic values, etc.
- 1965–1968: Research Associate with Prof. F. Andrews. Work carried out: sound radiation of vibrating bodies and plate vibrations.
- 1969–1970: Technical Director and Program Manager of the Themis Project, on cable system dynamics, sponsored by the US Department of Defense (Fig. 1).

During his stay in the US, he was an Instructor, Assistant Professor, Associate Professor, and Full Professor successively in the Dept. of Mechanical Engineering at the Catholic University of America. He taught courses in Fluid Mechanics, Applied Mathematics, Strength of Materials, Applied Elasticity Theory, Plate Theory, and Vibrations of Mechanical Systems. Patricio Laura was also responsible for supervising several graduate students and their theses (Laura 1987; p. 7).

Fig. 1 Distinction obtained in the US in 1968. Note: Picture belonging to Dr. Laura's family archive

2.4 Social and Family Life in the US Until 1969

The 1960s were years of great social and political upheaval throughout the world, and Latin America was no stranger to that. In the case of Argentina, a country marked by continuous military dictatorships, its brief democratic periods served to achieve some scientific advances in the universities. In 1962, the Government underwrote a loan with the Inter-American Development Bank (IDB) for US$ 5 million for the "Technical Re-equipment of National Universities," (Montano 2016) and thanks to this the period, 1965–1968 is known as the 'Golden Years of the Argentine University.' During this period, many research universities laboratories were created and strengthened, and the Scientific and Technical National Research Council (Consejo Nacional de Investigaciones Científicas y Técnicas–CONICET) tried to keep as many researchers as possible in Argentina and invited many expatriates to return to the country.

Patricio Laura remained in the US throughout the 60 s, but he was never "isolated" in his research; on the contrary, he was always concerned about the development of science in Argentina and the Latin American social issues. Yiyi testifies to this by mentioning that on May 7, 1969, in an open letter to President Richard Nixon, published by the prestigious The Washington Post, Patricio Laura wrote a reply to the erroneous assessments regarding the political situation in Latin America (Laura, 1987; p. 89) made by Nixon; his letter was answered by the White House, by Nixon's Chief of Staff John R. Brown, saying that "*they will be more careful in the future when issuing opinions of that kind.*"

In 1968, the Argentine Nobel Prize winner Dr. Bernardo Houssay (then president of the CONICET), began to contact all Argentine scientists abroad, who personally visited Patricio Laura at his home in the US. As Yiyi tells, after evaluating many factors for pros and cons, they decided to return to Argentina in 1970 with his wife and his first three children; he had only provisional teaching proposal and all desire to develop science in Argentina.

2.5 The Return to Argentina in 1970

Patricio Laura returned to Argentina with his family at the beginning of 1970 (he had requested a leave of absence in the US) to join the Catholic University of Salta (1500 km north of Buenos Aires City) as a professor; this university was run by the Jesuits and was only three years old. The presence of Patricio Laura did not go unnoticed in Salta City; he was interviewed by NORTE newspaper, and the journalist took the opportunity to ask him about the arrival of man on the Moon, since Patricio Laura had worked for NASA, and the question "*Are you opposed to the space race?*" he answered, "*I am not opposed. I am objective. On Earth there are many problems to be solved*..." Then, Dr. Laura mentioned the ravages of cancer, stressing that there is no research on that. The article ends with a paragraph that defines his life, "*There is a latent enthusiastic human vocation for science –science and conscience– reflected in all its definitions. The word -humanity- pops up with happy frequency... something undoubtedly comforting.*"

In 1970, science was only just becoming important at that university. The academic environment of the city was still very small, compared to what he was used to in Washington, he returned briefly to the US, where the Universidad Nacional del Sur (National University of the South) offered him a teaching position.

Patricio Laura accepted the offer to join the Universidad Nacional del Sur (UNS), established at Bahía Blanca City (635 km south of Buenos Aires City), as a full Professor; he and his family packed their bags again and moved to this city. He joined the Stability area of the Engineering Department on July 1, 1970 and was soon appointed Coordinator of the same field. That date is mentioned exactly, because that is when he began his tireless scientific work in Argentina until he left us forever one afternoon in November 2006.

Dr. Laura inaugurated a tradition of scientific research at the UNS that continues today. It was not easy at the beginning: he insisted that everyone had to be fluent in languages, especially English; encouraged the writing and publication of articles and encouraged full-time professors to be trained in research.

Despite moving and new experiences, Patricio had time to continue his research and, already as a professor at the UNS, he collaborated in the first publication of Mario Maurizi entitled "Eigenvalues for a Uniform Fluid Waveguide with an Eccentric-Annulus Cross Section," published in the Journal of Sound and Vibration Vol. 18(3) 1971 pp. 445–447.

2.6 The First Years Working at UNS

The year 1972 is established a priori as a time window, as the 'starting year' of the immeasurable scientific work of Patricio A. A. Laura, because two emblematic events (in this author's opinion) took place that year since his return to Argentina:

a. The Junior Chamber of Buenos Aires distinguished him along with nine other people as "Outstanding Young Person of the Year 1972," with a Jury presided by the Nobel Prize winner Dr. Luis F. Leloir, composed by Jorge L. Borges, Domingo Liotta, among other great Argentinian persons.
b. In April, his work done with his colleagues, "Torsion of bars of regular polygonal cross section," is published in the Journal of the Engineering Mechanics Division.

An event which highlights the dedication, open heart and total humility toward his coworkers, was manifested when he received the acceptance of the publication "Torsion of bars of regular polygonal cross section [...] According to his colleagues, "*with the letter-acceptance of the Editor of the journal in his right hand, he goes through the corridor of our work area and moves to the offices of other neighboring areas, and there he comments to the other coordinators the achievement reached.*" For the

Fig. 2 Patricio Laura at his home office (circa 1995). Note: Photo belonging to Dr. Laura's family archive

first time, since the creation of the UNS in 1956, a professor at that university had reached an international level with any of the projects and results of their scientific work.

In February 1975, Patricio Laura created the Institute of Applied Mechanics (Instituto de Mecánica Aplicada–IMA). A small group of professors from the UNS and the National Technological University (Universidad Tecnológica Nacional–UTN), together with four laboratory technicians who made up the support staff, were included. From its original headquarters, located at the Puerto Belgrano Naval Base, a more than fruitful development has begun that goes beyond the stability and dynamics of machine structures.

A very important explanation must be made here: Some publications state that "Patricio Laura worked for the Argentine Navy," which is false. The years 1974–1975 were very difficult political times, and anyone who had studied in the US and returned to Argentina was immediately branded as a "spy for Yankee capitalism" by extremist groups. Unfortunately, Dr. Laura's first office was set on fire in a terrorist attack[1]; this happened in the early hours of the morning and there was no loss of life. As a result of this, for the protection of the instruments and people, his research work was carried out in buildings within the facilities of the Argentine Navy (hence the misinterpretation).

[1] This attack is not properly documented; some of Laura's collaborators are pained to recall the cowardly and ridiculous fire. With the due respect that the subject deserves, his family (and those that lived under death threats) wants to share this inexplicable event, because the only thing Patricio Laura defended was freedom of expression and opinion.

In December 1975, Patricio Laura became IMA's Director, and many activities were developed in different scientific disciplines there, from nuclear physics to biomechanics, including acoustics and the manufacture of medical devices; they dealt with oncological issues, pediatrics, neonatology, etc.

2.7 IMA Director Until 2001

The creation of the Institute of Applied Mechanics (Instituto de Mecánica Aplicada–IMA) had international repercussions; the publication Mechanics, in June 1976 comments on this fact; also, the magazine *Experimental Mechanics* makes praising comments for this laboratory. It is clear that Dr. Laura was already a recognized scientist of world-wide reach. In 1977, the publication *Foreign Noise Research* of the US EPA, commented on the work done on vibrations (it was the only time a study carried out in Latin America was mentioned). In 1980, the Spanish version of the medical journal MD, mentioned IMA's work in bioengineering and the invention of medical equipment. The *Journal of the Acoustical Society of America*, in 1982, highlighted the activities of the IMA.

Throughout the existence of the IMA, many foreign specialists shared research without traveling to Argentina, and this fact is highlighted because it is difficult to imagine sharing work by mail during the 1970s and 1980s decades -without the benefit of the Internet—and correspondence took weeks to travel back-and-forth.

Not only did researchers work at the IMA, but Patricio Laura encouraged the participation of college and high school students, something unprecedented in Argentina dozens of researchers were trained, and postgraduate students carried out their masters or doctoral theses; specialization courses were given in many areas, tailored classes were given, etc.; in short, the IMA developed countless scientific activities.

Patricio Laura and his collaborators presented more than 400 scientific papers (it is impossible to reproduce their titles; they occupy 35 pages), published about 12 books. From this Institute, they participated in dozens of world congresses and organized seminars, conferences, courses, not only of a scientific nature, but also of divulgation. It is worth mentioning that the prestigious publishing house MIR, of the former Soviet Union, requested permission to include a work by the IMA in one of its books.

The IMA ended the first part of its life and momentarily closed in 2000, then was re-founded and strengthened in 2001, when it reopened with new agreements between CONICET and different faculties of the UNS; it remains open and maintains the academic tradition and pedagogical work initiated by Patricio Laura to this day.

2.8 *The Social, Cultural, and Religious Life of Patricio Laura*

Since his return from the US, Patricio Laura was an enthusiastic citizen who participated in various cultural, social, educational, and religious activities, the latter being the most publicly known, because he kept close ties with John XXIII Salesian School of Bahía Blanca city (Fig. 3).

Dr. Laura has about 50 lay publications, including "Science and Faith" of 1987 and "The Catholic Church and the Sciences" (see Fig. 3a) of 2000, which had great international affect (in contrast, in the same year he wrote one of his most abstract works on mechanics, see Fig. 3b). Both were commented on by many newspapers, and even had a mention by the Vatican Nuncio.

In "The Catholic Church and the Sciences," he rescues 46 eminent scientists who never left the Catholic religion, highlighting the importance of Jesuit education and defining humankind as a "religious animal and scientific animal;" he criticizes the fact that history only remembers the years of Darkness imposed by Catholicism and forgets to highlight the science it produced (even if it is minimal).

As an example of his social concern over scientific education, Patricio Laura was the 'architect' in introducing scientific computing in high school and tertiary education in Argentina (in this case at the John XXIII Salesian School of Bahía Blanca City). Due to his diligence, he obtained part of the money for the installation of a complete PDP–11 computer system, which was put into operation in 1979

Fig. 3 Books front covers **a** *The Catholic Church and Sciences*; **b** *Approximate methods in applied mechanics*. Note: Pictures belonging to Dr. Laura's family archive

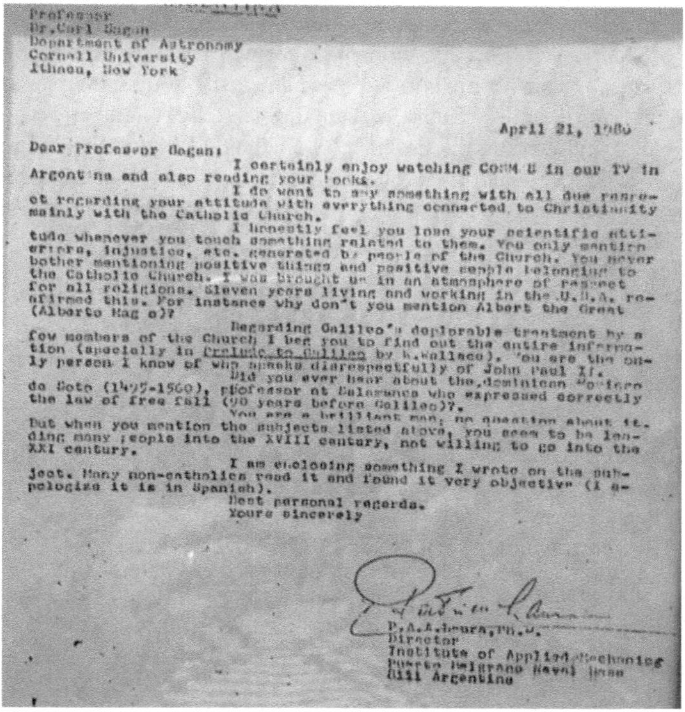

Professor
Dr. Carl Sagan
Department of Astronomy
Cornell University
Ithaca, New York

April 21, 1980

Dear Professor Sagan:

I certainly enjoy watching COSMOS in our TV in Argentina and also reading your books.

I do want to say something with all due respect regarding your attitude with everything connected to Christianity mainly with the Catholic Church.

I honestly feel you lose your scientific attitude whenever you touch something related to them. You only mention errors, injustice, etc. generated by people of the Church. You never bother mentioning positive things and positive people belonging to the Catholic Church. I was brought up in an atmosphere of respect for all religions. Eleven years living and working in the U.S.A. reafirmed this. For instance why don't you mention Albert the Great (Alberto Magno)?

Regarding Galileo's deplorable treatment by a few members of the Church I beg you to find out the entire information (specially in Prelude to Galileo by W. Wallace). You are the only person I know of who speaks disrespectfully of John Paul II.

Did you ever hear about the dominican Domingo de Soto (1495-1560), professor at Salamanca who expressed correctly the law of free fall (90 years before Galileo)?

You are a brilliant man, no question about it. But when you mention the subjects listed above, you seem to be leading many people into the XVIII century, not willing to go into the XXI century.

I am enclosing something I wrote on the subject. Many non-catholics read it and found it very objective (I apologize it is in Spanish).

Best personal regards.
Yours sincerely

P.A.A. Laura, Ph.D.
Director
Institute of Applied Mechanics
Puerto Belgrano Naval Base
8111 Argentina

Fig. 4 Dr. Laura's letter to Carl Sagan. Note: Picture belonging to Laura's family archive, reproduced with permission

and the "Don Bosco Computer Center" was created. Dr. Laura was an ad honorem professor at its outset, teaching Differential Equations; some mathematical methods of his Papers were solved using that advanced calculation system.

On an episode of COSMOS TV series in 1986, Patricio Laura had an epistolary exchange with Carl Sagan himself (See Fig. 4), about the fact that the Catholic Church was not always responsible for religious obscurantism, and that, on the contrary, it had outstanding scientists (Fig. 4).

2.9 *Scientific Family Tree of Patricio A. A. Laura*

Something that caught the attention of the authors of this article, is that in 2016 when he started researching Dr. Laura's family background, they found on the Internet that the great Italian mathematician Ernesto Laura (1879–1949) was assistant to the eminent Giuseppe Peano (1858–1932)—he is Patricio's great-uncle, and even more surprising is the fact that Ernesto Laura specialized in Analytical Mechanics and Rational Mechanics (Saltini, 2013) and wrote about mechanical waves in "Sulla propagazione delle onde in un mezzo indefinito" (On the propagation of waves in an indefinite medium) and "Sopra le vibrazioni normali di un corpo elastico immerso in un fluido" (Above the normal vibrations of an elastic body immersed in a fluid).

His father, Lauro Olimpio Laura, was an eminent civil engineer, responsible for most of the first highways in the Province of Buenos Aires. In 1959, he represented Argentina at a UNESCO world event, in a seminar on urbanization problems in Latin America, which was held in Santiago de Chile, and chaired the Commission, "Urbanization and economic development."

3 Dr. Laura's Immeasurable Scientific Work

Given the length of the original document, this section will comment on and extract paragraphs from Patricio Laura's personal CV, which he submitted in 1987 to the National Council for Scientific and Technical Research (CONICET) of Argentina. Patricio Laura's doctoral thesis, within the field of Applied Mathematics in Conformal Representation, consisted of the following:

> … in the derivation of two integral equations and the determination of functions that transform regions with a circular outer perimeter and a star-shaped inner contour into an annular region. This problem is of interest for the solution of problems of stress, temperature, and dynamics of solid fuel rockets. (Laura 1987; p. 20)

Other approximate techniques of conformal representation were also developed by Dr. Laura:

> … to be used in the calculation of stresses around mining galleries and exposed in the First International Congress on Rock Mechanics, 1966, also being cited by Keinosuke Cotoh: "A Numerical Method sor Determining the Mapping Functions of Some Simply Connected Regions," in *Theoretieal and Applied Mechanies*, Vol. 21, University of Tokyo Press, 1973. (Laura 1987; p. 20)

One of the most emblematic papers is "Dynamic Analysis of a Simplified Bone Model During the Process of Fracture Healing," published in the *Journal of Biomedical Engineering*, March 1990 Vol. 12(2), Mario Maurizi (one of his colleagues) completes the information commenting that "the work was requested by 26 specialists and institutes worldwide, mentioning the Department of Anatomy and Human Biology of England, Veteran Hospital of California USA, Laboratoire de Biologie du Tissu Osseux de Saint Etienne France."

3.1 Research on Approximate Methods

Regarding his work on the development of *Approximate Methods and their Application in the Determination of Characteristic Values*, Dr. Laura and his associates published several papers in the US and England, where "… they described research carried out using new criteria for the application of the weighted residual method. Such criteria were "Placement along arcs and Minimization of the residual function by successive derivatives" (Laura 1987; p. 21). Regarding the latter criterion, it is worth mentioning a laudatory review published in Applied Mechanics Reviews by R. B. Mc Calley in 1966, among four others received in the same journal.

3.2 *Research on Theoretical and Applied Mechanics*

Only the most outstanding papers that use Dr. Laura's original scientific contribution will be presented here; in his CV he points out many more research, and also doctoral papers that mention him. The problems studied by Dr. Laura and his associates, in the topic of *Vibrations in Continuous Media*, have been the following:

> Vibrations of plates, membranes, and bars of complex shape. In general, the solution method consisted of the conformal transformation of the given domain and then use of a variational method to solve the transformed differential equation (Laura, 1987; p. 21). The high accuracy achieved by this approach has been pointed out by the well-known vibration specialist J.C. Robson (Great Britain), in Applied Mechanics Reviews (1968), referring to the work: "An Application of Conformal Mapping to the Determination of the Natural Frequency of Membranes of Regular Polygonal Shape." Developments in Mechanics, Vol. 39 Part II, John Wiley & Sons, pp. 155-163, 1967. (Laura 1987; p. 22)

The mechanical-mathematical method developed by Laura and her collaborators has been referred to many times in scientific publications related to the subject, and in publications on microwaves, waves, electronics, and acoustics:

> B.H. Bates in "The Theory of the. Point - Matching Method for Perfectly Conducting waveguides and Transmission Lines," 1969. J. A. Fuller en "The Point - Matching Solution of Uniform Non-Symmetric Waveguides," 1969. B. H. T. Bates in "The Point - Matching Method for Interior and Exterior Two-Dimensional Boundary Value Problems," 1967. (Laura, 1987; p. 22)
>
> T.M. Bristol in "Waveguides of Arbitrary Cross Section by Moment Methods" (Doctoral Thesis) 1967. O.C. Zienkiewicz in "Application of Finite Elements to the Solution of Helmholtz's Equation," IEEE 1968. R.M. Bulley y J.B. Davies in "Computation of Approximate Polynomial Solutions to TE Modes in an Arbitrarily Shaped Waveguide," IEEE 1969. A. K. Bahrani in "Solution of Laplace's Equation in a Region with Dielectric Interfaces of Arbitrary Shape," IEEE 1970. M.J. Hine in "Eigenvalues for a Uniform Fluid Waveguide with an Eccentric Annulus Cross Section," Journal S&V 1971. M.J. Hine y F.J. Fahy in "A Membrane Analogy to an Acoustic Duct" Journal S&V 1971. J.B. Davies in "Review of Methods for Numerical Solution of the Hollow Waveguide Problem," IEEE 1972. B.E. Splelman in "Waveguides of Arbitrary Cross Section by Solution of a Nonlinear Integral Eigenvalues Equation," 1971. B.H. Bates in "Some Reviews in Radio Science," 1972. L. Ng in "Tabulation of Methods for the Numerical Solution of the Hollow Waveguide Problem," IEEE 1974. (Laura, 1987; p. 23)

3.3 Research in Underwater Acoustics

In Dr. Laura's own words:

> This topic is related to vibration investigations of continuous media, but with emphasis on the sound radiation problem. Although the problems analyzed are of a basic nature, they constitute a good foundation for the rational design of acoustic transducers and have been favorably commented in the publication *Applied Mechanics Review* in 1968. The paper "Directional Characteristics of Vibrating Plates and Membranes" JASA Vol. 40 1968, has been referred to by the distinguished acoustician G. Maidanik, in his paper: "Operational Method for Evaluating the Farfield Radiation from Plane Boundaries," JASA Vol.45 1969 (Laura 1987; p. 25).

3.4 Research in Fluid Mechanics and Heat Transfer

His work in this area is in the field of nuclear energy, Dr. Laura in his CV mentions the research that used his original work, the main ones are A. L. London from Stanford University; B. K. Shah and K. P. Johansen (Technical University Berlin); Professor Bruno Boley (the most cited name in Thermal Stress Theory); Professor B. A. Finlayson in his book: "The Method of Weighted Residuals and Variational Principles" Academic Press, 1972 (Laura 1987; p. 26). And also in Dr. Laura's own words:

> The problems studied belong to two basic categories: (a) determination of non-stationary thermal fields in regions of complex shape, (b) study of problems where it is necessary to analyze flows with heat transfer in ducts of arbitrary cross-section. In both cases the solution has been generalized and at the same time unified using the conformal representation method in an original way, since in classical applications this method is used, but only to solve the Laplace equation. The work published in the British Journal of Applied Physics, 1967, has aroused interest in the chemical industry. (Laura 1987; p. 25)

3.5 Research on the Representation of Vibration Modes of Structural Elements by Polynomials

Throughout several pages of his CV, Dr. Laura comments on the many publications that mention the use of his innovative mathematical method. It is impossible to transcribe them all here, but in its own words:

> Using polynomials, in 1967 Dr. Laura and one of his disciples (B. F. Saffel) published a paper: "Study of Small Amplitude Vibrations of Clamped Rectangular Plates Using Polynomial Approximation" (JASA Vol. 41, pp.836-839, 1967); this modest paper shows that it is possible to calculate frequencies for various modes of vibration of rectangular plates by approximating them with simple polynomials. The paper presents a large number of numerical results and simple formulas for various frequencies that can be used with minimal effort by a design engineer. It is of interest to note the fact that at the *Third Canadian Congress of Applied*

Mechanics, 1971, Ch. Bert of the School of Aerospace, Mechanical and Nuclear Engineering, University of Oklahoma, presented a NASA–sponsored research paper using the method of Laura and Saffel (Laura 1987; p. 30)

4 Dr. Laura Legacy in the Research on Oceanographic Cable System Dynamics and Acoustics

Under the direction of Dr. Laura and the sponsorship of the US Department of Defense the "Cable System Dynamics Program," began in 1968 at Catholic University in Washington; this program has been unique in the US in the problem of oceanographic cables "…proof of the interest it aroused was the First National Symposium held in Washington in August 1970, sponsored by the Office of Naval Research and held at Catholic University" (Laura 1987; p. 26).

Dr. Laura was directly responsible for studies related to dynamic behavior of cables under periodic excitation and shock loading, elastic detection of cable failures, wave propagation velocity as a function of initial cable prestress, etc. In the work "On the Dynamic Behavior of a Cable System in a Recovery Operation" by Dr. Laura and Goeller (JASA Vol. 49 pp. 615–621), for the first time in the technical and scientific literature, the dynamic problem of a salvage operation in the ocean is introduced; both the analytical and experimental models developed are a first approximation to the real problem, allowing to draw several conclusions of practical interest (Fig. 5).

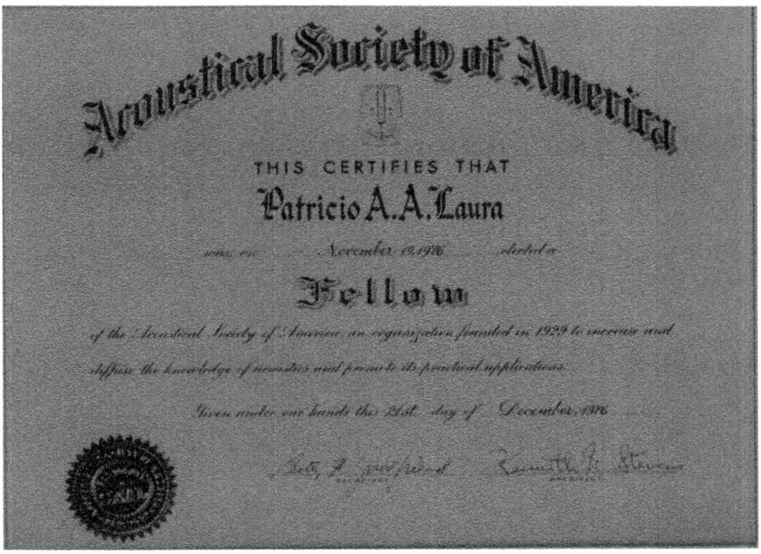

Fig. 5 Presentation of Fellowship to Dr. Laura, for his contributions to science. Note: Picture belonging to Laura's family archive, reproduced with permission

Dr. Laura in his CV mentions the papers he presented in different journals, among them: the *Journal of the Acoustical Society of America* (JASA), and the *Journal of Sound & Vibration* (JSV) of Elsevier. About his pioneering work in Applied Mechanics in the field of Acoustics, where he deals with topics such as torsion and other mechanical effects in cables, he explains one of these:

> Given the following problem "Let there be a cable carrying a payload at one end, while being sinusoidally excited at its upper end": We vary the excitation frequency slowly and monothematically. If the net force acting on the cable (resultant of static and dynamic forces) is positive (tensile), the cable behaves like a solid bar, and it is possible to predict stresses and strains in the system as a function of the spatial and temporal variables. If the excitation frequency increases, a "critical frequency" value is reached, for which the total force changes sign and becomes compressive. Since the cable stiffness is practically zero the cable takes an unstable configuration. Therefore, the problem belongs to the general theory of dynamic stability. This situation often occurs in oceanographic applications and its effects can have serious practical consequences, since after reaching this unstable configuration the suspended weight falls abruptly, producing an impact phenomenon (Laura 1987; p. 28).

This phenomenon is of dynamic instability, consisting of a new contribution to the Theory of Dynamic Stability; he comments that "in experiments carried out by the US Navy in the ocean (1971), values of impact loads were obtained that differed by only 5% from those obtained by means of the mathematical model" (Laura 1987, p. 28).

It is important to mention that the possibility of detecting faults in oceanographic cables using the acoustic emission method was used for the first time in these studies carried out by Dr. Laura and collaborators; according to their own words "It is interesting the fact that already in 1981 the system is operational with Canadian warships using it in the towing of sonars" (Laura 1987; p. 29).

5 Awards and Distinctions to Dr. Laura

Throughout his life he obtained a great number of appointments, awards, distinctions, etc. Here are the most important ones (it is impossible to mention them all).

5.1 Academic Distinctions

a. Certificate for outstanding work (US Navy, Naval Ship Research and Development Center), 1968. Washington, DC
b. Chairman of different Sessions of many International Congress on Acoustics congresses.
c. Member of Tau Beta Pi and Sigma XI.
d. Reviewer of papers for the following magazines: Electrical and Electronic; Applied Mechanics Reviews: Journal of Hydronautics; Journal of Real and Mass

Transfer, Journal of Sound and Vibration, Journal of Applied Mechanics, Journal of the Acoustical Society of America, Microwave Theory and Techniques, etc.

e. Member of the Editorial Board of many journals: "Experimental Mechanics;" Associate of Ocean Engineering; etc.
f. Editorial Consultant for MECHANICS (annual publication of the American Academy of Mechanics).
g. Member of the Board of Directors of the Scientific Research Commission of the Province of Buenos Aires,
h. Member of the Editorial Committee of the Journal of the Argentine Mathematical Union.
i. Second National Technology Award, Ministry of Culture and Education, 1975.
j. First Prize, Bonaerense Region, Ministry of Culture and Education, for the work: "Introduction to the Theory of Vibrations of Discrete and Continuous Systems."
k. Special Mention at the XXI Annual National Meeting of the Argentine Federation of Gynecology and Obstetrics Societies (1977).
l. Member of the International Committee of Composite Materials.
m. Distinguished Lecturer, Ohio Aerospace Institute USA (1991).
n. Annual Recognition of the American Academy of Mechanics USA (1995).

The specialized press in the US closely followed Dr. Laura's work, it is impossible to list here the newspaper articles highlighting his research; however, here it is highlighting two Argentinian emblematic distinctions:

- Second National Engineering Award 1978, together with R. E. Rossi and J. A. Reyes, Argentina.
- First National Engineering Award 1985, together with Dr. G. Sánchez Sarmiento, for a paper on mechanics applied to nuclear energy.

5.2 *The Most Important Recognitions to His Career*

a. Outstanding *Young Man* of the Year 1972, by the Junior Chamber of Buenos Aires, Argentina.

About the personalities that integrated the Jury that anointed him as one of the 10 outstanding young people of Argentina, there were the writer Jorge Luis Borges and Dr. Domingo Liotta, and it was presided by the Nobel Prize winner Luis Leloir.

b. FELLOW: *Acoustical Society of America* "For contributions to structural acoustics and for leadership in acoustical research."
c. FELLOW: *American Academy of Mechanics*, (US).
d. Corresponding Member of the *National Academy of Sciences*; Córdoba, Argentina.
e. Corresponding Member of the *National Academy of Exact and Natural Sciences*; Argentina.
f. Full Academician of the *Argentine Academy of Engineering*.

g. Honorary Professor of the *National University of Mar del Plata.*
h. KONEX 1983, one of the hundred best figures in the history of Argentine science and technology.

Dr. Laura was unanimously elected as *Corresponding Academician* in Bahía Blanca, by the *National Academies of Sciences*, of Córdoba and of *Exact, Physical and Natural Sciences*, of Buenos Aires. On September 17, 1976, he was appointed Full Academician of the Argentine Academy of Engineering, which in 1982 published his work "Recent applications of the conformal transformation method," written jointly with Gustavo S. Sarmiento (Laura, 1982).

5.3 Dr. Laura and His Participation in Scientific Associations

Only strictly scientific ones are listed here (Laura, 1987; p. 18), not Fraternities:

a. American Society for Engineering Education
b. American Institute of Aeronautics and Astronautics
c. Founding member of the American Academy of Mechanics
d. Group of Latin American Acousticians
e. Argentine Association of Applied Mathematics
f. Argentine Mathematical Union
g. Member of the International Committee of Composite Materials.
h. Association of Argentine Acousticians (AdAA)
i. Argentine Association of Computational Mechanics.

Regarding the Association of Argentine Acousticians (AdAA), which was created in 1976, Patricio Laura had great participation in different events in the years prior to its foundation, since through CONICET he was in contact with all university researchers related to mechanical vibration issues; he participated in many of its Congresses and left in his disciples the seed planted that, today, integrate the Board of Directors of the AdAA.

6 Patricio Laura: An Enlightenment Character in the Twentieth Century

The Enlightenment figures of the seventeenth century excelled in science, literature, religiosity, construction of devices, etc., and Dr. Laura in the twentieth century was in his own way an enlightened man out of time. Here are some of his achievements.

Biomechanics. Among the many disciplines he covered throughout his life, Dr. Laura, in addition to writing books on biomechanics, together with his disciples was dedicated to Applied Biomechanics, designing devices, and solving problems in the Hospitals of the city of Bahia Blanca. They built the first artificial elbow in Argentina, which was manufactured in series. At the IMA, they rehabilitated crippled people at work by means of: alignment coupler for leg amputees, abduction devices for newborns with hip dysplasia, phototherapy devices, lymphography pump, intrauterine infusion pump, portable rope injector for the treatment of thalassemic children, etc. Perhaps the most important development was that of an incubator with very low noise level with a project financed by the IDB, a design that had repercussions in the main medical journals.

Invention patents. Achieved two technological developments, in 1979, which were registered (Patricio Laura assigned his Rights) by the National Atomic Energy Commission of Argentina: **(a)** Damping device at the bottom of the operating pit of a nuclear fuel container (Patent No. 216612). **(b)** Flexible assembly for the transport of nuclear fuels (Patent No. 216613).

Science versus Religion. They were mentioned before.

Quotes to his work. Dr. Laura scrutinized all scientific publications and kept a careful record of research mentioning his work and that of the IMA which, up to the year 1988 he identified more than 90 citations. It is worth mentioning that the prestigious publishing house MIR of the former Soviet Union requested permission to include a work of the IMA in one of its books (Fig. 6).

Reviewer of scientific publications. At the time of his death, Dr. Laura was Associate and Emeritus Editor of *Ocean Engineering* and a member of the Editorial Board of the *Journal of Sound and Vibration, Acta Mechanica, International Journal of Mechanical Sciences, Structural Engineering and Mechanics, and Structural Health Monitoring*.

Fig. 6 Dr. Laura on his last visit to NASA. Note: Picture belonging to Laura's family archive, reproduced with permission

The authors highlight two facts that demonstrate the immense personality of Dr. Laura:

- In 2001, he was the first Argentine to be awarded the honorary degree of Professor Emeritus, from the National University of the South (UNS).
- In 2011, the *Academy of Exact, Physical and Natural Sciences* (ANCEFN) of Argentina, awarded the "Patricio Laura" prize to Mechanical Engineering.

The authors had receiving many anecdotes about Dr. Laura, and one of them is remarkable as a review of his sense of humor, Juan Cruz Giménez de Paz (member of the AdAA) share that in the occasion of the oral presentation of one of his Papers, at the *II Interamerican Conference on Noise and Community* in 1977, Dr. Laura said: "God, make this Mechanical Law true, I am in charge of linearizing it."

7 Conclusions

Dr. Laura always taught his family and colleagues, citing the example of many scientists who worked humbly in their places, without all the "pomp" that many bureaucrats at different levels in Argentina tried to impose, that academic and scientific achievements must be based on hard and constant work, and that often large buildings and large investments were not necessary. His education at the Catholic University in Washington D.C. had also been a very good school for him, because of the Jesuit educational line. All of this led him to work simply and with a very small budget in his beloved IMA, to publish and to contribute great scientific achievements with a small budget that was admirable in terms of its scarcity, as Dr. Laura always showed with great pride.

He trained many university students in various disciplines and always generously gave them his free time for their help and scientific guidance in whatever way he could. Today, many of them have fond memories of his generosity, both as a person and as a scientist. Not only researchers worked at the IMA, but Dr. Laura also encouraged the participation of university and high school students, something unprecedented in Argentina, an activity that began before the IMA was created, and his legacy of scientific work continues to this day. Fifty years ago, Patricio Laura initiated the academic tradition that the IMA continues today.

Dr. Laura's departure from this world did not go unnoticed in the scientific community or in society; obituaries were published in the local media of the city of Bahía Blanca and also throughout Argentina (such as La Nación newspaper); the Acoustical Society of America and the rest of the associations to which he belonged remembered his life; even the prestigious ELSEVIER published a special edition saying: "It is with great sadness that we inform you of the death of Professor Patricio A. A. Laura".

Acknowledgements The author would like to thank Rafael López García of Jaen University (Spain) for inviting the authors to write this chapter. The collaboration and patience of Yiyi, wife of Patricio Laura for sharing Dr. Laura's archive, and their sons and daughters Patricio, Paul, Mónica, Silvina and Diego, for sharing dozens of mails telling the authors personal aspects about their father. Also, to Noel Martínez–Pascal for revising the original version of this manuscript in Spanish.

References

C. Hernán, Redes académicas transnacionales y la física argentina en las décadas de 1940 y 1950. Publicación electrónica biblioteca científica SciELO. Chile, 2013. Disponible en Internet

L. Patricio, Conformal Mapping of a Class of Doubly Connected Regions. Catholic Univ. of America 1965 Washington, DC. Disponible en Internet

A.A. Laura, Patricio, Sarmiento Gustavo. Aplicaciones recientes del método de transformación conforme. Academia Nacional de Ciencias. 1982, Argentina. Disponible en Internet

L. Patricio. Curriculum Vitae. CONICET. 1987, Argentina

Puglisi; Alfio A. La enseñanza de la física en la escuela naval militar. Armada de la República Argentina. Buenos Aires, 2011. Disponible en Internet

S. Paola T. Il fondo gian Antonio Maggi (1885–1937). Publicación disponible en Internet

Domingo Santo Liotta (1924–2022)

Manuel Esperon Miguez, Víctor Rodríguez de la Cruz, Daniel Fernández Caballero, and Julián Martín Jarillo

Abstract Domingo Santo Liotta was born in Argentina where he studied medicine and specialized in cardiovascular surgery, although his professional career stands out for his numerous inventions of technology applied to medicine. Undoubtedly, his best-known achievement is the invention of the first complete artificial heart that was successfully implanted in 1969. In his long career as a surgeon and inventor he developed prostheses, surgical instruments and diagnostic methods. Domingo Liotta lived and worked for years in Latin America, Europe and the United States. Back in Argentina his successes in the field of medicine catapulted him into the political sphere, being appointed Secretary of Public Health of the Nation, Secretary of State for Science and Technology, and President of the National Council of Scientific and Technical Research. Domingo Liotta combined these positions with his vocation as an inventor and continued to patent new designs until the twenty-first century. This chapter discusses his career path and his inventions, which have influenced the lives of patients around the world.

M. Esperon Miguez (✉)
Aerospace Engineer, London, UK
e-mail: manuelesperon@hotmail.com

M. Esperon Miguez · V. Rodríguez de la Cruz · D. Fernández Caballero · J. Martín Jarillo
Mechanical Engineer, Madrid, Spain
e-mail: v.rodriguez.delacruz@gmail.com

D. Fernández Caballero
e-mail: daniel.fernandez.caballero@gmail.com

J. Martín Jarillo
e-mail: julianmjarillo@gmail.com

R. López-García and M. Ceccarelli (eds.), *Distinguished Figures in Mechanical Engineering in Spain and Ibero-America*, History of Mechanism and Machine Science 43, https://doi.org/10.1007/978-3-031-31075-1_8

1 Introduction

When analyzing the biography of engineers who have made great contributions in the field of mechanical engineering, it is common to discover that they were multifaceted individuals, with interests that went beyond the purely technical and that, in many cases, contributed to the advancement in other areas of knowledge. Similarly, it is common to find individuals who, without having received training in engineering, have had a professional and personal career in which they have made great advances in this field.

Domingo Santo Liotta belongs to the second group. After studying medicine in Argentina, Liotta had an illustrious professional career as a cardiovascular surgeon, but above all he is known for his great contributions to health technology. Over the course of more than 60 years he invented numerous medical devices, the most famous of which was the first artificial heart implanted in a patient in 1969. Analyzing his many inventions, which include prosthetics, surgical instruments, or diagnostic methods, we can see that, in addition to being a great doctor, Liotta had in-depth knowledge of fluid dynamics, mechanisms, materials and manufacturing.

2 Early Life and Education

Domingo Santo Liotta was born on November 29, 1924 on the banks of the Paraná River in the town of Diamante in the province of Entre Ríos, Argentina, a small city with about 20,000 inhabitants. Domingo Liotta is the son of Italian parents, the demographic with the highest number of immigrants in Argentina. His mother was a primary school teacher. His father, a musician by profession and an amateur botanist, was the director of the music band of the Third Regiment of the Argentine army. This was the second army in which he served, having previously participated in the Russo-Japanese War with the British navy's Mediterranean fleet.

His primary school was called 'No. 1 Independencia', or School no. 1—Independence, located in Diamante. To pursue secondary education he moved 260 km to the town of Concepción del Uruguay, where he was a student of the Justo José de Urquiza School, which was the first secular and free school in Argentina when General Justo José de Urquiza inaugurated it in 1849. Among its alumni are three Presidents of Argentina and one of Paraguay; two Vice Presidents; seven Governors; and five Ministers. Domingo Liotta finished his secondary studies in 1942 and decided to pursue a carrier in medicine.

3 College Years

In 1943 he left Concepción del Uruguay and moved to the city of Córdoba to study medicine, finishing university on July 18, 1949. Between 1949 and 1953 Domingo Liotta did his general surgery residency under the tutelage of Professor Dr. Pablo L. Mirizzi. During this period he began his research work on the clinical anatomy of the bile duct, this being the subject of his thesis.

This was a politically turbulent time in Argentina, with a military coup perpetrated on June 4, 1943 known as the Revolution of '43. However, it is important to emphasize that, despite the instability of the country's internal politics during these years, successive governments maintained a neutral position during World War II, allowing Liotta and his contemporaries to focus on their studies without being involved in the conflict.

By remaining neutral, Argentina benefited from great economic growth and in 1943 industrial production surpassed the agricultural sector for the first time. To this we must add the arrival of Juan Domingo Perón to the government after the elections of 1946, who established a series of social policies that increased investment in public health. This increased flow of investment into the medical sector was very beneficial for Liotta in his formative years and early years of research.

4 First Steps as a Doctor, Researcher and Inventor

After completing his residency in 1953, Domingo Liotta continued to work in the provincial general surgery service of Nuestra Señora del Valle, where he had already done his residency. During his early years he combined his work as a surgeon with research on pancreatic and ampullary cancer. The result of this work was a new technique for the detection and early diagnosis of this disease that was published in 1955. This success put Liotta in contact with international medical teams, first in France and, later, in the United States (Adventures and of a Heart Surgeon 2007).

Between September 12 and 23, 1955, the Liberating Revolution took place in Argentina, in which the democratic government was supplanted by a military government. With the change of regime, investment in science and research suffered sharp cuts and university institutions were restructured with the aim of purging them of sympathizers of the democratic government. Domingo Liotta worked in a hospital attached to the University of Córdoba, which endangered his professional and personal situation.

In 1956 he moved to France where he completed his residency in the thoracic and cardiac service of the University of Lyon, which he completed in 1959 (The Courage to Fail 2002). It is here that he implements his method of diagnosing pancreatic and ampullary cancer with great success. It is in his last year of residence in Lyon when he begins to investigate hemodynamics and the requirements to implant an artificial heart.

5 Development of the Artificial Heart

Despite having moved to France, Liotta maintained his position as a surgeon in Córdoba and continued his collaboration with researchers in Argentina. One of his most important collaborators was Tomasso Taliani, an engineer who helped Liotta design and manufacture the first artificial hearts that Liotta invented. This experience was key for the young surgeon who took the opportunity to learn everything he could about mechanical engineering.

These early designs consisted of hemodynamic pumps that were implanted in dogs. The pumps were compact enough to be implanted in the animal's chest cavity, but the electric motors to which Liotta had access were not compact or reliable enough, so he opted for the use of a pneumatic system that supplied compressed air to the pump through an external module. With this design Liotta managed to replace the function of the left ventricle of the animal.

Domingo Liotta published his first designs in specialized magazines and presented them at several conferences in France and the United States. In May 1961, while Liotta was doing a stay at the Cleveland Clinic, he received a letter from Michael E DeBakey inviting him to participate in a research program in the department of cardiovascular surgery that DeBakey directed at Methodist Hospital in Huston (Liotta et al. 1960). What began as a one-year stay became a decade during which Domingo Liotta worked as an Advanced Research Fellow of the American Heart Association and, from 1964 onwards, as a professor of cardiovascular surgery at the Bayor College of Medicine.

DeBakey and Liotta organized a research group at Baylor University dedicated to the development of the artificial heart. However, despite having generous financial resources, Liotta was surprised to discover the lack of technical means of the laboratory in which he had to work. While the University of Córdoba had had teams of engineers and access to the university's workshops, Baylor College lacked an engineering department. In 1961 Liotta began working in a laboratory that had a few machine tools. The first prototypes were designed and manufactured by Liotta and laboratory technician Louis Feldman (Foundation and Biography) (Fig. 1).

By December 1961 Liotta had already developed a first prototype based on his animal-tested designs and over the next two years Liotta and DeBakey expanded the laboratory's team of researchers and technicians. Although both were surgeons of recognized prestige and both had great technical knowledge about the design of assisted circulation systems, Liotta was the one who focused on the design and improvement of the mechanism and DeBakey in the development of surgical techniques to implant them.

The first intrathoracic artificial left ventricle, or LVAD (originally called Left Ventricular Assist Device) was an incredibly simple and effective system (Fig. 3). An artificial duct made of Dacron, a special reinforced tissue for surgical use, connects the left atrium directly to the aorta, surrounding the left ventricle. At each end of the Dacron duct two non-return ball valves are placed, and the Dacron duct is connected

Fig. 1 Escuela N° 1 Independencia, Diamante, Entre Ríos, Argentina (left). Colegio Superior del Uruguay Justo José de Urquiza, Concepción del Uruguay, Entre Ríos, Argentina (right)

to an air pumping system that varies the pressure in the tube cyclically, pumping blood from the atrium to the aorta.

The intention of this device, like the rest of Liotta's designs of the 1960s, was to keep the patient alive long enough to find a donor or for the heart to heal and recover its function after an operation or cardiac arrest.

With a design ready and prototypes tested in the laboratory, Domingo Liotta and his team were waiting for a patient. On July 18, 1963, George Washington (no relation to the General) underwent an operation to implant an artificial heart valve. The morning after the operation, the patient suffered a cardiac arrest and, after being resuscitated, fell into a coma. The patient's heart was not pumping enough blood, leading to a number of other complications, including pulmonary edema. That same

Fig. 2 Faculty of Medical Sciences, National University of Córdoba, Argentina

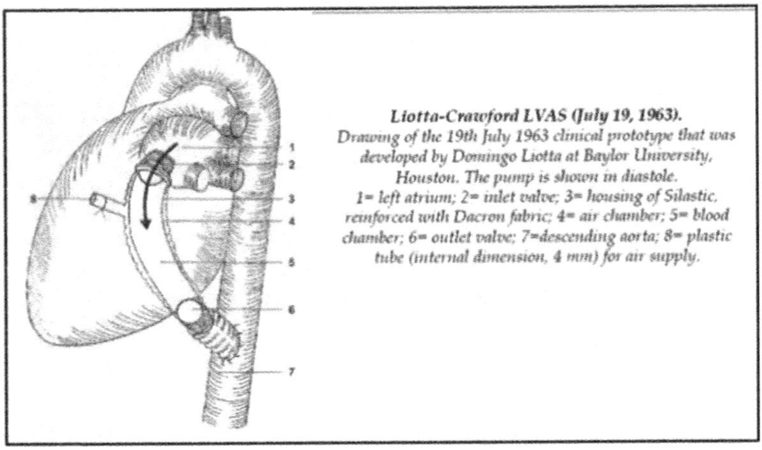

Fig. 3 First intrathoracic artificial left ventricle, or LVAD implanted in a patient on July 18, 1968

day Domingo Liotta and Stanley Crawford successfully implanted the first LVAD in a patient. The pulmonary edema subsided, but the patient remained in a coma due to brain damage caused by the cardiac arrest and died within a few days.

With this first success Liotta and DeBakey began work on a second LVAD model (Liotta et al. 1964). This second model was implemented with most of its components extracorporealally, offering better control of the state of the LVAD. In this second generation the LVAD already had a compact pump that adhered to the patient's skin.

The first patient to receive an extracorporeal LVAD was Mr. DeRudder on April 21, 1966 (Fig. 4). This operation allowed Liotta to further study the connections between the various LVAD ducts and improve their design. A few months later Esperanza del Valle received a double implant of artificial heart valves, but after the operation her heart could not pump enough blood, so an improved version of extracorporeal LVAD was implanted. Once the heart recovered from the operation, the LVAD was disconnected from the patient. Esperanza del Valle lived another ten years and died in a traffic accident.

The success of the first LVAD implanted in 1963 allowed Liotta and DeBakey to obtain funding from the U.S. government for the development of an artificial heart. The project was awarded to Baylor and Rice universities. Rice University provided a team of internationally renowned engineers: John H. Manes (electronics), William Akers (biomaterials), and John Jurgens (electric motors), all working under Liotta and DeBakey.

The first conceptual design of a complete artificial heart was published in 1964, but it was not until December 1968 that the first prototype was ready to begin laboratory testing. Animal experiments began in January 1969. This prototype used two state-of-the-art electric pumps controlled by variable frequency drives specifically designed for this application that would be controlled by the medical team based on the patient's response (Fig. 5).

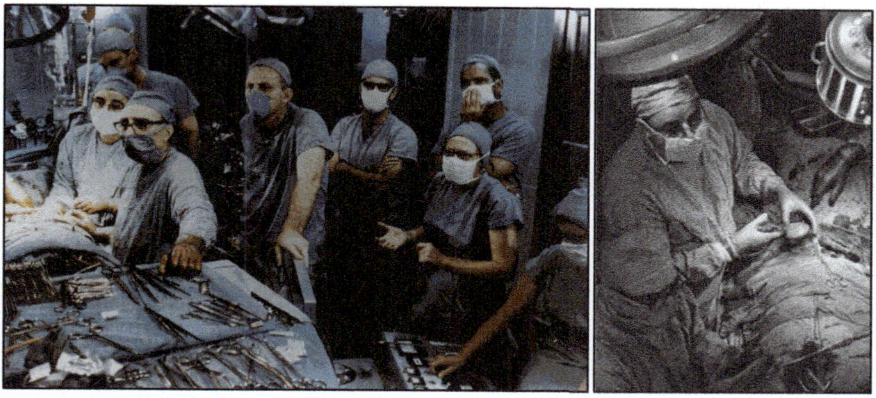

Fig. 4 Left: Dr. Michael E. DeBakey with patient and Dr. Domingo Liotta on his right during operation to implant the Liotta-DeBakey Left Ventricular Assist Device (LVAD) in Extracorporeal Position, April 21, 1966. Methodist Hospital, Houston, Texas, USA. Right: Dr. DeBakey (foreground) and Dr. Liotta (holds the LVAD) in the same operation

Fig. 5 First complete artificial heart designed by Domnigo Liotta and implanted together with Denton Cooley on April 4, 1969. The specimen is part of the collection of the Smithsonian Museum in Washington D.C., United States

In April 1969, Denton Cooley had a patient on the waiting list to receive a heart that could be transplanted. His health was deteriorating rapidly and, to gain time, Cooley proposed to carry out an operation to try to repair the heart. Cooley also proposed that, if during the intervention the medical team believed that the heart could not be repaired, they would try to implant the artificial heart invented by Liotta. The patient agreed.

The operation was performed at St. Luke's Episcopal Hospital and on April 4, 1969, Haskell Karp became the first person to have an artificial heart implanted

Fig. 6 Left: Front page of the Houston Chronicle on April 5, 1969 with the news of the first artificial heart to replace the patient's heart. Right: Dr Dento Cooley and Dr Domingo Liotta at the press conference explaining the operation of the artificial heart invented by Liotta

(Fig. 6). The patient successfully recovered from the operation and remained conscious and lucid until doctors located a viable heart to be transplanted. Unfortunately, the transplant was unsuccessful and Haskell Karp passed away due to a post-operative infection. Liotta's artificial heart had kept him alive for 65 h.

6 Other Relevant Inventions

Undoubtedly, his contributions to the development of the artificial heart are Domingo Liotta's best-known achievements. However, it is important to remember that, when we analyze the creation of great inventions, the process of innovation and invention is not linear and therefore gives rise to many other new inventions.

Engineering work always involves creating new ideas that must be analyzed, tested and discarded or improved with each new prototype. This can be seen in the career path of every great inventor. This experience gained from years of research and development of the artificial heart allowed Liotta to continue contributing to the field of engineering applied to cardiovascular surgery for decades.

In this section we will explore some of the most important inventions of Liotta and his collaborators from the 1970s to the twenty-first century.

6.1 Low Profile Aortic and Mitral Prosthesis—1979

In the late 1970s Liotta presented a new design of an aortic prosthesis (Patent of the United States of America). The purpose of these prostheses is to provide structural reinforcement around the heart valve in patients in whom the tissue surrounding the valve is damaged or weakened. This can result in valves that deform irregularly when opened and closed which can generate disturbances in the flow and even result in valves that do not close or open properly.

This invention was part of a line of research that Domingo Liotta and his collaborators carried out in pigs. Prototypes and patents mention applications in animals, but the revolutionary design of this aortic prosthesis was very influential in the development of other cardiovascular prostheses for humans. Proof of this is that this prosthesis has been cited in more than eighty patents registered in the United States since 1979.

The main innovation of this invention came from the extensive study and improvement of the geometry of the prosthesis. After analyzing the design of other valves Liotta concluded that the rest of the designs used walls that were too tall. Other prostheses tried to gain rigidity by making the cylindrical frame larger, so the height of the prosthesis usually exceeded 14 mm. This presented two types of problems: first mechanical, since this generated deformations and tensions in the aorta to accommodate the prosthesis, even producing perforations over time; The second was related fluid dynamics, since larger prostheses tended to offer greater resistance to blood flow.

The research team analyzed more than 200 specimens and conducted statistical analysis to determine the space available to insert an aortic valve prosthesis without producing additional stresses on the cavity around the valve. The team concluded that the prosthesis should not exceed 11 mm in height. With this premise, Liotta designed a more compact prosthesis that also featured a cylindrical frame with a wavy lower edge that allowed a more natural movement of the valve membranes and, at the same time, increasing the rigidity of the frame (Fig. 7).

The frame is surrounded by a suture ring allows the prosthesis to be fixed. This ring is soaked in a glutaraldehyde solution to reduce the risk of infection. Liotta and his team conducted several tests to analyze the interaction between the tissue used for this ring and the polymer of the cylinder to ensure that the prosthesis was able to survive the large number of cycles to which it would be subjected.

6.2 Artificial Two-Gate Valve—1987

Domingo Liotta continued working on the design of prostheses and artificial valves for several years, combining his work as an inventor and researcher with that of Chief of Surgery, university professor and public servant. The end of the 1980s and

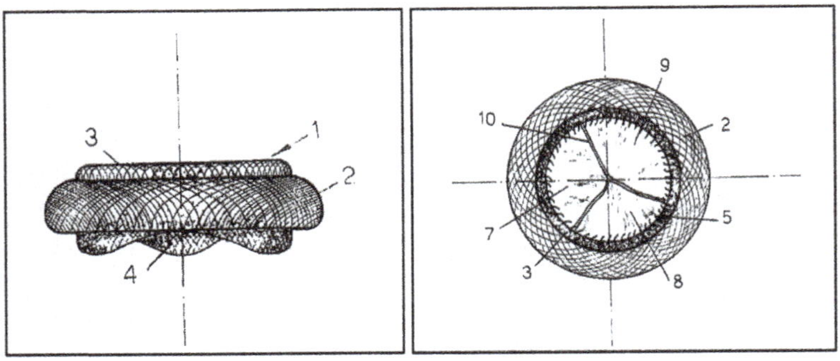

Fig. 7 Low profile porcine aortic prosthesis fixed with glutaraldehyde (United States Patent 4,079,468). (1) semi-rigid cylindrical frame, (2) suture ring, (3) upper and (4) lower edge of cylindrical frame, (5) suture, (7, 8, 9) aortic valve membranes

the beginning of the 1990s were particularly prolific for Liotta, who introduced the world to four new inventions.

In 1987 Liotta patented a new articulated artificial valve to replace heart valves that are damaged to such an extent that they cannot recover their function even with the help of surgery. At this time the vast majority of artificial valves were based on a spherical element retained inside a cage that allows and interrupts blood flow according to the systole and diastole cycle of the heart (Fig. 8, left). While these valves fulfill the function of controlling the flow with the pulsations of the heart, the flow distribution and the flow-time profile resulting from the opening and closing of the valve is markedly different from that of a natural heart valve.

The first disc valves, released in 1969, were more similar to the geometry and operation of a natural valve (Fig. 8, right). These valves used a circular element held by at least two metal supports that allow it to pivot without requiring a through-shaft for the gate. However, this design is still far from replicating the operation

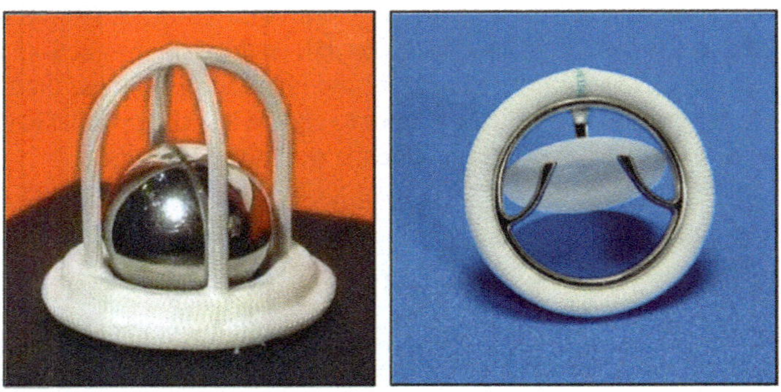

Fig. 8 Starr-Edwards artificial mitral valve (left) and Björk–Shiley oscillating disc valve (right)

of biological valves: the mitral valve is formed by two membranes and the aortic, tricuspid and pulmonary valves have three.

The first valves using two gates appeared in 1979 and in 1987 the Hemex, Duromedics or St. Jude type valves had already been used successfully. However, these models placed the gate joint near the center, which did not prevent another big problem of artificial valves: turbulence.

A key principle in hemodynamics is that the viscosity of the blood increases markedly when the flow velocity decreases. Therefore, a design that results in low-speed zones increases the risk of thrombosis (one of the main risk factors and cause of most complications after implanting artificial valves is thromboembolism). The artificial valves used until then used parts located in the center or near the center of the flow, creating turbulence with the consequent risk of clot formation.

Once again Liotta demonstrated his skills as an inventor and his knowledge of fluid dynamics when he turned his attention to the design of a new artificial valve (Patent of the United States of America) (Fig. 9). This valve placed the center of rotation of the gates away from both the central valve axis and the walls of the suture ring. This design drastically decreases the turbulence created by the gates.

The valve invented by Liotta was the result of a careful analysis that contemplated not only the characteristics of blood flow when the gates were fully open, but also the dynamic effects on the flow when the valve opened and closed. This was also one of the first prostheses that attempted to reduce the water hammer effect by careful analysis of the angle of the gates in their closed position.

This prosthesis not only represented an important qualitative leap compared to other contemporary valves, but also introduced a whole series of improvements that are still used in valves that have been designed more recently.

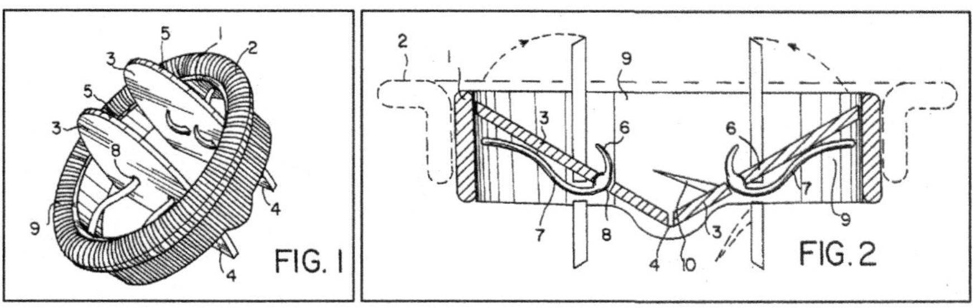

Fig. 9 Prosthetic Heart Valve (United States Patent 4,655,772). (1) Bracket Regulated Ring, (2) Suture Ring, (3) Gates, (4) Gate Contact Zone, (5) Gate Top Edge, (6) Virtual Gate Shaft, (7) Gate Pivot Bracket, (8) Gate Hole for Support, (9) Rigid Support Ring Inner Surface, (10) Rod to ensure simultaneous opening of both gates

6.3 Diaphragm for Aortic Occlusion—1987

Nineteen eighty-seven was a prolific year for Domingo Liotta in which, in addition to presenting to the world his artificial valve with two gates, he patented a new method to block the thoracic aorta after a bypass operation (Patent of the United States of America). This type of intervention is performed to treat aneurysms by diverting or *bypassing* the aorta by inserting an artificial duct to surround the damaged segment of the aorta (Fig. 10). Once the bypass has been inserted, surgeons may attempt to remove the aneurysm or, depending on the risk to the patient, insert a block that prevents blood flow to the aneurysm, leaving the bypass permanently in the patient.

Liotta took advantage of technological improvements in fluoroscopy and radioscopy that allowed better visualization of aneurysms in real time with equipment installed in the operating room (Fig. 10). This helped to develop new operating techniques that opt for occlusion of the aorta. His invention uses a catheter to insert a diaphragm into the aorta (Fig. 11) which not only simplifies this last step considerably, but also reduces the trauma suffered by the patient. The result is an intervention with a lower risk to the patient and a shorter recovery period.

The diaphragm, made of surgical plastic, contracts inside the head of the catheter before the catheter is inserted into the patient. During the operation, the surgeon guides the catheter with the help of a fluoroscope to the point where the aorta must be blocked. At that point the surgeon pushes the diaphragm with a concentric stylet to the catheter removing the diaphragm from the catheter.

Liotta's design pays great attention to the geometry and materials of the diaphragm. The block is provided by a sheet of 22 to 40 mm in diameter made of surgical polyester like the Dacron. This sheet is attached to radial reinforcements made of Elgiloy, a special alloy for surgical elements that had recently been marketed by the American Edwards Laboratory when Liotta designed his prototype. These reinforcements expand the diaphragm when it is removed from the catheter and press

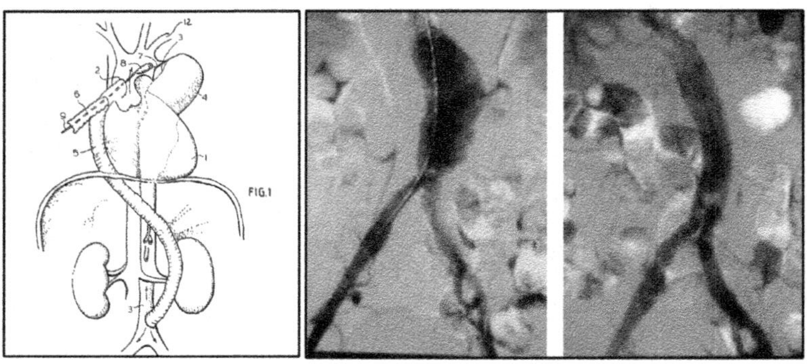

Fig. 10 Left: Thoracic aorta bypass with artificial aorta (5) and catheter (9) to insert the occlusion diaphragm (8) of the blocked aorta (3). Right: example of abdominal aortic aneurysm seen with a fluoroscope before and after surgery

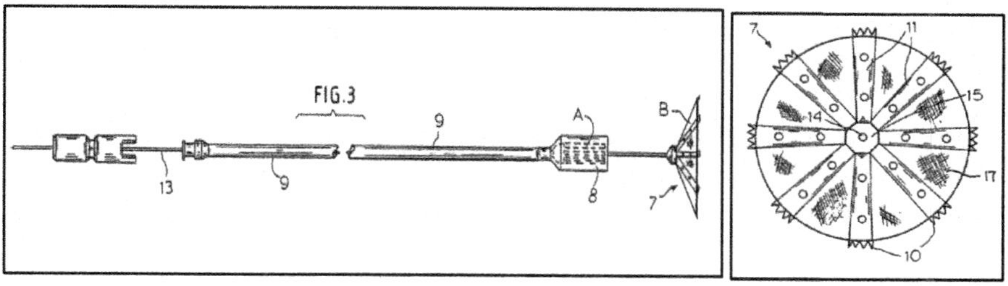

Fig. 11 Left: Detail of the catheter with the stylet to control the position of the diaphragm (13) and the diaphragm contracted for insertion (A) and expanded after implantation (B). Right: Front view of the diaphragm with radial reinforcements (11), anchor points to the aorta (10), and diaphragm membrane (17) (United States Patent 4,710,192)

the edge against the walls of the aorta. To prevent the diaphragm from moving, the radial reinforcements protrude a few millimeters and are finished in the shape of a sawtooth, allowing it to be permanently fixed.

6.4 Heart Valve Bioprosthesis—1992

Over the years, the advancement in the technology of plastics used in medicine allowed several research groups to develop artificial valves in the early 1990s that tried to faithfully reproduce the geometry and operation of natural valves. In this new generation of artificial valves the gates are replaced by membranes made of polymers that deform with each systole and diastole, opening and closing the valve.

Once again, Liotta was at the forefront of this new era in the design of artificial valves with a prosthesis for aortic valves (Stentless bioprosthetic cardiac valve) (Fig. 12).

As he did with previous inventions, Domingo Liotta based his design on a careful study of the geometry of valves and their dynamic behavior. Thanks to the study of

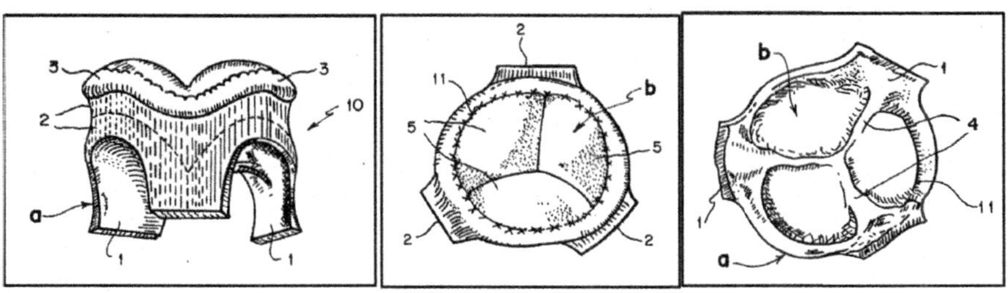

Fig. 12 Heart valve bioprosthesis. Tubular body to fix the valve (a) attached to the membrane body or valve plug (b) by means of a suture ring (3). Side view (left), top (center) and bottom (right) (Patent of the United States of America 5.156.621)

multiple angiograms, his artificial valve faithfully reproduces the deformations of the biological valve it supplants.

Additionally, this new type of prosthesis eliminated many of the problems generated by mechanical artificial valves that increased the risk of clot formation. Thanks to the improvements of Liotta and his contemporaries, patients did not need permanent treatments with anticoagulants, with the risk that they entail.

6.5 Mechanical System to Support Blood Circulation—1995

One of the main limitations of artificial hearts is the need for an external energy source to power the pumping mechanism, which is usually electric. One of the most radical ideas that has been explored by various teams for several decades has been to use the relative movement of the patient's muscles to activate the pumping mechanism. The idea is to stimulate the contraction of the muscle externally through electrical pulses. While this solution still requires an external source of energy to generate the electrical pulses that stimulate muscle, most of the energy is generated by muscle contraction and total energy consumption of the device is lower.

In 1995 he presented one of his most groundbreaking ideas since his first designs of the artificial heart (Patent of the United States of America) (Fig. 13).

In this invention Liotta proposes to use part of the latissimus dorsi muscle (15), the strongest muscle in the entire trunk, as a source of linear movement. This linear motion drives a transmission (6) which is in turn connected to the blood pumping unit (11). This pumping unit uses a cam mechanism (23, 24) to drive a hydraulic piston responsible for forcing blood circulation.

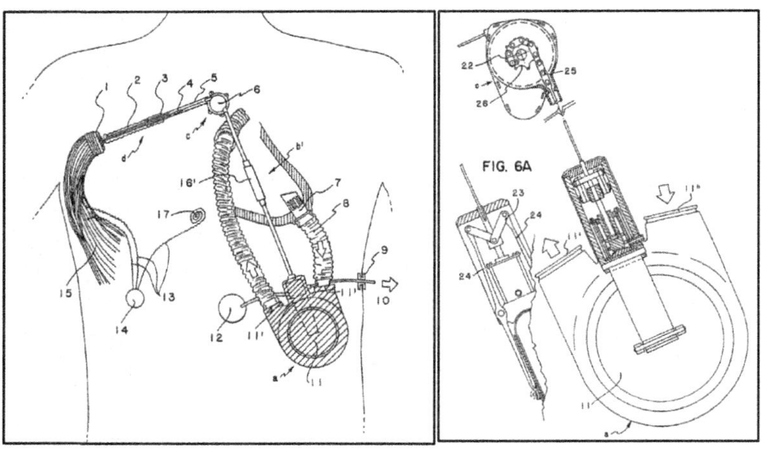

Fig. 13 Mechanical system for the aid to blood circulation. (Patent of the United States of America 5.456.715)

The system includes a programmable electro-stimulator (14) that controls the rate of muscle contraction through a series of electrodes (13). In the event of a failure, the pumping system can be connected to an external system via an external connection (10).

Obviously, this is a temporary solution that allows to prolong the patient's life until a viable donor is found. Implanting this device also requires that the muscle is preconditioned by electrostimulation for several weeks. This allows for the preparation of the the muscle to develop strength and endurance, but also to make adjustments to the electro-stimulator and make sure that it works reliably. At the time of writing, no clinical case has been found that has used this device.

6.6 *Intracorporeal Implantation Device to Assist Ventricular Circulation—2005*

Domingo Liotta continued to innovate in the twenty-first century and in 2005 introduced a pneumatically operated heart pump (Patent of the United States of America). This unidirectional pumping device, like other similar systems, has the function of assisting the functioning of the heart while the patient recovers from surgery or as a bridge to a heart transplant.

The patent for this invention focuses on pump design as an innovation and assumes that an external system provides the pneumatic pressure needed to drive the pump in a cyclical and controlled manner (Fig. 14, left).

The device operates on a principle similar to diaphragm pumps. The image in the center of Fig. 14 shows how the chambers (11) expand when the pneumatic system increases the pressure, reducing the volume of the inner chamber of the pump (12). When the pressure is lower than blood pressure at the pump inlet the

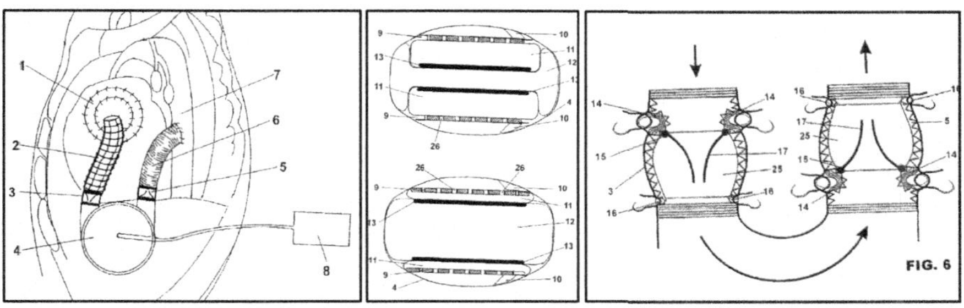

Fig. 14 Intracorporeal device to assist ventricular circulation. Left: external view of the pump (4) with the inlet (2) and outlet (6) ducts driven by an external pneumatic system (8). Center: pneumatic chambers (11) withtrolls due to their connection to the pneumatic system (10) regulate the volume of the pumping chamber (12). Right: inlet and outlet valves (17) oriented to ensure unidirectional flow (Patent of the United States of America 6,945,998)

inner chamber expands forcing the entry of blood. The pumping direction is ensured by two oppositely oriented diaphragm valves (Fig. 14, right).

7 Return to Argentina and Public Work

Due to professional, personal and political circumstances Domingo Liotta spent most of the 1960s outside Argentina. However, he maintained professional ties with his native country during the years he spent in France and the United States.

His successes in developing the first artificial heart brought him international fame and prestige. This allowed him to return to Argentina in 1971. His first assignment was to take charge of cardiovascular surgery services at the Italian Hospital and the Durand Hospital. Both hospitals are located in Buenos Aires.

During the 1970s Liotta combined his work as an inventor and researcher with that of several public offices. While his professional career and reputation were clear evidence of his ability to hold these positions, it is also important to analyze his political connections. Particularly his personal relationship with Juan Domingo Perón, who was President of Argentina on three occasions.

After the Liberating Revolution of September 1955, Perón was forced into exile. Between 1955 and 1960 he took refuge in Paraguay, Panama, Venezuela, and the Dominican Republic, moving repeatedly due to frequent regime changes in each of these countries. He finally settled in Madrid in 1960, where he and Liotta met.

Domingo Liotta and Denton Cooley had been hired by the Spanish government in 1971 to train surgeons in the operating techniques they had developed at the Texas Heart Institute. Over the next seven years Liotta and Cooley traveled frequently to Spain as part of this agreement.

Salvador Liotta, Domingo's brother, worked at the time in the Department of Hemodynamicsof the Giménez Díaz Foundation and at the Cardiovascular Center of La Paz in Madrid. Salvador was also Perón's personal physician. It was through this connection that Liotta and Perón began a relationship that would last for years.

7.1 Secretary of Public Health of the Nation

On March 11, 1973, the military leadership, which had remained in government since the 1966 revolution, agreed to organize democratic elections. After a series of constitutional reforms, Juan Domingo Perón won the elections onSeptember 23, 1973, becoming president of Argentina for the third time.

That same year Perón appointed Domingo Liotta Secretary of State for Public Health. In this position Liotta fulfilled a double function: to reform the public health system promoted by the new government and to act as a scientific ambassador and promoter of collaborations with international medical and research centers.

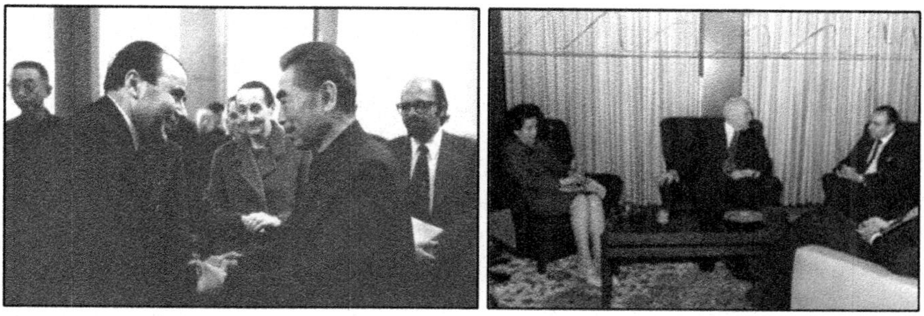

Fig. 15 Right: Domingo S Liotta with Chinese Premier Chou-en.Lai on November 20, 1973. Left: Israeli President Ephrain Katzir and Dominic S Liotta inJerusalem in 1974

Liotta was the author of Law 20 748, which created a "National Public Health System". With this law, numerous reforms were implemented that made the state responsible for guaranteeing the right to health of all inhabitants. An important point of this law is that it made the state responsible for guaranteeing the financing of these services. In addition, the law defined a series of functions of the state, control and supervision of the production, distribution, and consumption of medical consumables.

As Secretary of Health, he visited China and Israel, where he signed historic agreements with Premier Chou En-lai and President Ephraim Katzir, respectively (Fig. 15). His relationship with hospitals and research groups in China was very long and continued after leaving the post of Secretary of Health, proof of this is his appointment as honorary director of the Guanzhou cardiovascular center in 1990.

Juan Domingo Perón died on July 1, 1974. Domingo Liotta, who had also been Perón's personal cardiologist since his return to Argentina, left the post of Secretary of Health that same year.

7.2 *Secretary of State for Science and Technology and President of the National Council for Scientific and Technical Research*

In 1994 Liotta was appointed Secretary of State for Science and Technology. At the same time, he was made President of the National Council of Scientific and Technical Research (CONICET). Domingo Liotta remained in both positions until 1996.

CONICET was founded in February 1958, its first director being the Nobel Prize in Medicine Bernardo Houssay. CONICET is the largest scientific institution in Argentina and one of the most important in Latin America. As its president, Liotta was responsible for more than 700 establishments including research centers, laboratories and institutes.

7.3 Final Years

In 1997 he was appointed Dean of the School of Medicine of the University of Morón, located in Buenos Aires. In this position, Liotta promoted the construction of a new headquarters of the Faculty of Medicine located in the General Interzonal Hospital of Acute "Prof. Dr. Luis Güemes", which was inaugurated in 1999. He also promoted the expansion of the radius of action of the faculty of medicine with the creation of 11 annexes in various cities of Argentina including Córdoba and Rosario, and another nine distributed in the province of Buenos Aires.

In 2013, at seventy-two years of age, Domingo Liotta was appointed Vice-Chancellor Emeritus of the University of Morón.

Domingo Liotta died on the 31st of August 2022 in the Italian Hospital of Buenos Aires, Argentina.

8 Awards and Recognitions

1962 Finalist for the Young Investigator Award, Competition (Denver) of the American College of Cardiology.

1964 Annual Award, Southwester Surgery Society, USA Shared with William C. Hall and Michael E. DeBakey.

1968 Golden Eagle Award, International Events Council, USA, shared with Denton A. Cooley, Robert Bloodwell, and Grady Hallman.

1969 American Medical Association Award of Merit shares with Denton A. Cooley, MD.

1969 Decoration of the Institute of Spanish Culture, Madrid, Spain. Shared with Jesús Zerbini (Brazil).

1970 Decoration, Grand Cross of Alfonso X El Sabio. Shared with Denton A. Cooley.

1971 Decoration, Order of the Legion of Merit, Government of the Province of Entre Ríos, Argentina.

1973 Professor of Surgery (Honorary), National University of Córdoba, Argentina.

1973 Decorated by the National State Council of the People's Republic of China.

1974 Decoration, Grand Cavallieri d'Onore of Humanitarian Services, Rome, Italy, the highest Italian civilian decoration.

1990 Award, Director of the Cardiovascular Center, Guangzhou, China. Honorary Professor, Beijing Friendship Hospital of Capital University of Medical Sciences, China.

1997 Science Award. Assisted Circulation for Chronic Heart Failure, International Society of Artificial Organs (ISAO), USA

2005 Grand Council Knight of the Sovereign Hospitaller Order of St. John of Jerusalem and Rhodes, Cyprus, Rhodes.

References

Amazing Adventures of a Heart Surgeon: The Artificial Heart: The Frontiers of Human Life, Domingo Liotta, iUniverse books (2007)

Cardiac valvular prosthesis - Patent of the United States of America N°4.655.772

Corporal Implantation device for assisting blood and heart ventricular circulation- Patent of the United States of America N°6,945,998

Diaphragm and method for occlusion of the descending thoracic aorta - Patent of the United States of America N°4.710.192

Emilio Liotta Foundation—Biography, http://www.fdliotta.org

Implantable mechanical system for assisting blood circulation- Patent of the United States of America N°5,456,715

Low profile cluteraldehyde-fixed porcine aortic proscetic device - Patent of the United States of America N°4.079.468

D. Liotta, C. W. Hall, D. A. Cooley and M. E. DeBakey, Prolonged Ventricular Bypass with Intrathoracic Pumps, Transactions—American Society for Artificial Internal Organs, Vol. 10, No. 10, 1964, pp. 154–156

Stentless bioprosthetic cardiac valve - Patent of the United States of America No. 5,156,621

The Courage to Fail, *A Social View of Organ Transplants and Dialysis, Judith P* (Taylor & Francis, Swazey, 2002)

D. S. Liotta, M. E. DeBakey, A. Denton, A. Cooley—Mike, The Master Assembler; Denton, the Courageous Fighter: A Personal Overview Unforgettable Past Remembrances in the 1960s, Open Journal of Thoracic Surgery, 2, 37–45 (2012). https://doi.org/10.4236/ojts.2012.23010 Published Online September 2012 http://www.SciRP.org/journal/ojts

Cipriano Segundo Montesino y Estrada (1817–1901)

J. Echávarri Otero, E. de la Guerra Ochoa, E. Chacón Tanarro, E. Bautista Paz, and J. L. Muñoz Sanz

Abstract Cipriano Montesino, a multifaceted engineer who developed his activity in the second half of the nineteenth century, is the initiator of the academic study of machines in Spain, throughout his teaching activity at the Real Instituto Industrial in Madrid. His book on Machines Construction, a work of ambitious approach, content and length, not only compiled and imported ideas from other foreign texts, but also contributed to the methodology for teaching this subject and to the process of reflection upon machines that was taking place in the European framework at the time. This chapter summarises his biography, including both his teaching work and his activity as a representative, senator, academician, director of a private company, etc., and then continues with a description of Montesino's contribution to mechanical engineering and the content of the book he developed for his course on Machines Construction.

Keywords Reflection · Machine · Treatise · Machines construction · Steam engine

1 Biographical Notes

Cipriano Segundo Montesino y Estrada was born on the 26th of September 1817 in the town of Valencia de Alcántara in the province of Cáceres (Spain). At the age of six he had to go into exile with his family in England at the end of the Liberal Triennium and the beginning of the Ominous Decade, during the reign of Ferdinand VII, as his father was a doctor and liberal politician who became a representative in the Courts of 1812.

J. Echávarri Otero (✉) · E. de la Guerra Ochoa · E. Chacón Tanarro · E. Bautista Paz · J. L. Muñoz Sanz
Universidad Politécnica de Madrid, Madrid, Spain
e-mail: javier.echavarri@upm.es

R. López-García and M. Ceccarelli (eds.), *Distinguished Figures in Mechanical Engineering in Spain and Ibero-America*, History of Mechanism and Machine Science 43, https://doi.org/10.1007/978-3-031-31075-1_9

(a) (b)

Fig. 1 **a** Portrait of Montesino as General Director of Public Works, by José Vallejo y Galeazo (1855); **b** Portrait of the National Constituent Assembly, by José Suárez (1869)

Cipriano Segundo Montesino remained in England until he began his university studies in London, but was able to return to Madrid with his family after the death of Ferdinand VII in 1833. (Martínez-Val 2001).

Figure 1 shows two portraits of Cipriano Segundo Montesino, available on the website of the Spanish National Library (Biblioteca Nacional de España: www.bne.es).

In 1834 he received a scholarship to study in France and travelled to Paris, where he was graduated as an engineer in 1837 at the École Centrale des Arts et Manufactures (Asociación de Ingenieros Industriales de Madrid 2008). He validated this degree at the Real Instituto Industrial in Madrid in 1856, within the first graduating class of this pioneer higher technical school of engineering in Spain. Due to his youth, as he was 20 years old when he finished his studies in Paris, his scholarship was extended for two more years, so that he could continue his education in London and increase his knowledge about construction and design of machines.

Once he had finished his studies, he took part in the revolution of the Progressive Party (1840), which led General Espartero to take power and with whom he would later become related by marrying his niece Eladia Fernández Espartero y Blanco (second Duchess of La Victoria and second Duchess of Luchana), thus becoming Duke of La Victoria Consort. They had nine children, some of them with a brilliant career, such as Pablo Montesino-Espartero, third Duke of La Victoria and third Duke of Luchana (military officer and diplomat) and his brother Luis (industrial engineer and pioneer of aviation in Spain).

On the 17th of October 1842 he was awarded the chair of Mechanics of the Royal Conservatory of Arts and Crafts and in 1851 he became professor of Machines Construction of the Real Instituto Industrial in Madrid, where he was interim director during the period 1853–1854. His main didactic work was the book on Machines Construction entitled *Resumen de las lecciones del curso de Construcción de Máquinas* (Montesino 1853–54), a work of ambitious approach, content and length, which imported some ideas from other previous foreign texts. The content of this text will be described in depth in Sect. 3.

In addition to his teaching activity, Cipriano Segundo Montesino performed important public and political work throughout his life, becoming a representative, senator and member of the Royal Academy of Sciences. He stood in different elections for Cáceres to the Congress of Deputies, being elected a representative in the elections of 1843, 1854, 1858, 1869, 1871 and 1872. He became an Official in the Ministry of the Governance and in 1854 Director of Public Works (Congreso de los Diputados 2022).

During his period as Director of Public Works, the Railway Law of 1855 (which had been under discussion since 1850) was approved, establishing the Iberian track gauge (six Castilian feet or 1671 mm) as reference for the Spanish railway line, being Cipriano himself the only technician who was in favour of the standard track gauge during the study performed by the permanent commission (García Álvarez 2010).

Moreover, during his second period as a representative, he developed one of his most important works, a report on the situation of public works in Spain (Montesino 1856). This report was published in 1856 and consists of two main parts. The first one is divided into seven chapters, dedicated to ordinary ways, railways, water exploitation, ports, lighthouses, buoys and beacons, telegraphs and the organisation of public works. The second part includes a total of ninety-four appendixes related to the contents of the chapters of the previous part.

Additionally, he was appointed academician of the Royal Academy of Sciences (the 3rd of April 1847). Subsequently, he was Vice-Secretary of this Academy from 1848 to 1861, later becoming Vice-President from 1875 to 1882 and President from 1882 until his death (Real Academia de Ciencias 2022). As an academician, he was assigned to be part of the international commission that studied the Suez Canal, the characteristics of the project and the implications of its creation, which led to the publication of his book *Rompimiento del Istmo de Suez* (Montesino 1857). He also participated in other commissions and positions, and was shareholder of the so-called Free Educational Institution.

Besides being a member of the Royal Academy of Sciences and representative in the Congress, he was Senator for several periods, being elected by the province of Cáceres (1872–1873) and by the Royal Academy of Exact, Physics and Natural Sciences (1881–1901). He became Second Vice-President of the Senate during the Legislature of 1881–1882 (Senado de España 2022) and was awarded with the Grand Cross of Charles III.

Furthermore, he was also active in the private sector (De la Torre 1901; Peris Torner 2022). In 1868 he was appointed President of the Railway Company from Madrid to Zaragoza and Alicante (MZA) and remained linked to his board of directors until his death on the 27th of August 1901 in Madrid.

2 Reflection Upon Machines

In order to understand the importance of the contribution of this illustrious person towards the reflection upon machines, it seems appropriate to analyse the Theory of Machines and Mechanisms within the European context. It can be said that during the teaching period of Montesino, a unification of two different visions upon machines, initiated since the Renaissance, came to an end. The practice of machines construction (understood as manufacturing) and the theory of mechanics (understood as design) converged from two very different lines of development. Treatises on machines were developed, in the format of a collection of rationally ordered machines, along with studies on machines, as an application of mechanical physics (Bautista et al. 2010).

From the Renaissance period onwards, the interest for the theoretical aspects of machines (physics, geometry and mechanics involved in them, simplifying hypotheses, etc.) increased, and Ancient knowledge was recovered. Thus, the works of Greek mechanics and the machines created by Roman engineers were rediscovered and studied in depth. However, due to the enormous development of machines during the Industrial Revolution, both in terms of capacity and diversity of usage, the creation and grouping of knowledge and its implementation were necessary for the design, manufacture, operation and reparation of these machines. After the treatises on machines of the European Renaissance, during the Industrial Revolution, treatises on mechanisms appeared that combined graphics and texts in accordance with the evolution of that time, and which allowed the training of professionals working on machines.

A further evolution led to focus on two separate and complementary approaches: one concentrated on the transformation of movement, without considering the causes that originate the movement; and the other one focused on the functionality of machines. In the first of these, Betancourt and Lanz wrote in 1808 their *Essai sur la composition des machines* (Lanz and Betancourt 1808), in which they compiled the types of movement and a classification based on the transformation of movement with a similar vision to the one of Monge (Monge 1805): to use the theoretical knowledge of geometry in the professional field of machines. This is how Lanz and Betancourt created a systematic treatise on machines, whose study would continue throughout all the nineteenth century. The classification was structured in ten divisions, starting from four types of movement: uniform rectilinear, alternative rectilinear, uniform circular and alternative circular. The relationships existing between these four starting movements and their transformation into any of the four types led to the ten divisions mentioned above.

Following the steps of Lanz and Betancourt, Hachette (Hachette 1811), in his book *Traite Élémentaire des Machines* of 1811, continued the same guidelines as his predecessors and divided the movements into the same ten types. This book, written for the course on machines of the École Polytechnique of Paris, includes, in addition to brief descriptions, calculations of different machines.

Since these dates, and over a period of a few decades, several authors approached the systematic study of machines, such as Borgnis *Traité complet de mécanique appliquée aux arts* in 1818 (Borgnis 1818), Willis *Principles of Mechanism* in 1841 (Willis 1870) or Weisbach *Principles of the mechanics of machinery and engineering* in 1848 (Weisbach 1848). All of them developed technical compendia of machines, with theoretical explanations of their operation, different from a more informative purpose of the previous treatises.

Borgnis' study included a classification of machines, dividing them into six types: receptors, communicators, modifiers, supports, regulators and operators, that is to say, he adopted a functional approach of the machines. This division did not please his contemporaries, who continued supporting the "kinematic" division of Betancourt and Lanz.

It is noteworthy to mention that, after the edition of Montesino's book in 1853–54, this difference of approaches continued in the following decades. Labouyale in his *Traité de cinématique o Théorie des mécanismes* (Laboulaye 1861) of 1861 and Redtenbacher in his *Die Bewegungs-Mechanismen* (Redtenbacher 1866) of 1866 likewise summarised the types and scope of each one, although Labouyale preferred to continue with Betancourt and Lanz's division, and Redtenbacher chose to classify machines by their usage, in the same line as Borgnis.

A new approach followed when Reuleaux published six books about machinery, the two most important being *Lehrbuch der Kinematik, V.1—Theoretische Kinematik* (Reuleaux 1875) of 1875 and *Kinematics of Machinery* (Reuleaux 1876) of 1876, where he considers the relative movement of the constituent elements of the machine as a basis. Reuleaux's contributions are characterised by the excellent symbology employed in the identification of the kinematic parameters, which is still used today.

Schröder, in his *Catalog of Reuleaux Models* (Schröder 1899) of 1899, and Voight, in his work *Kinematische Modelle nach Prof Reuleaux* (Voight 1907) of 1907, masterfully gathered the teachings of their predecessor.

Thus, during these decades, the fundamental bases for the study of machines and mechanisms were completed. Figure 2 shows the covers of some of the main books mentioned above.

Within the Spanish environment, there were few contributions, besides the work of Lanz and Betancourt mentioned above. Spain was in a context of destruction of industry, seizure, etc. At that time, education in Spain was mainly influenced by the French model.

One of the main contributions in Spain was carried out by Azofra (Azofra 1838) with his *Curso industrial o lecciones de aritmética, geometría y mecánica aplicadas a las artes* (1838), where he collected a series of rules for machines construction based on the contemporary treatises he was aware of (Carnot, Borgnis, Poncelet, Coriolis…). Other important works were published by Odriozola (Odriozola 1839)

ECOLE IMPÉRIALE POLYTECHNIQUE.

PROGRAMME
DU
COURS ÉLÉMENTAIRE DES MACHINES,
POUR L'AN 1808,
Par M. HACHETTE.

ESSAI
SUR
LA COMPOSITION DES MACHINES,
Par MM. LANZ et BÉTANCOURT.

A PARIS,
DE L'IMPRIMERIE IMPÉRIALE.
1808.

TRAITÉ ÉLÉMENTAIRE
DES MACHINES,
PAR M. HACHETTE,
INSTITUTEUR DE L'ÉCOLE IMPÉRIALE POLYTECHNIQUE.

PARIS,
J. KLOSTERMANN fils, Libraire de l'École Impériale Polytechnique, rue du Jardinet, nº. 13, quartier St.-André-des-Arts;
SAINT-PÉTERSBOURG,
KLOSTERMANN père et fils, Libraires.
M. DCCC. XI.

THE
KINEMATICS OF MACHINERY.
OUTLINES
OF A
THEORY OF MACHINES.
BY
F. REULEAUX,
TRANSLATED AND EDITED BY
ALEX. B. W. KENNEDY, C.E.
WITH NUMEROUS ILLUSTRATIONS.
London:
MACMILLAN AND CO.
1876.

LEHRBUCH
DER
KINEMATIK
VON
PROF. F. REULEAUX
ERSTER BAND
THEORETISCHE KINEMATIK
BRAUNSCHWEIG
DRUCK UND VERLAG VON FRIEDRICH VIEWEG UND SOHN
1875

PRINCIPLES
OF
MECHANISM,
DESIGNED FOR THE USE OF STUDENTS
IN THE UNIVERSITIES,
AND FOR
ENGINEERING STUDENTS GENERALLY.
BY
ROBERT WILLIS, M.A., F.R.S., &c.
LONDON:
JOHN W. PARKER, WEST STRAND.
CAMBRIDGE: J. & J. J. DEIGHTON.
M.DCCC.XLI.

TRAITÉ COMPLET
DE MÉCANIQUE
APPLIQUÉE AUX ARTS,
CONTENANT l'Exposition méthodique des théories et des expériences les plus utiles pour diriger le choix, l'invention, la construction et l'emploi de toutes les espèces de machines;
PAR M. J.-A. BORGNIS,
Composition des Machines
PARIS,
BACHELIER, LIBRAIRE, QUAI DES AUGUSTINS
1818.

Fig. 2 Covers of treatises on machines

with *Mecánica de las máquinas operando, o tratado teórico y experimental sobre el trabajo de las fuerzas* (1839) and Arau (Arau 1848) with *Tratado completo de maquinaria teórico y práctico que comprende: los principios de esta ciencia, cálculo práctico de los mismos, su aplicación a toda clase de máquinas como son los relojes, bombas para elevar el agua y otras máquinas hidráulicas, molinos de viento, de agua, de vapor, filatura, tejidos,* etc. (1848). This latter is a treatise on machinery which includes the principles, the practical calculation and the application to machines such as clocks, pumps for raising water and other hydraulic machines, windmills, watermills, steam mills, textile machinery, etc.

The covers of these books are shown in Fig. 3 and represent the main contributions in the Spanish framework, prior to Montesino's book on Machines Construction.

The reflection described above covers the whole nineteenth century, at least as far as a systematic and scientifically based treatment is concerned. This reflection is exclusively European and it is divided into two periods of similar length. The first of

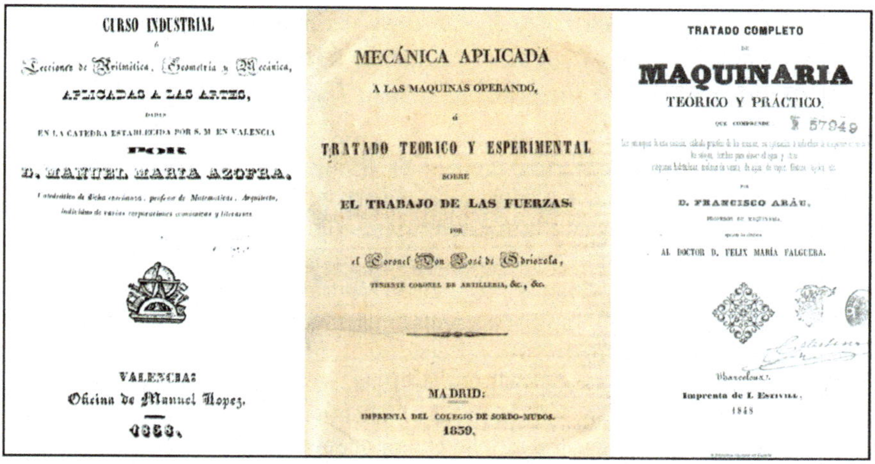

Fig. 3 Covers of treatises on machines in Spain

them starts with a work produced by Spanish authors (Lanz and Betancourt 1808), written in French, in 1808, and continues until 1848 with French (Hachette 1811; Borgnis 1818), English (Willis 1870), German (Weisbach 1848) and also Spanish authors (Azofra 1838; Odriozola 1839; Arau 1848). The second period covers from 1861 (Laboulaye 1861) to 1907 with prevalence of German authors (Redtenbacher 1866; Reuleaux 1875, 1876; Schröder 1899; Voight 1907). There is an interval of thirteen years between the two periods, in the middle of which Montesino's work appears, as a transition between two lines of reflection, and to a certain extent closing the Spanish participation in the subject.

From this brief summary of the historical background regarding the European reflection upon machines, it can be concluded that Montesino's work, published in 1853–54, represents a turning point in this subject, in terms of both Spanish and foreign authors.

3 Montesino's Contribution to Mechanical Engineering

The context analysed in the previous section allows a proper evaluation of the reason for including a chapter about this illustrious person due to his study of machines.

The pedagogical aim of his book is clearly stated in its own title. It is also an essentially practical approach to train engineers who can develop their activity in the industrial field.

3.1 Summary of lessons of the course on Machines Construction by Cipriano Segundo Montesino, professor of this subject at the Real Instituto Industrial in Madrid

There are fourteen lithographed copies of Montesino's treatise on machines (Montesino 1853–54), Fig. 4. The treatise consists of two volumes with more than a thousand pages and seventy illustrated prints. The volumes with the content of the course, without including the prints, are available in the "Colección Digital Politécnica" of the Universidad Politécnica de Madrid: http://cdp.upm.es.

The modesty in which the author presents himself is noteworthy, not giving importance to his extensive professional experience as an engineer, nor to the extraordinary work of synthesis that is reflected in his book. The text contemplates several aspects of mechanical engineering, not only in terms of machinery construction, but moreover in the systematisation of its structure and components.

In the introduction of the book, Professor Montesino begins by presenting the difficulties he has encountered when he took responsibility for the new subject of Machines Construction, which was introduced in the framework of a period of changes in the university. As it is a new subject, he has very few specific reference works, meaning that the contents of the course have to be compiled from a wide variety of sources.

RESUMEN
DE
LAS LECCIONES
DEL CURSO
DE CONSTRUCCION DE MAQUINAS
por
D. C. S. Montesino
profesor de dicha asignatura
en el
REAL INSTITUTO INDUSTRIAL
TOMO I.

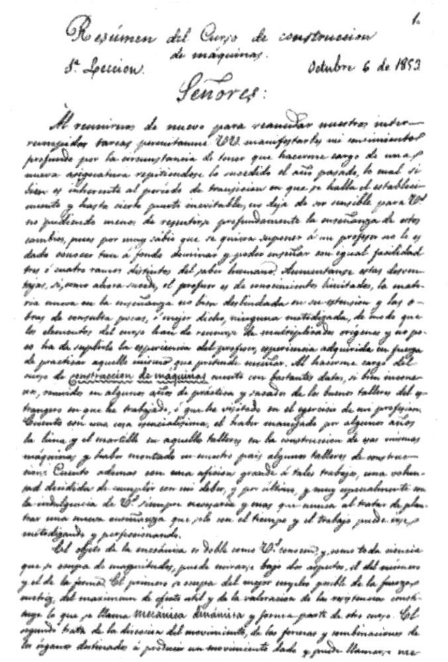

Resúmen del Curso de construccion de máquinas.
1.
1.ª Leccion. Octubre 6 de 1853
Señores:

Fig. 4 Cover of Montesino's treatise on machines and his introduction to his work

However, Montesino has on his side the experience gathered in the practice of the very thing he intends to teach, which will undoubtedly be decisive in achieving an adequate planning and programming to successfully develop the new course. In Montesino's own words: "At the moment of taking over the course on machines construction, I can rely on enough data, despite it being disconnected, collected during several years of practice and obtained from the excellent workshops abroad in which I have worked, or that I have visited during the exercise of my profession. I count on one very essential thing, that is, having handled for some years the file and the hammer in those workshops in the construction of those same machines and having set up in our country several construction workshops".

Subsequently, he discusses two possible points of view in the study of mechanics. The first approach covers dynamic mechanics and would be beyond the proper content of the machine construction course. The second one includes the direction of movement, the shapes and combinations of the machine elements intended to produce a given movement. He considers this second point of view to be very essential for the machines construction course.

He reflects on the fact that most of the treatises on industrial mechanics published to date deal with dynamic mechanics and theoretical knowledge, which, although essential and indispensable, he considers to be insufficient for the adequate knowledge on machines. It further requires knowledge oriented to practical application, which is the result of experience and which is difficult to obtain in technical schools. This perspective is always found in Montesino's book, as in many nineteenth century engineers (Silva Suárez 2011), whose activity is structured through their intuition, their experience and their art, together with their rigorous knowledge, which differentiates them from artisans.

Montesino himself considers that a more practical perspective is behind the great inventions and advances that have taken place in industry since the beginning of the so-called Industrial Revolution, and he quotes, as an example of this, distinguished figures like Watt, Arkwright and Stephenson. The principal icon of the Industrial Revolution is the steam engine (Silva Suárez 2011), about which Montesino writes in reference (Montesino 1861): "Almost up to recent times everything was done by the muscular strength of man and animals, with the help of some of the most elementary machines. Today, improved machines, and in particular the steam engine, which only seems to be lacking in intelligence to be considered as an iron man, replace animal strength with immense advantage regarding every concept, thus gaining in terms of work perfection, the economy of constructions and even in human dignity".

Montesino was training engineers who would play an essential role in the development of his country. Spain had the previous seed of the Royal Manufactures promoted by the Crown, a failed industrial revolution, which nevertheless laid the bases for the industrial boom that was to come in the following centuries in very different regions (Comín and Martín 1991).

In the introduction, Montesino also refers to the importance of studying the best design and arrangement of the elements involved in machines construction, as was initiated by Lanz and Betancourt (Lanz and Betancourt 1808), which is a work of great significance considering the time in which it was published, despite the fact that some of its aspects have already been overcome.

In accordance with the above, he divides the content of the course into ten sections, dedicated to the following subjects:

1. Study of the nature and properties of the materials that are used in machines construction.
2. Receptors. Shapes of their elements and nature of the movement produced according to the mode of action of the driving force.
3. Elements of communication, of transformation of movement from one part of the machine to another.
4. Elements of the machines that are used to modify the movement and to arrange the components in a particular order.
5. Operators. Elements that allow to overcome the resistances and differ according to the nature of the resistances and the product to be obtained.
6. Ways of assembling.
7. Layouts, shapes and usage of the different parts of the machines.
8. Construction and setting of machines: steam engines, water wheels, cranes, etc.
9. Installation and assembly of machines.
10. Construction workshops.

It can be remarked that, in addition to establishing criteria for the design and construction of machinery, aspects linked to its manufacturing, and to its different industrial applications, are addressed. Moreover, the last two sections contain guidelines for installation and assembly, as well as for the layout and organisation of workshops. Although the book is essentially a reflection upon machines in line with the theoretical approaches of his time in Europe, the whole text is imbued with a practical concern regarding what an engineer must take into consideration so that his projects are adapted to the industrial reality, and to specific objectives.

The most relevant aspects of each section of the book are summarised below.

3.2 First Section. Materials Used for Machines Construction

In this section, the author indicates the importance of having an exact knowledge of the materials that are employed in the machines, as well as their properties and usages. Among them, he includes information on the following: metals, woods, flexible materials (leather, hemp, tow, rope, cotton, etc.), greasy materials (vegetable and animal oils, greases, etc.), bitumen, protective coatings and materials for abrading and polishing.

Regarding metals, he describes the origin and properties of wrought iron, cast iron, steel, copper, lead, tin, zinc and some alloys. He emphasises on the necessary tests to be performed on irons to verify their good or bad quality, describing different types of tests to achieve this purpose.

The composition of steel is also described, and the most commonly used ones are listed. The different steel tempering processes are presented in detail, as well as the different properties obtained according to the process applied. Likewise, the properties of copper and lead and the different types of alloys obtained from these are also covered.

Within woods, he describes the classification between hardwoods, softwoods and resinous woods. He explains the application of each of them, obtained from different varieties of trees for hard wood (oak, holm oak, elm, beech…), soft wood (lime, birch, maple…) and resinous wood (fir, cedar, juniper, pine…).

Lastly, a brief description of other materials is given, considering their origin, their mechanical properties and their application in machines construction.

3.3 Second Section. Receptors. Shapes of Their Elements and Nature of the Movement Produced According to the Mode of Action of the Driving Force

This part of the book initiates the study of the constituent parts of machines, defines receptors, and remarks the nature of the movement produced according to the mode of action of the driving force. It classifies engines into three main classes: animated engines, gravity, and the heat and chemical actions. Each of these classes is described in depth.

Man's force and some means for using this force (different types of levers, pulleys and ropes), the use of body weight and the force of animals are all considered as animated engines. Within the group of gravity engines, the weight of solids and liquids, their movements, wind and air pressure and elasticity are the ones considered. In the last group, different means of using the dilation force of heated bodies are presented, along with chemical, electrical and magnetic actions.

An example of the use of atmospheric air as an engine in industry is the Ericsson caloric machine (Colegio de Ingenieros de Caminos, Canales y Puertos 1853), also featured in one of the prints in Montesino's book and shown in Fig. 5. The sources of that time (Colegio de Ingenieros de Caminos, Canales y Puertos 1853) contain information about its existence in New York and mention that the machine presented sixty horsepower.

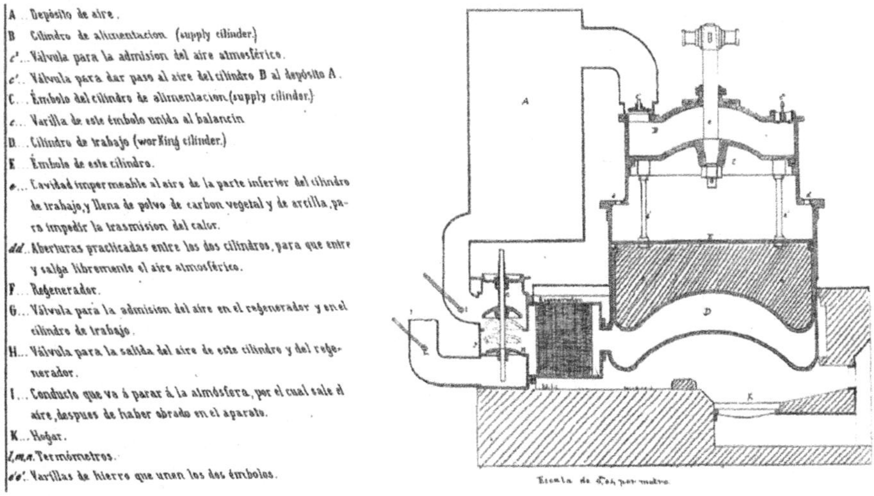

Fig. 5 Ericsson caloric engine (Montesino 1853–54; Colegio de Ingenieros de Caminos, Canales y Puertos 1853)

3.4 Third Section. Elements for Communicating the Movement from One Part of a Machine to Another, in Order to Obtain it in a Predetermined Direction and Speed

This is one of the most extensive sections, with more than a hundred pages, where the possible transformations between the different types of movements are studied, considering the different forces to be overcome. Within the chapter, the transformation of movement with mechanisms using different elements is explained. This includes levers, cranks, connecting rods, rockers, revolute joints, belts, chains, etc.

Gears, both ordinary and planetary, are described in wider detail and the necessary recommendations for their design are listed. Strength formulae are provided to select appropriate dimensions, along with practical tables, such as the one shown in Fig. 6a for gears of common use. In addition, Fig. 6b–d shows some examples of transmissions included in the book.

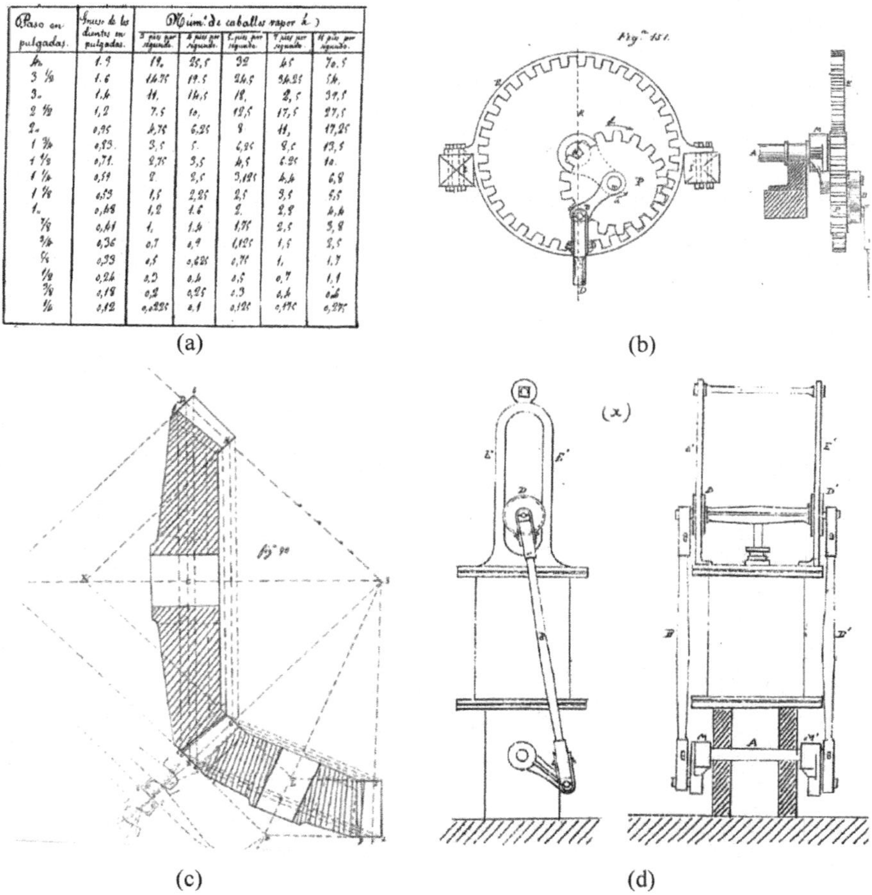

Paso en pulgadas.	Grueso de los dientes en pulgadas.	Núm.º de caballos vapor h ;				
		3 pies por segundo.	4 pies por segundo.	5 pies por segundo.	7 pies por segundo.	11 pies por segundo.
4.	1.9	19.	25,5	32	45	70.5
3 1/2	1.6	14.75	19.5	24.5	34.25	54.
3.	1,4	11.	14,5	18.	2,5	39,5
2 1/2	1,2	7.5	10,	12,5	17,5	27,5
2.	0,95	4,75	6,25	8	11,	17,25
1 3/4	0,83.	3,5	5.	6,25	8,5	13,5
1 1/2	0,71.	2,75	3,5	4,5	6.25	10.
1 1/4	0,59	2.	2,5	3,125	4,4	6,8
1 1/8	0,53	1,5	2,25	2,5	3,5	5,5
1.	0,48	1,2	1.6	2.	2,8	4,4
7/8	0,41	1,	1.4	1,75	2,5	3,8
3/4	0,36	0,7	0,9	1,125	1,5	2,5
5/8	0,33	0,5	0,625	0,75	1,	1,7
1/2	0,24	0,3	0,4	0,5	0,7	1,1
3/8	0,18	0,2	0,25	0,3	0,4	0,6
1/4	0,12	0,0225	0,1	0,125	0,175	0,275

Fig. 6 **a** Table for the design of gears that can safely transmit a force expressed in horsepower at a given speed; **b–d** Mechanical transmissions (Montesino 1853–54)

3.5 Fourth Section. Elements that Are Used to Modify the Movement and to Arrange the Components in a Particular Order

This section includes the elements that provide the movement under appropriate conditions to the proper operation of the machines, in order to perform the work for which they are intended. This includes clutches, sleeves, wheels, pawls, etc. As an example, Fig. 7a shows some speed modification systems on mechanical looms, which include flywheels. Figure 7b shows a braking system. All of them are described in detail in the book and are illustrated with examples of applications in machines.

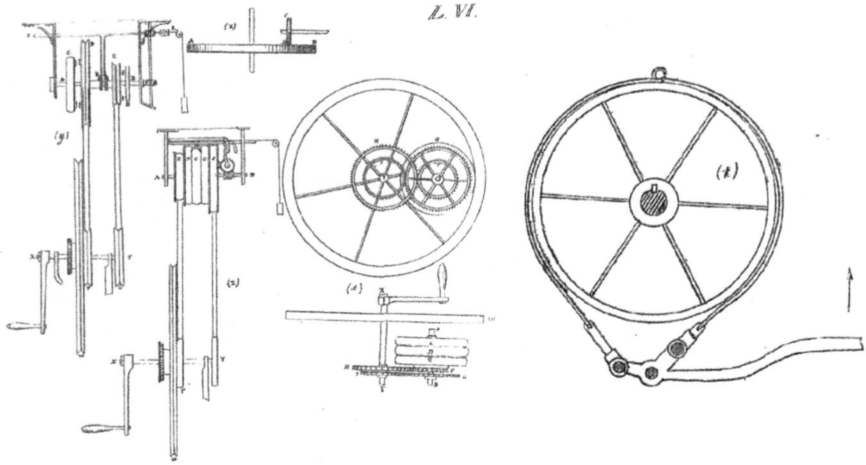

(a) (b)

Fig. 7 Examples of elements to modify movement (Montesino 1853–54): **a** Systems to modify speed; **b** Braking system

3.6 Fifth Section. Operators. Elements that Allow to Overcome the Resistances and Differ According to the Nature of the Resistances and the Product to be Obtained

This section of the book describes the resistances that arise when specific shapes are to be given to bodies through the use of machines, working with materials both in hot and cold conditions. The different tools and machines used are presented in depth, distinguishing those that operate by pressure, those that operate by wear, and those that operate by dividing. Among all of them, several examples are provided, such as the steam hammer or trim hammer, the press, the lathes, the slotting machines, the drills, the threading machines, etc.

Some of them are illustrated with highly detailed figures to fully describe their main components and their operation. Moreover, some important features to be considered for their correct operation are indicated.

An example of the different machines described in this section is the slotting machine shown in Fig. 8. Another outstanding example is the machine for making mortise holes, which is shown in Fig. 9.

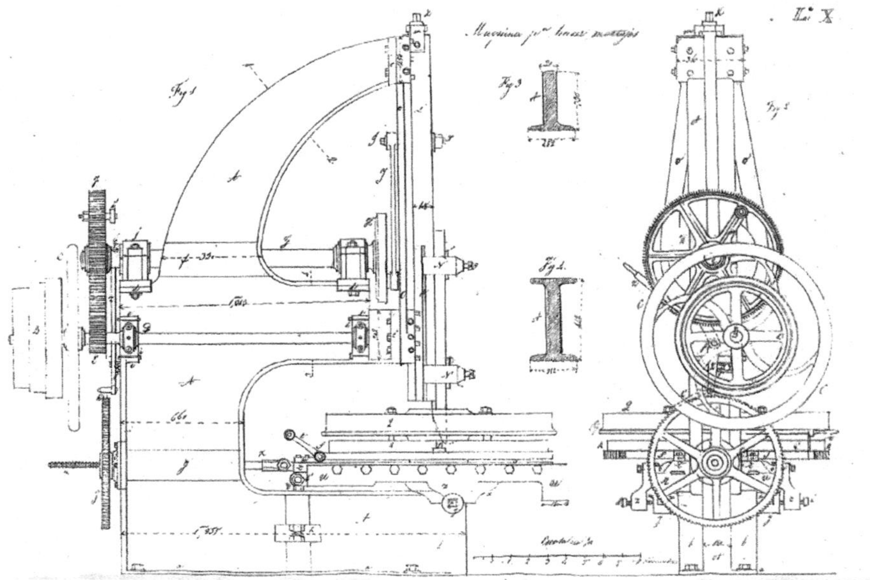

Fig. 8 Slotting machine (Montesino 1853–54)

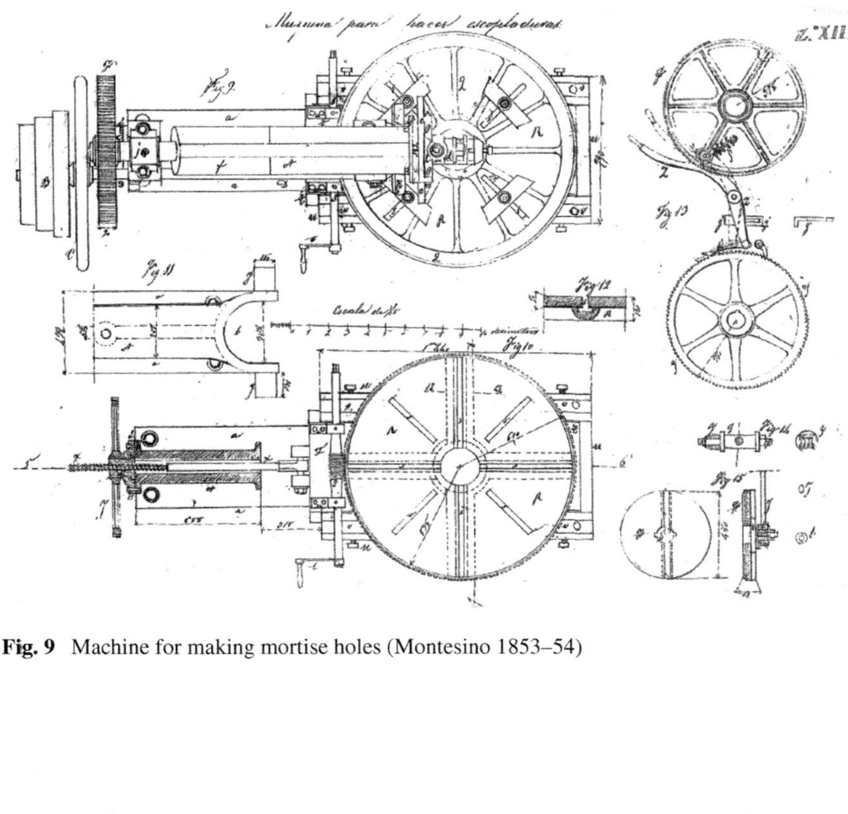

Fig. 9 Machine for making mortise holes (Montesino 1853–54)

3.7 Sixth Section. Ways of Assembling

This is a short chapter, where the ways of joining machine parts without the use of welding are presented. It describes and considers the applications of rivets, threaded joints, shaft keys, and connections of tubes and hollow parts.

Montesino also indicates in his book the materials commonly used to carry out the assemblies, the strict procedures to perform them correctly and the advantages and disadvantages of one technique or another according to each application.

3.8 Seventh Section. Layouts, Shapes and Usage of the Different Parts of the Machines

In this section it is explained that every machine, whether it is an engine or an operator, has some principal machine elements and other secondary ones, and a series of connections between elements. An exhaustive overview of the elements used for the transformation of movement and the assemblies is presented. The first group mainly considers pulleys, wheels, pinions, racks, shafts, connecting rods, parallelograms, rockers, screws, eccentrics and levers. The second group includes rivets, joints, shaft keys, hubs, stuffing boxes, brackets, etc.

The author focuses on the description of the resistance of the materials used in the machines and lists the different types of stresses to which an element can be subjected, with the aim of providing the engineer with criteria for designing machines of appropriate dimensions, which can operate correctly and avoid breakages during service.

3.9 Eighth Section. Construction and Setting of Machines. Steam Engines, Water Wheels, Cranes, etc.

Firstly, he considers steam engines and their classification: of high and low pressure, of simple or double effect, without expansion or with expansion, and within these, with fixed or variable expansion. He also presents a classification of machines according to the mode of application of the driving force in the following types: hydraulic machines for raising and pumping water, blowers with bellows to generate air flow, rotating machines, marine engines, locomotives and hydraulic turbines.

At the same time as he classifies them, he lists and describes their main parts, such as the generator, the steam distribution apparatus, the steam cylinder, the transmission and regulation apparatus of the piston movement, the steam condensation apparatus, the generator feeding apparatus and finally the machine frame. He also details their operation within the machine as a whole, applying his experience and knowledge to compile different dimensional and functional values representative of such components in machines of the time. For many of these parameters, he establishes comparisons between the different types of machines.

Once the main components have been presented and described, Montesino introduces a collection of different expressions and adjustments for their calculation and dimensioning, establishing the general operating conditions of these machines (power, consumption, etc.). These calculation procedures are complemented with graphics that help the engineer when performing first approximations to design these elements. As an example, Fig. 10 shows one of the tables for dimensioning a cylinder of a high pressure, expansion and condensing steam engine.

Once the operation and calculation procedure of each of the components has been analysed, a series of recommendations are performed regarding their position within the machine. To this end, Montesino uses different prints of real machines and mechanisms, which are included in the volume of illustrated prints. Regarding these prints, he makes several comments about the integration and interaction between the components and their effect on the functionality of the machines.

Some of these machines are shown in Figs. 11 and 12. The print in Fig. 11 represents a steam engine with Watt regulator. This machine in particular was designed for raising water in Paris and Montesino made use of it, among other things, to present and analyse several regulation systems existing in the steam engines of the time, such as the well-known Watt regulator. The steam engine represented in Fig. 12 is the Maudslay type, commonly used at that time as marine engines, as they were very popular on paddle steamers in the nineteenth century.

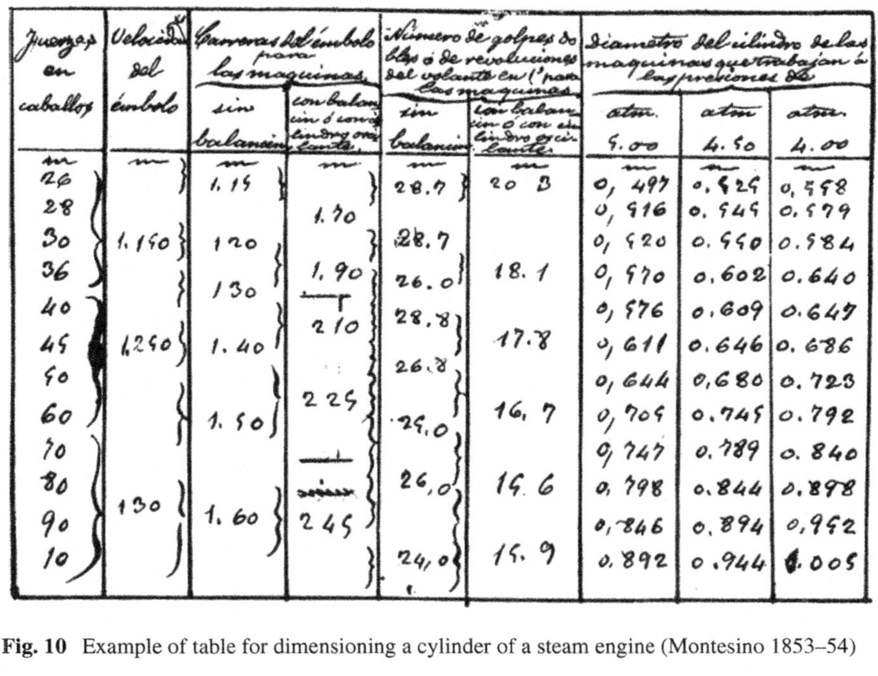

Fuerzas en caballos	Velocidad del émbolo	Carreras del émbolo para las maquinas		Número de golpes dobles ó de revoluciones del volante en 1' para las maquinas		Diámetro del cilindro de las maquinas que trabajan á las presiones de		
		sin balancín	con balancín ó con cilindro oscilante	sin balancín	con balancín ó con cilindro oscilante	atm. 5.00	atm 4.50	atm. 4.00
m	m	m	m	m	m	m	m	m
26		1.15		28.7	20 B	0, 497	0.525	0, 558
28			1.70			0, 516	0. 545	0.579
30	1.150	120		28.7		0, 520	0. 550	0.584
36		130	1.90	26.0	18.1	0, 570	0.602	0.640
40				28.8		0, 576	0.609	0.647
45	1,250	1.40	2 10		17.8	0, 611	0.646	0.686
50				26.8		0, 644	0,680	0.723
60		1.50	2 25	25.0	16, 7	0, 705	0.745	0.792
70						0, 747	0.789	0.840
80				26,0	15.6	0, 798	0.844	0.898
90	130	1.60	2 45			0, 846	0.894	0,952
10				24,0	15.9	0.892	0.944	1.005

Fig. 10 Example of table for dimensioning a cylinder of a steam engine (Montesino 1853–54)

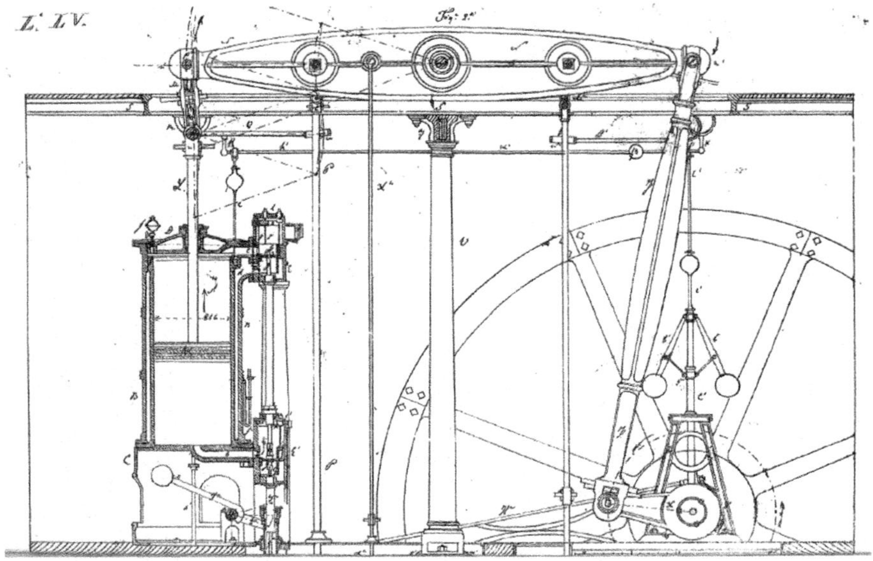

Fig. 11 Example of a steam engine with Watt regulator (Montesino 1853–54)

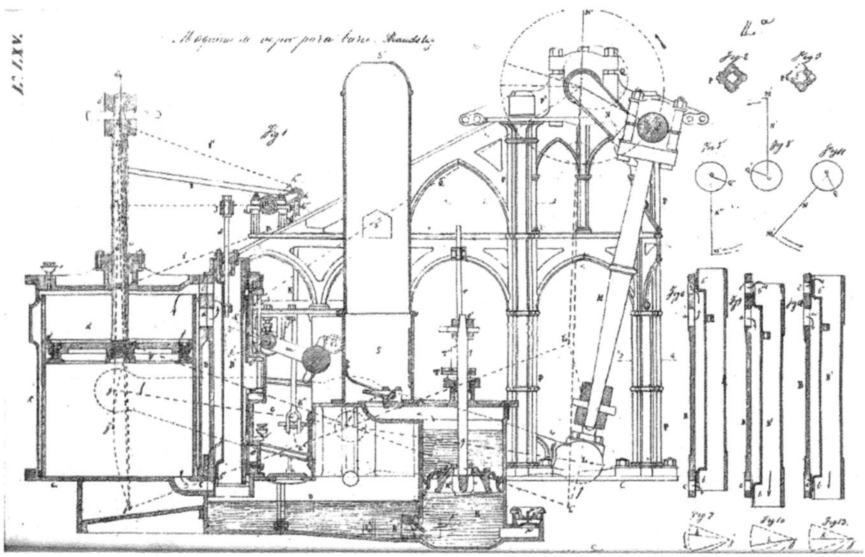

Fig. 12 Example of a marine steam engine (Montesino 1853–54)

As he describes each of the types of machines, collected in one or several prints, he also analyses their differences in the operation of each of their components (boilers, cylinders, condensers, valves, pumps, etc.), as well as their mechanisms (rockers, connecting rods, regulators, etc.).

For example, in the case of the Maudslay engine in Fig. 12, an extensive analysis is dedicated to the different technical solutions of the time that enabled to achieve reductions in the weight and volume of the machine, as well as to lower its gravity centre, which were very important aspects in marine applications, steam locomotives, etc.

Within the field of steam engines, the text dedicates considerable attention to the machines that Montesino was most closely involved with during his professional life, namely railway locomotives. This extensive section analyses the different systems that comprise them, based on a large number of concepts already seen in previous sections: vaporisation in the boiler, the piston system and valves, transmission systems and frame, always highlighting their particularities regarding the steam engines seen up to this point.

Moreover, the book makes a clear distinction between locomotive designs intended for freight transport, with speeds ranging between 20 and 30 km/h, and those intended for passenger transport, with speeds of up to 100 km/h. Figure 13 shows a locomotive of the time.

In this same section, Montesino also presents water wheels, analysing the design of the turbines of the time. Figure 14 shows a Fountaine water turbine with a regulator that adjusts the water inlet by the use of cylinders, which move a leather belt that

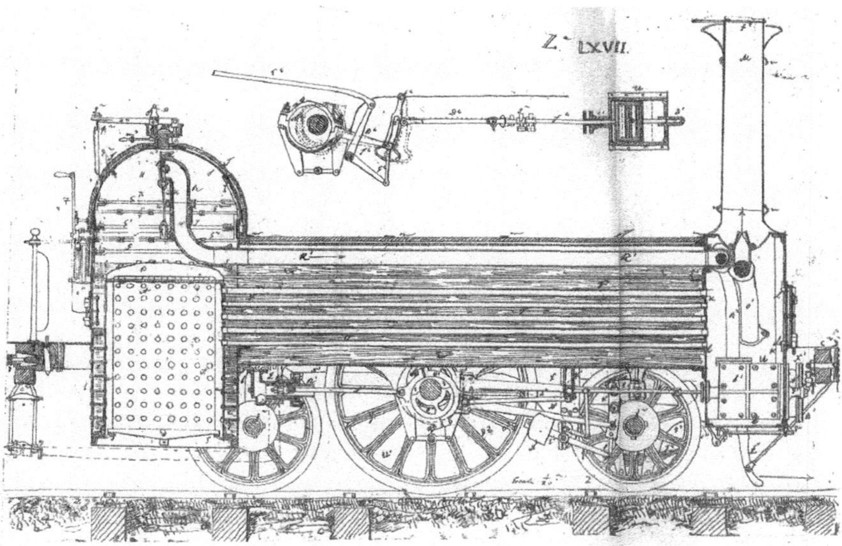

Fig. 13 Example of a locomotive (Montesino 1853–54)

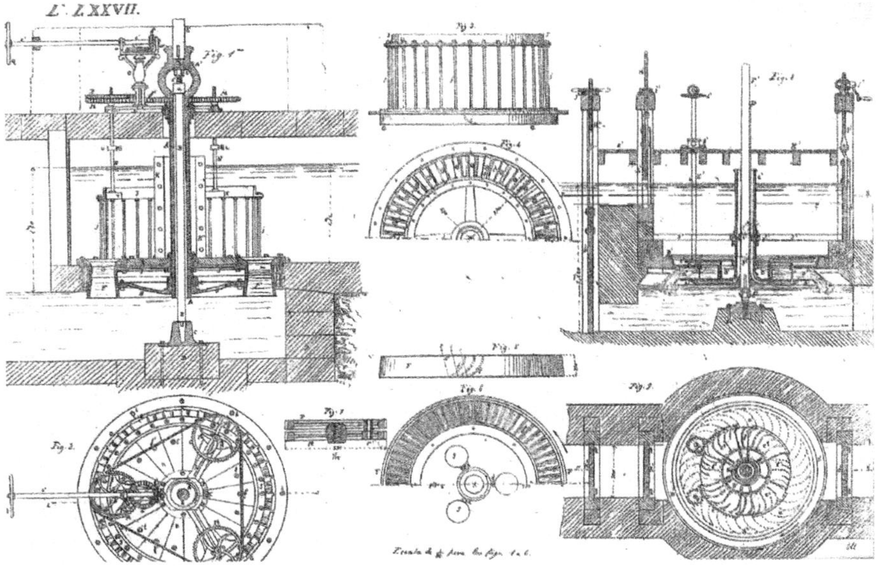

Fig. 14 Fountaine water turbine (Montesino 1853–54)

governs the water flow. This type of adjustable turbines was very well adapted to the applications in Mediterranean rivers, of a very variable flow rate.

3.10 Ninth Section. Installation and Assembly of Machines

This section provides notions about foundations and the subsequent process of installing and assembling the machines over them. It also dedicates part of the chapter to present knowledge on the construction of water wheels and canals.

3.11 Tenth Section. Construction Workshops

This last section presents various layouts and ways of organising some different typologies of construction workshops, such as a foundry, forge, locomotive workshop, gear workshop, etc. Tables are also provided for calculating the diameters of the shafts obtained by different manufacturing processes.

4 Concluding Remarks

Montesino's book belongs to a period of important change in the academic study of machines in Europe in general, and in Spain in particular, which seeks to go beyond providing theoretical knowledge. This latter is logically necessary, but the aim is to complement the engineer's training with a practical vision. Montesino developed part of his professional activity in machine workshops and was aware that he was training engineers who would play an essential role in the country's industrial progress, so the practical perspective was clearly emphasised in his course on machines. In this sense, his work is less "academic" than other books of the time, meaning that it is less addressed to the small number of theoreticians who were concerned with defining concepts about machines, since the ultimate aim was to train professionals for the industry.

Although most of the machines and mechanisms in the book do not represent significant innovations for the time, the depth of the analyses performed, their vast length and practical approach are noteworthy. Montesino himself considers his work imperfect, as he had very limited time for its preparation. He wrote it as a reference text for the engineers he trained, limiting the number of copies published to one for each student of the course, and this undoubtedly made its dissemination difficult.

The work written by Montesino is not limited to the book mentioned above, which is obviously the most significant in relation to the title given to this chapter. The multifaceted activity shown in his biography (Sect. 1) enabled that he, throughout his life, wrote studies, reports, memoirs, etc., of different natures, whose exhaustive analysis is out of the scope of this chapter, in the same way that his character goes far beyond his distinguished role in the field of mechanical engineering. Among these other publications, the following remarkable examples can be highlighted: a study on the characteristics of the project of the Suez Canal and the implications of its creation (Montesino 1857), a report on the situation of public works in Spain (Montesino 1856), and the speech given before the Royal Academy of Sciences in 1861 (Montesino 1861).

The bibliography of Montesino is mainly in Spanish, with very few references to his work by foreign authors, which, in the opinion of the authors of this chapter, is not in line with the fact that his contributions were at the forefront of what existed at the time. This situation is not due to an attitude of exclusivist nationalism, but to reasons attributable to his own country, and to the personality of the character himself, who had to live through a particularly complicated period in the political and social environment in Spain, in which he was actively involved from a very early age. His activity as a machine theorist was marginal and sporadic, as he himself explains in the introduction to his book, which was not widely distributed due to these same circumstances.

His personality led him to devote his efforts primarily to the development of his country, both through his political and business activities, without paying proper attention to international academic relationships in the field of machines reflection. To this must be added the modesty with which he himself considers his work.

Montesino is, on the contrary, a mandatory reference for any author who approaches the History of Engineering in Spain, such as Martínez-Val (Martínez-Val 2001), García Álvarez (García Álvarez 2010), De la Torre (De la Torre 1901), Silva Suárez (Silva Suárez 2011), Comín and Martín (Comín and Martín 1991), and he even deserves to be included in this publication about the History of Mechanical Engineering.

References

F. Arau, *Tratado completo de maquinaria teórico y práctico que comprende: los principios de esta ciencia, cálculo práctico de los mismos, su aplicación a toda clase de máquinas como son los relojes, bombas para elevar el agua y otras máquinas hidráulicas, molinos de viento, de agua, de vapor, filatura, tejidos, etc* (Imprenta I. Estivill, Barcelona, 1848)

Asociación de Ingenieros Industriales de Madrid: Boletín informativo n.13, May 2008.

M.M. Azofra, Curso Industrial o Lecciones de Aritmética, Geometría y Mecánica aplicadas a las Artes. Oficina de Manuel López, Valencia (1838)

E. Bautista, M. Ceccarelli, J. Echávarri, J.L. Muñoz, *A Brief Illustrated History of Machines and Mechanisms* (History of Mechanism and Machine Science Series. Springer, Dordrecht, 2010)

F.A. Borgnis, *Traité complet de mécanique appliquée aux arts* (Bachelier, Paris, 1818)

F. Comín, P. Martín, *Historia de la empresa pública en España* (Espasa Calpe, Madrid, 1991)

Colegio de Ingenieros de Caminos, Canales y Puertos. Revista de Obras Públicas 1, I(6): 76–77, Madrid (1853)

Congreso de los Diputados: Index of past members, https://www.congreso.es, last accessed 2022/11/21.

E. De la Torre, *Anuario de ferrocarriles españoles 1901* (Imprenta Central de los Ferrocarriles, Madrid, 1901)

A. García Álvarez, Cambio automático de ancho de vía de los trenes en España. Documento de explotación económica y técnica del Ferrocarril. Fundación de los Ferrocarriles Españoles, Madrid (2010)

M. Hachette, *Traite élémentaire des machines* (Klostermann, Paris, 1811)

J. Lanz, A. Betancourt, *Essai sur la composition des machines* (Bernard, Paris, 1808)

Ch. Laboulaye, Traité de Cinématique, ou Théorie des Mécanismes. E. Lacroix, Paris (1861).

J.M. Martínez-Val, *Un empeño Industrial que cambió a España 1850–2000: siglo y medio de Ingeniería Industrial* (Síntesis, Madrid, 2001)

C.S. Montesino, Resumen de las lecciones del curso de Construcción de Máquinas, Madrid (1853–54). Two volumes and seventy prints. The volumes (without the prints) are available in: "Colección Digital Politécnica" of the Universidad Politécnica de Madrid: http://cdp.upm.es, last accessed 2022/11/21.

C.S. Montesino, *Memoria sobre el Estado de las Obras Públicas en España* (Imprenta Nacional, Madrid, 1856)

C.S. Montesino, *Rompimiento del Istmo de Suez* (Imprenta Nacional, Madrid, 1857)

C.S. Montesino, *Discurso en contestación al ingeniero de caminos Lucio del Valle: Influencia que han tenido los progresos de las ciencias exactas en las artes de construcción, y más especialmente en las que entra el hierro por principal elemento* (Speech before the Royal Academy de Sciences, Madrid, 1861)

G. Monge, *Application de l'analyse à la géométrie* (Bernard, Paris, 1805)

J. Odriozola, *Mecánica de las máquinas operando, o Tratado teórico y experimental sobre el trabajo de las fuerzas* (Imprenta del colegio de sordomudos, Madrid, 1839)

J. Peris Torner, Website on Spanish railways: http://www.spanishrailway.com, last accessed 2022/11/21
F. Redtenbacher, *Die Bewegungs-Mechanismen* (F. Bassermann, Heidelberg, 1866)
F. Reuleaux, *Lehrbuch der Kinematik* (Verlag von Friedrich Vieweg und Sohn, Braunschweig, 1875)
F. Reuleaux, *Kinematics of Machinery* (MacMillan and co, London, 1876)
Real Academia de Ciencias: Galery of presidents, http://www.rac.es, last accessed 2022/11/21.
Senado de España: http://www.senado.es/web/conocersenado/senadohistoria/senado18341923/senadores/fichasenador/index.html?id1=3114, last accessed 2022/11/21.
J. Schröder, Catalog of Reuleaux Models: Polytechnisches Arbeits-Institut. Illustrationen von Unterrichts-Modellen und Apparateni. Polytechnisches Arbeits-Institut, Darmstadt (1899).
M. Silva Suárez, El Ochocientos. De los lenguajes al patrimonio. Técnica e Ingeniería en España. Institución "Fernando el Católico", Zaragoza (2011)
G. Voight, *Kinematische Modelle nach Prof* (Reuleaux. Voigt Mechanische Werkstatt, Berlin, 1907)
R. Willis, *Principles of Mechanism*, 2nd edn. (Green and co., London, Longmans, 1870)
J. Weisbach, Principles of the Mechanics of Machinery and Engineering. Lea and Blanchard, Philadelphia (1848)

Tomás de Morla y Pacheco (1747–1811)

I. Durán Montero, R. López-García, and G. Medina-Sánchez

Abstract Tomás de Morla y Pacheco was a military engineer who made notable contributions to the development and dissemination of metallurgical techniques and military strategies of his time. Andalusian born in Jerez de la Frontera, Spain, in 1747, was part of the first promotion of the Artillery Academy of Segovia. At the end of his studies, he remained there as a professor of the Tactics subject. He wrote *"Tratado de Artillería para el uso de Caballeros Cadetes del Real Cuerpo de Artillería"*, (Artillery Treatise for the use of Gentlemen Cadets of the Royal Corps of Artillery), an extensive work that contains the knowledge that was available in his time on metals, alloys, smelting furnaces and cannon manufacturing, with a collection of prints with detailed designs of machines and mechanisms. This work was translated into several languages and used as a text for European students. Morla was a prolific author who wrote on topics as diverse as military strategies, the manufacture of gunpowder, or ways to end epidemics. He was commissioned to make a trip with the aim of reporting on the latest technological advances that were taking place in Europe. For four years, he visited the industrial facilities of the most advanced countries, and collected the information in a work called *"Apuntes autógrafos"* (Autograph Notes). As a soldier, he participated in numerous battles in Gibraltar, the Pyrenees, Portugal, Cádiz and Madrid. He came to occupy high positions in the military structure and played a decisive and controversial role during the Spanish War of Independence.

Keywords History of MMS · Metallurgical techniques · Artillery equipment · Military engineering

I. D. Montero (✉)
Fundación Patrimonio Industrial de Andalucía, Sevilla, Spain
e-mail: inesdumo@gmail.com

R. López-García · G. Medina-Sánchez
University of Jaen, Jaen, Spain
e-mail: rlgarcia@ujaen.es

G. Medina-Sánchez
e-mail: gmedina@ujaen.es

R. López-García and M. Ceccarelli (eds.), *Distinguished Figures in Mechanical Engineering in Spain and Ibero-America*, History of Mechanism and Machine Science 43, https://doi.org/10.1007/978-3-031-31075-1_10

1 Introduction (Biographical Notes)

1.1 *Historic Context*

In the middle of the eighteenth century, Tomás de Morla was born in Andalusia, under the reign of Fernando VI, and died at the age of 64, during the Spanish War of Independence.

The principles of the Enlightenment, promoted in Spain by King Carlos III, marked his academic training. Later, he had numerous opportunities to develop his military knowledge in the battles resulting from the different alliances between European countries, which were hatched under the mandate of Carlos IV and his favorite Manuel Godoy. Finally, he played a controversial diplomatic role during the Napoleonic invasion of Spain.

His life was initially linked to the study and teaching of artillery-related sciences, to experimentation with machines and materials, to research on metallurgical techniques, and ultimately to developing as an engineer who applied his knowledge to strategies military. But in his last years, the great and profound changes that shook Spain produced a radical change in his life, dragging him to play various roles of great political significance, reaping numerous victories and some failures that unfortunately hid the brilliant career of this character, who dedicated mainly his life to the study and knowledge (Ruiz Lagos 1972).

1.2 *Personality*

Tomás de Morla was a person with an enormous capacity for work, a scientist, studious, patriotic and brilliant. He was the prototype of an eighteenth-century man, individualistic, open-minded, insatiable curiosity and eager for knowledge. On his personal side, it was said that he was a misogynist. The truth is that he remained single throughout his life.

His enemies, who were numerous, accused him of being intolerant, ambitious, and treasonous. It was also said that he was Frenchified, perhaps because he had a large library with texts in several languages, including French, and he said he admired the figures of the Illustration (Fig. 1).

From the graphological study done on his manuscripts, we know that he was a balanced person, who knew what he wanted and that to achieve it he did not care about the means. He attached great importance to social conventions, although he was capable of breaking with them at any given moment. He was suspicious and an enemy of confidences. He also suffered from depression (Vega Viguera 1995).

Fig. 1 Portrait of Tomás de Morla y Pacheco

1.3 His Early Years in Cádiz

Tomás was born in Jerez de la Frontera on July 9, 1747, into a noble family, but which only enjoyed a modest position. His father was a numerary public scrivener from Jerez. He was son of his father's second marriage, becoming eight siblings.

He began his studies at the Convent of the Padres Dominicos in Jerez, an institution created with the aim of combating the existing neglect of culture and education in the area. There he received a magnificent initial training in the humanities, Latin and Philosophy, which made it easier for him to excel in his later studies. He was from the beginning a very diligent student who obtained the best grades in all subjects.

Tomás had a brother Guardia de Corps and another clergyman brother, but he inclined to instruct himself in artillery sciences, since he never showed sympathy for the clergy.

Being very young, he moved to Cádiz to enter the Provincial Academy of Artillery. This institution was initiating a new approach in military education. There he studied Mathematics and Drawing, subjects in which he caught the attention of his teachers because of the ease he showed to assimilate them. This institution also taught the necessary knowledge for the training of an officer, such as the use of gunpowder and the assembly of cannons (Ruiz Lagos 1972).

2 Main Works

Tomás de Morla left an abundant written work consisting of treaties, speeches, proclamations and correspondence.

His most important technical works are:

- *"Tratado de Artillería para el uso de Caballeros Cadetes del Real Cuerpo de Artillería"* (Artillery Treatise for the use of Gentlemen Cadets of the Royal Artillery Corps) (1784) [3, 4]
- *"Apuntes autógrafos"* (Autograph notes) (1789–1790) [5]
- *"El arte de fabricar pólvora"* (The art of making gunpowder) (1800) [6]:
  Volume I: About the collection of saltpeter
  Volume II: About the recognition of saltpeter, its refining and preparation
  Volume III: About the gunpowder factory

Other works related to his professional experience were:

- Recognition of the Pyrenees
- Report on the invalidity of the fortifications of the Castle of Figueras
- News of the Prussian military constitution (1790)
- News of the artillery pieces called carronades with plans and figures
- Notions of fortification
- Report on the method that should be established in the Royal Gunpowder Factories to test nitrates
- Report on the campaign of the Count of the Union in 1794
- Reflections on the causes of spreading the contagion of yellow fever.

2.1 Tratado De Artillería Para El Uso De Caballeros Cadetes Del Real Cuerpo De Artillería

Morla's main work is an artillery treatise, called "*Tratado de Artillería para el uso de Caballeros Cadetes del Real Cuerpo de Artillería*" (Artillery Treatise for the use of Gentlemen Cadets of the Royal Artillery Corps), written during his stay at the Royal College of Artillery of Segovia between 1784 and 1786 (Morla 2022) (Fig. 2).

The treatise is made up of four volumes. In the first two volumes, he exposes the notions and doctrines on the functions of the officers of the Corps in peace periods. The index of that section is:

Article I: About gunpowder
Article II: About casting of bronze artillery pieces
Article III: About iron and the foundry and factory of the pieces and ammunition composed by it
Article IV: About construction of the carriage, tools and machines for the service of the artillery and the wood more suitable for them
Article V: About military bridges

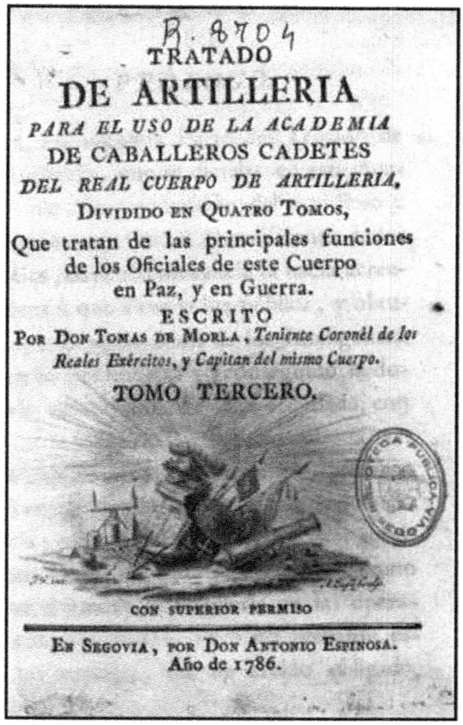

Fig. 2 Cover of "*Tratado de Artillería para el uso de Caballeros Cadetes del Real Cuerpo de Artillería*" (Morla 2022)

Article VI: About cordage and rope-wick
Article VII: About reconnaissance and inventories of the artillery
Article VIII: About firearms, point and cut
Article IX: About fireworks
Article X: About Practical Schools of Artillery
Article XI: About ranges and charges of firearms
Article XII: About mines

The third volume develop the knowledge on the use of artillery in military operations in wartime, both in the campaign, attacking the places and defending them.

The following articles develop the matter:

Article I: About campaign trains.
Article II: About use of field artillery.
Article III: About military trains.
Article IV: About attack of the places.
Article V: About endowment of the places.
Article VI: About defense of the places.

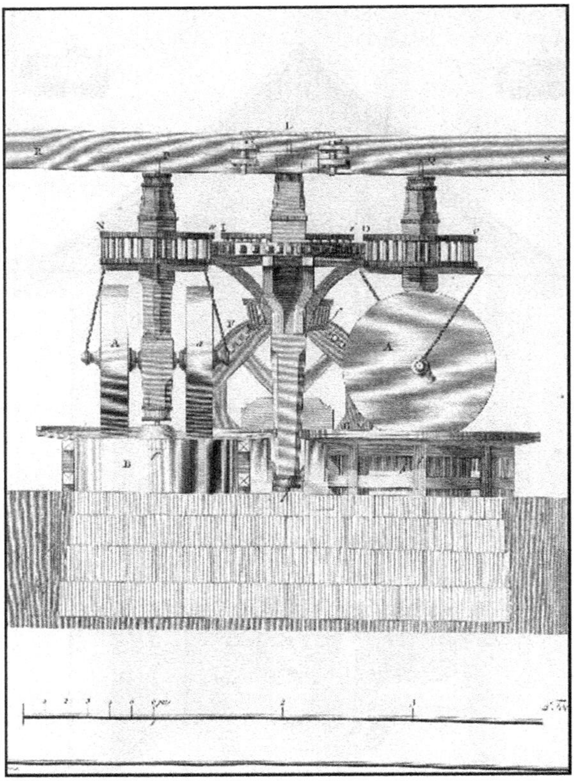

Fig. 3 Sheet with a design of a mechanism from the IV volume of the Treatise (Morla 1803)

The fourth volume is a beautiful collection of copper-engraved sheets displaying the knowledge explained in the previous volumes (Morla 1803). This volume was published years later than the previous ones, due to the high cost of printing (Fig. 3).

3 Review of Main Works and Events of His Life

3.1 The Royal College of Artillery of Segovia (1764–1787)

When King Carlos III arrived in Spain, he verified the disorder that existed in the artillery and, aware of the boom that it had been taking since the seventeenth century, he initiated important reforms in the army. He saw that scientific and metallurgical knowledge was necessary in the training of cadets, and to achieve this he founded the Royal College of Artillery of Segovia, located in the Alcázar of this city and directed by Felix Gazzola, who had been his collaborator in Naples (Artola 1968, 1983).

Fig. 4 Drawing of the Alcazar of Segovia

The creation of the Academy marked a new era in the history of Spanish artillery. The purpose of this institution was to provide scientific-technical-military training to the cadets to turn them into fundamental pieces of the technological and cultural development of enlightened Spain (Fig. 4).

Carlos III wanted to attract students from the nobility, and that they feel honored to contribute to the service of the King. For this, he would create a body characterized by its excellent training and its high social origin. After his training, in addition to being an artilleryman, he should be a mathematician, philosopher, politician and hero.

The College operated on a boarding school basis, both for the students and for the majority of the teachers, who also had to remain single. The day was full of activities, both theoretical and practical. Only one day a week, they walked through the countryside and on Sunday, they could visit the city.

Tomás de Morla was admitted to this College, after presenting the corresponding nobility tests, together with sixty other students, when he was 17. In the first course only fifteen passed, Morla obtaining one of the best grades, possibly due to the fact that he was one of the oldest students and the good preparation he had acquire.

Teaching activity. When Morla finished his studies, at just 21, he was offered, like other outstanding students, the opportunity to do teaching work at the Academy and became an assistant to Captain Vicente Gutiérrez de los Ríos, a tenured professor of the subject of Tactics. He immediately devoted himself to the task of writing notes to facilitate student learning and to continue training as a teacher. In the following academic year, he took over the secretariat of the Board of Professors, which led him to acquire a wide knowledge of the College. In the minutes that he drew up, he detailed the pedagogical work of the Academy, showing signs of rigor and application.

He was promoted to artillery lieutenant in 1773, and continued his work as an assistant professor, but he began to feel uncomfortable living in the Alcázar, requesting permission to live in an apartment in Segovia. He gave several reasons for the moving, among others he complained about the poor quality of the food that came out of their kitchens and that they shared with the students of the Academy. There is a lot of correspondence addressed to the director of the school in which he expresses his discomfort for various issues, such as the refusal to grant him his promotion as a teacher or to send him to other destinations that he repeatedly requested.

Morla was a great scholar who began his study day at five in the morning to leave at nine to teach his classes. As the main professor of the Tactics subject was an academic from several institutions and had family affairs that separated him from Segovia for a long time, Morla was in charge of the Tactics subject and the preparation of the study material. The academy had been operating for a short time and the study plan had not yet been fully defined, so his methodical work was a very interesting contribution.

However, Morla's dissatisfaction was increasing, he considers himself a theoretical officer who had not been able to demonstrate his knowledge in practice, and this caused him some frustration. He also saw how some of his classmates were already actively serving in the military and how others had become full professors, even with worse records than his. This made him impatient, he was uncomfortable in Segovia, and he suffered when, after spending rest periods in the warm Andalusia, he had to endure the cold and harsh Castilian winters again. At the age of 29, he wrote a letter to his superiors requesting his transfer to America, but once again the answer was the same as always: they assure him that instructing and educating cadets is the most important destination for an artilleryman (M. D. Herrero Fernández –Quesada 1992).

The Artillery Treatise. Considering the great theoretical preparation, work capacity and pedagogical qualities of Morla, in 1779 the direction of the Academy decided to make him a commission that would be transcendental in his life, the drafting of an Artillery Treatise. This work would be the continuation of some annotations that his teacher Gutiérrez de los Ríos had started, but that he had not had time to write. Morla dedicated himself to that task with all his efforts.

They provided him with a financial allowance so that he could dedicate himself fully to the writing of the work that he titled "*Tratado de Artillería para el uso de Caballeros Cadetes del Real Cuerpo de Artillería*" (Morla 2023). In this work, it became clear that its author was an engineer who had done important research work and who thoroughly studied the different opinions of foreign scholars.

Some of Morla's enemies questioned the authorship of the work and he was accused of having appropriated the work of his teacher Ríos and publishing it after making some enlargements. However, a thorough study of his work has shown that said accusation is unfounded. Morla himself justifies in the prologue that the treaty ends because the occupations and the early death of his teacher prevented him from finishing it, and he also humbly affirms that his intervention will never be up to the level of Gutiérrez de los Ríos (Herrero Fernández–Quesada 1992).

The Siege of Gibraltar (1782). The Great Siege of Gibraltar was one of Spain's attempts to recover the British colony. For four years, the rock was subjected to

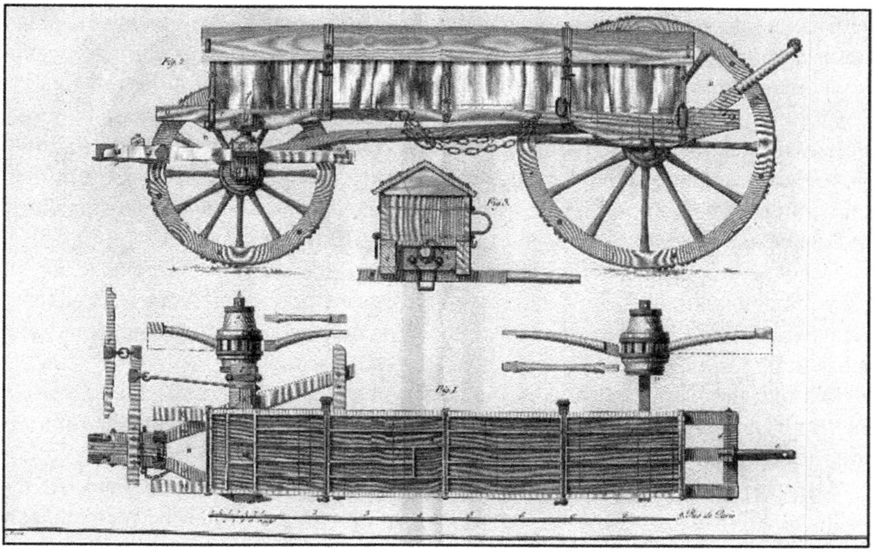

Fig. 5 Sheet with a mechanical design from the IV volume of the Treatise (Morla 1803)

a naval blockade, intense bombardments and the action of the innovative floating batteries.

In 1782, Morla finally managed to live the experience of war when he was assigned to the Siege of Gibraltar. There, he was assigned a place in a floating battery, which consisted of a boat with two rows of cannons. In the middle of the battle, a bullet fired from Gibraltar set his boat on fire and sank it, but Morla managed to survive. When he recovered, he was commissioned to prepare an underground mine to open a way overland to Gibraltar. The mine worked as he had anticipated, but the explosion seriously injured him. When he recovered, he returned to Segovia, holding the rank of Graduate Captain, and with a service record stating that his behavior had been courageous and effective.

Already at the Academy, he was satisfied to have practiced with artillery in war, and to have used gunpowder for the use of underground mines.

A few years later, he was appointed professor of Tactics and Artillery, replacing his teacher, already deceased.

3.2 *The Scientific Journey (1787–1791)*

When Felipe V came to the throne, he began to promote the steel industry, so that Spain would be self-sufficient in the production of weapons, which was essential for the defense of his interests. One of the initiatives to achieve these objectives was the realization of scientific trips. To make these trips, the government selected the officers with the best technical training, who were capable of capturing the advances in manufacturing engineering and had the ability to establish relationships with technicians from other countries (Fontana 1983).

Continuing this policy, the government of Carlos III proposed Morla, who had already turned 40, to form part of a technical commission that would make a trip to those European nations that had the best and most up-to-date industries for military applications. The aim of this trip was to know the advances in weapons, metallurgy and chemistry that would be useful to Spanish industry.

Jorge Guillelmi, a former professor at the Royal Academy, was selected to accompany him. This kind of trip lasted several years and were carried out with advantageous conditions, since they traveled accompanied by several assistants and with substantial allowances.

In 1787, they left Barcelona in the direction of Paris. He reported the details of the trip promptly, but unfortunately a large part of the documentation has been lost and little is known about the artillerymen's stay in France.

After spending two years there, they traveled to England and Ireland. At that moment, Morla began to write the "Autograph Notes" in which he recounted the details of the visits he made, describing the machines, inventions and industrial techniques he found, as well as the impressions these places produced on him (Fig. 6).

The travelers sometimes had serious difficulties to enter some facilities, and even more to learn the secrets of manufacturing, since really their task had very much of industrial espionage.

Near London, he visited Herschel, the famous English astronomer who had discovered the planet Uranus. He admired the 13-m long telescope that he had just built and saw how he corrected the curvature of the mirrors to get a sharper image. A few years later, Godoy commissioned a telescope from Herschel for the Spanish Astronomical Observatory.

Fig. 6 Drawing of an English industrial town in the eighteenth century

In Birmingham, a city with great industrial activity, he had the opportunity to visit factories making horse harnesses, metal buttons, steel watch chains, buckles, candlesticks and other household items. The one that most attracted him was that of harquebuses, shotguns and blunderbuss. It was in this city where he saw for the first time what he called firebombs, which were the primitive steam engines, and his astonishment was enormous because he immediately became aware of the infinite advantages that these machines were going to bring to the world in the future. Throughout the trip he did not stop making references to them, every time he discovered their uses in the facilities he visited.

He came to visit the steam engine manufacturing company owned by Mr. Boulton and Mr. Watt. He thought that thanks to them, the problems of limiting available energy would end, and that from now, the factories could be located next to the places where extract the raw materials.

He went to Bradley, where he visited the ironworks, and there he saw that charcoal was being replaced by mineral coal, a very abundant natural resource in that region. That would solve the deforestation problem that England was suffering from.

At Sheffield, he learned about the research that was being done on iron alloys. At Newcastle, he went down into a colliery, passed through the galleries supported by props, and studied the mechanical mechanisms used to transport ore within the mine. There, he criticized the scarce safety measures for the workers.

He took advantage of his time in Wakefields and Leeds to get to know the manufacture of wool and the subsidiary chemical industry dedicated to detergents, bleaches and acids for dyes. He also observed the looms, drawing his attention to the flying shuttle, which allowed weaving pieces of any width.

At Glasgow, he was able to learn the technique of printing cloth with copper plates and presses, and at Manchester, where silk and cotton were worked, he saw wool carding machines, similar to the ones he studied in Rouan (Fig. 7).

On the way, he praised the perfect organization of Blackfriend's flourmill, and at Greenwich, he visited the residence for invalid sailors and the zero meridian, however what he liked best was its park, for the varied species it contained.

He had a logical interest in getting to know the Wolwich Artillery Department, and there he criticized the lack of regulations and organization, as well as the low level of teaching given, the poor installations of the shooting range, and even that the foundry was without activity.

He does not like Oxford University either, saying that the so-called interns taught the students, while the teachers did not attend the classrooms.

However, he admired English hospitals, where the sick were not only cared for but cured as well. He also liked Liverpool jail, for being safe and ventilated at the same time.

He ended his tour of England by seeing the Tower of London, which he described as the safe for English royalty, and the menagerie, where he was able to observe tigers, a lion and a white bear.

He went over to the mainland and headed for Flanders; there the bridges that rotated over the canals caught his attention. In the Cabinet of Natural History in The

Fig. 7 Drawing of a weaving factory in England

Hague he had the opportunity to see the mechanism of a canal lock and a machine that lifted large weights and deposited them exactly at the chosen height.

In Amsterdam, he was able to enter several mills, but he could not pass to the gunpowder mills, which were of greatest interest to him.

He then went to Saxony to the Military College, of which he got a good impression, especially because of its extensive cavalries. He also visited the linen factories, where they made canvases and tablecloths. Saxony had an abundant production of minerals; they extracted copper, tin, lead, iron and cobalt. There, he also had the opportunity to see the Freiberg mines and the School of Mining and Metallurgy.

He tried to visit the Rotterdam Military Arsenal, but his entry was denied. From there he went to The Hague Foundry, whose director was the nephew of the famous foundryman Jean Maritz, who had reformed and updated the Royal Artillery Factory in Seville a few years before. Morla did not totally agree with the methods of casting solid barrels that Maritz had implemented in Spain, and he recognized that better techniques were applied in the Hague factory than those considered unbeatable in Seville.

He went on to Prussia where he thoroughly studied its military strength. There he marveled at his organization, that the sovereign Frederick, called the soldier-king, had implanted years before. He liked it so much that he wrote a very long monographic book that he titled "Prussian Military Constitution", where he described the composition of the army and the tactics used. In the eight months that he remained in Prussia, he had the opportunity to visit barracks, study their weapons, and praised everything, from the discipline applied to the elegant dress of the officers.

From Prusia he traveled to Brussels, a city in which he observed the social customs and clothing of the population. In Antwerp he was in the strongholds built by the Spanish; in Brunswick he saw a chicory production industry, and described how chicory was consumed in the region. Later he passed through Liepzig and went to the Dresden Public Library, which struck him for the beauty of the building and for containing more than four hundred thousand books.

He finally moved to Prague, to the Bohemian glass factories and finished his trip in Vienna.

Finding himself exhausted, he decided to spend some time at a spa in Pisa, before returning to Barcelona.

Upon his return, he convinced the management of the Segovia Academy of the convenience of artillerymen knowing languages, and managed to make this institution the first in Spain where the students took English classes (Vega Viguera 1995; Herrero Fernández–Quesada 1992).

3.3 *Factory Work (1791–1800)*

The Barcelona Artillery Factory. He returned to Spain in 1791 and found a very different country from the one he had left four years earlier. During his absence Carlos III had died and his son Carlos IV had inherited the throne; on the other hand, the French Revolution had convulsed all of Europe.

Morla was assigned to the Barcelona Artillery and Maestranza Foundry, and there he had the opportunity to apply the knowledge acquired during the trip. He perfected the manufacturing processes used up to that time. He launched the theories of the Frenchman Gribeauval. This technician had carried out the renewal of the French artillery, endowing it with great mobility, designing much lighter carriages and changing the layout of the horse teams. His cannons had carefully designed ammunition, elevation screws and shot measurements. This improved aiming and shortened the rate of fire, made the reloading much faster, and also standardized the shape and weight of the bullets. He achieved the cannon was the greatest destroyer of the battles.

In Barcelona he also built mortars that had a cone-shaped chamber that proved to be more effective.

The Roussillon War (1793). When the year 1793 arrived, Spain entered into a war against France, due to the discrepancies that arose against the French revolutionaries. They had led their kings, related to Spanish monarchs from the house of the Bourbons, to the guillotine, threatening the political balance in Europe. Godoy had established himself as a defender of the monarchical cause and prepared an offensive in the Pyrenees.

Spain wanted to quickly resolve the conflict and placed three fronts along the border to defend the territory and invade the Roussillon region, bordering Catalonia, which a century earlier had belonged to Spain. He sent General Ricardos along with 32,000 soldiers, who entered France and fought in the battle of Saint Laurent de

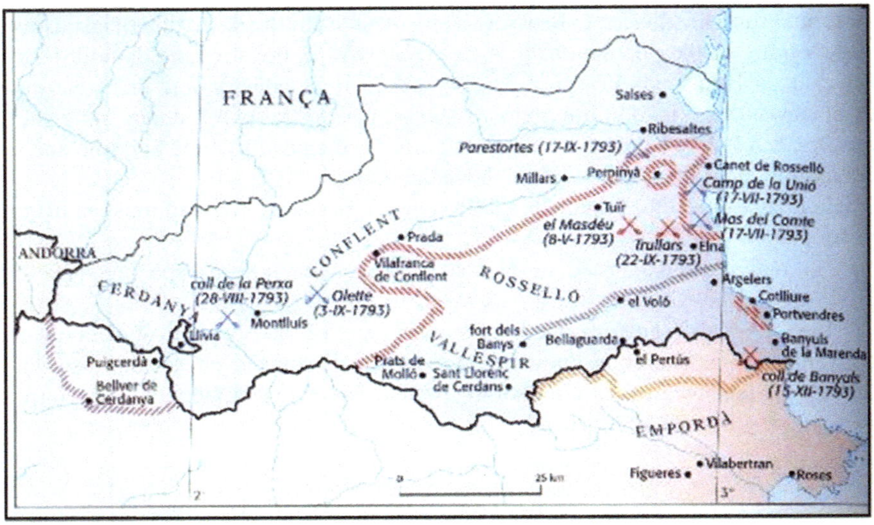

Fig. 8 Phases of the Roussillon War

Cerdá, where the Spanish army was victorious. But the initial euphoria turned into one defeat after another, which finally cost Spain the loss of the lands of Roussillon and the Island of Santo Domingo, and they ended up capitulating in the castle of Figueras two years later (Cuenca Toribio 2006).

Morla was required to attend the Roussillon campaigns, which forced him to put an end to his activity at the Barcelona factory. He was appointed Master Headquarters General of the army of operations, a position equivalent to the current Chief of Staff. He wanted the artillery used in the campaign to be fundamentally pieces of the Griveauval model, which fitted perfectly to the artillerymen's plans, demonstrating their ease of use and its accuracy.

During the war there was no good understanding between General Ricardos and Morla. He thought that the orders given to him by his superior were confusing and difficult to comply with. He also did not agree with the choice of the most strategic place to establish the main battle effort.

Although the Roussillon campaigns were unsuccessful for the Spaniards, the figure of Morla began to be renowned and he came to be awarded by Godoy, along with other prominent soldiers during the campaign.

From that moment on, Morla's military life became increasingly active, forcing him to abandon the intense work as an engineer and scientist that he had carried out up to then, both at the Academy and at the cannon factory.

Regulations and factories. Godoy had created a Board of Generals whose mission was to raise proposals to make a more educated and operative army, and he chose Morla to work on the artillery corps. In just six months, he wrote four regulations where he ordered the staff and their organization, functions and promotions of the military, study of weapons, uniformity and accruals.

At the age of 48, he was promoted to Lieutenant General of the Army. Godoy placed great trust in him that later became a strong friendship, and commissioned him to reform and update the El Salitre factory in Murcia and the El Farge factory in Granada. This mission took him two years, and at the end, he returned to Madrid where he dedicated himself to writing another of his great works, "The art of making gunpowder", which was declared of compulsory use at the Academy of Segovia and sent for its application to all the artillery garrisons (Ruiz Lagos 1972).

3.4 Political Commissions (1800–1808)

Governor of Cádiz (1800). At the age of 53, he was promoted to Captain General of Andalusia and settled in Cádiz. He liked this city more than any other did, since he sympathized with the character of the population and enjoyed its climate. When he arrived at the city, he found himself in an extremely difficult situation, as an epidemic of yellow fever had started.

He found a city invaded by patients with colic, diarrhea and high fevers, devastated by an epidemic that did not subside, as the citizens fed on products from gardens irrigated with non-potable water, and they fished on nearby rocks next to the city sewer (Fig. 9).

The town was disconcerted and launching prayers and invocations to the saints, while resorting to healers, but the disease spread without stopping through lower Andalusia. Without showing any fear, Morla visited all the neighborhoods to find

Fig. 9 Drawing of Cádiz in the eighteenth century

out the extent of the epidemic, and ordered to clean everything with lime, sulfur and boiled water, providing these products to the population. Such was his dedication that he came to catch the disease, fortunately without seriousness.

The epidemic was not the only problem he encountered in Cádiz. A powerful English fleet, made up of one hundred and forty-eight ships and commanded by Admiral Keith, threatened to attack at any moment. Morla opted for the diplomatic route and communicated with the english admiral, asking him to abandon the plans to occupy Cádiz, since the landing could cause his soldiers to become infected, and he warned that useless bloodshed would take place, since the people of Cádiz, before their desperate situation, they were willing to fight to the death. The admiral replied that he was willing to capitulate in exchange for the delivery of the Spanish ships anchored in the bay. Morla decided to write a letter to Godoy where he related the facts and the position of the Englishman, and he distributed this letter by the foreign chancelleries, causing great indignation against the English admiral due to his inhumane attitude. In those days, there were strong storms at sea, and the situation of the English deteriorated in such a way that led Admiral Keith to give up the idea of invasion. Morla's courageous and determined attitude helped the people of Cádiz to increase their appreciation and admiration for him.

Once the dangers were over, Morla focused on definitively banishing the epidemic and improving the fortifications. He proposed to build a new fortress, restructuring the area of Isla del León, Puntales and La Cortadura. He wanted to expand the perimeter of Cádiz and build new walls to prevent invasion of the city. He met with engineers and drew up a project that he sent to Madrid, but nothing could be carried out, because at that time, when he had only been in Cádiz for a year, he was called to the front and had to leave the city (Vega Viguera 1995; Herrero Fernández–Quesada 1992).

The War of the Oranges (1801). In 1801, the international situation was marked by the dispute between France and England to hold power in Europe. Napoleon wanted Portugal to break its traditional alliance with England and close its ports to English shipping, and he committed Spain, by treaty, to declare war on Portugal.

Godoy knew that Morla knew artillery in depth and entrusted him the organization and the strategic and logistical approach of the attack. This war only lasted two weeks, as the Portuguese army put up little resistance, and there were hardly any casualties. It took place on the border with Badajoz and the victory went to Spain, which won the border city of Olivenza.

It was called the War of the Oranges because Godoy sent the queen a bouquet of that fruit collected in Elvas as a symbol of victory.

Collaboration with Godoy (1801–1805). After the intervention in the war, Morla stayed in Madrid together with Godoy, as a technician in charge of studying the problems of the army and proposing solutions, and updating the Ordinances of Carlos III. They had an expert team of eleven chiefs and six officers of different arms. In those years, both made up the military power of Spain (Fig. 10).

Nevertheless, Morla's presence in Madrid and his growing friendship with Godoy annoyed other soldiers, who felt relegated to the increasing attributions that he received. Many envied his knowledge and industriousness; others disliked his firm

Fig. 10 Portrait of Manuel Godoy, by Francisco de Goya

and undiplomatic character. Among the people that Morla disliked was Queen María Luisa de Parma herself, perhaps because he had made the decision to reorganize and reduce the staff of the Corps Guards, which the Queen understood as a contempt for her person. Therefore, she wanted to remove the artilleryman from the Court and that he returned to his place in Cádiz. This decision pleased him because he did not like the capital, because of his customs and frivolity. He openly criticized Madrid and stated that he preferred the societies of Seville or Cádiz.

Stay in Granada (1805). He arrived in Cádiz convinced that this would be his last destination, since he was 57 years old and wanted to live in peace for the rest of his days. Nevertheless, when he was spending a few days off in El Puerto de Santa María, he was urgently called to join to Granada, because a yellow fever epidemic had been declared there and they asked him to apply the strategies that he had implemented with such good results in Cádiz years before.

Once in Granada he took charge of the military, judicial and health command of the province. His actions were very effective, he prepared brochures, which were distributed among the population, explaining the measures that needed to be followed to prevent the spread of the plague. He also wrote letters encouraging judges and magistrates to improve their performance of their duties.

After six months in Granada, and sanitary order restored, he requested to return to Cádiz, but since his place was occupied, he was assigned to Puerto de Santa María without specific attributions, and he lived there quietly for three years (Herrero Fernández–Quesada 1992) (Fig. 11).

Fig. 11 Illustration of El Puerto de Santa María in the eighteenth century

3.5 The French Invasion and Period of Decadence (1808–1812)

At the beginning of the year 1808, the presence of French troops in Spain was increasingly threatening as they occupied various Spanish towns. The total number of French soldiers stationed in Spain amounted to about 65,000, who controlled communications with Portugal and Madrid.

On March 17, a small crowd gathered in front of Godoy's palace in Aranjuez, burning and looting their belongings. The riot sought the dismissal of Godoy and the abdication of Carlos IV in his son Fernando. At that time, the lower classes were suffering the consequences of the defeat at Trafalgar, the nobility was dissatisfied with the situation of the Court and the clergy feared the confiscation measures (Cuenca Toribio 2006).

Morla, already 61 years old and in somewhat delicate health, resentful of old war wounds and periodic gout attacks, requested incorporation into active service.

After the events of May 2 in Madrid, the Board of Officers in Andalucía was summoned to order the multiple patriotic movements that were emerging throughout Spain.

Evaluating the few military means available, the military governor of Cádiz considered reckless to declare war on the French, since the best of the Spanish army was outside of Spain. However, the exalted people did not share that opinion and asked for arms and that the French ships anchored in the bay of Cádiz be destroyed. The exaltation of the population was such that a group of citizens went to look for the governor, who tried to flee through the roofs, but they captured and lynched him in the middle of the street accused of being a traitor to the country. This entire episode took place so quickly that Morla and the rest of the military did not arrive in time to prevent it. Morla felt angry and hurt while he was acclaimed by the crowd to once again take over the command of the military government of Cádiz.

The battle of the Poza de Santa Isabel (1808). The Political Board of Seville did not have full confidence in Morla and sent a representative to monitor all his actions and make sure that he did not agree with the French. He reported that Morla had appeased the people, that he was following the orders issued from Seville and was working on the measures that he had already designed a few years ago to increase the defense of Cádiz. He also created a body of volunteers, made up of three thousand men, which was called "Militias for the defense of Cádiz".

A French fleet commanded by Admiral Rosily was anchored in the bay, constituting a threat to the area. The Spanish admiral Ruiz de Apodaca prepared an attack strategy together with Morla. They pretended that they were capitulating, which suited the French as they expected reinforcements by land shortly. So, they withdrew through the La Carraca channel, but the Spanish attack unexpectedly took place on June 9, 1808.

After two days of hard battle, the French, seeing the attack device and the profusion of artillery that Morla had arranged, offered to capitulate, which was very opportune, since the Spaniards were running out of gunpowder stocks.

Nor did the French realize that many of the fire hydrants they reported from the ships were simulated. The capture of the French squadron had a favorable repercussion throughout the country and the Junta de Sevilla awarded distinctions to the most distinguished participants in the contest, including Morla (Fig. 12).

Before the attack, the English offered to support the battle with an infantry division that could land in the bay, but Morla kindly declined the offer, as he considered that accepting the help of England could lead to great inconvenience later on; he feared that they would make a new Gibraltar of Cádiz.

The victory over the French navy in Cádiz was the first defeat of the French army during the War of Independence and allowed it to gain booty consisting of six ships

Fig. 12 Drawing of a naval combat

and numerous weapons and gunpowder. The people of Cádiz devised numerous verses related to the War of Independence and the French. One of them spoke of Morla and said thus:

In Cádiz a squad.

we saw her deliver.

to Morla and Apodaca.

surrendered on land and sea.
Rosilly, who was the boss,

it fills you with regret.

see over your flag.

ours tremulous.

The people acclaimed Morla and he took the opportunity to publish an edict where he asked the people of Cádiz to comply with the laws and respect the adversary, since he feared that they would attack the French prisoners taking the law into their own hands. At the same time, he continued with the task of organizing the citizen militia and issued its ordinances, made up of ten articles.

After the victory in the battle of Bailén, he received the order to embark in Cádiz the more than ten thousand French prisoners bound for his country. Morla did not have the material means to carry out the mission, except using warships, but he refused to do so because he feared that once the ships arrived in France, it was very likely that they would be retained there. When the prisoners arrived at Puerto de Santa María the population assaulted them and dispossessed of their luggage, which in many cases was the product of looting in Spanish churches and palaces. Morla ordered to punish the assailants, since he wanted to maintain order and housed the prisoners in warehouses and pontoons, but the problem of feeding so many people was enormous and the English decided to send them to the Balearic Islands, with the approval of the governor. There, they disembarked on the island of Cabrera, where they remained in terrible conditions until the end of the War of Independence.

He had discrepancies with the Provincial Board of Seville and declared himself in favor of the creation of a single Central Board that would unify the criteria for action throughout the peninsula.

In October 1808, Morla was assigned to Madrid as General Director of the Artillery of Spain and the Indies.

General Director of the Artillery of Spain and the Indies (1808)

Although Morla was sorry to leave Cádiz, he was also excited by the idea of being able to reorganize the Artillery Corps with his own criteria. He passionately devoted himself to such a difficult task, because apart from the complexity, he would have to live with Napoleon's troops occupying the peninsula. It was time to measure himself with Napoleon, the European strategist. For this, he personally inspected the fortifications established in the mountains of Madrid and accelerated the production of artillery factories and foundries by ordering more cannons, light weapons and gunpowder.

When the French entered the Segovia Artillery Academy, they seized the library's collections and all the copies of the Artillery Treatise book. The book was translated into the most widely spoken languages in Europe and was highly praised, considering Morla one of the most important artillerymen of his time.

After the defeat of the French army at Bailén, Napoleon took command of his army in Spain, reinforcing it, and initiating attacks that opened the way for him to Madrid. The Central Board decided that Morla would take command of the army deployed in the north of Castilla leaving the coordination of the Artillery. These numerous changes of destination negatively affected Morla's spirit, since he was aware of the inexperience and mistakes made by the Central Board.

Meanwhile, Napoleon advanced on Madrid at the head of the Grande Armée, a veteran army made up of 250,000 soldiers, which quickly overwhelmed the resistance and after numerous battles left the Spanish army practically defeated (Vega Viguera 1995; Herrero Fernández –Quesada 1992).

The defense of Madrid (1808). Morla was appointed to be part of Madrid's defense team. He surveyed the city and proposed redeployment of the artillery. Morla's plan was not to prevent the entry of the French into the capital, but to present such resistance that it would wear them down and they would prefer to give up their efforts. For this, he had the entire people of Madrid, to whom he appealed to build barricades, dig trenches, raise stones to the roofs and open loopholes in the walls that surrounded the city, but deep down he was aware of the uselessness of the effort before the superiority of the French army.

At dawn on December 2, numerous French cavalry forces surrounded the city, and the French artillery opened fire against the walls, opening a wide gap through which the infantry entered. Everything happened so fast that the citizens of Madrid barely had time to react, although many resisted heroically. The action of the Spanish artillery was so effective that when Napoleon was already in the Castellana area, he had to withdraw due to the precision of the impacts. However, the strength of the French army was superior and a few hours later Napoleon sent an emissary demanding surrender within a maximum period of six hours in exchange for preserving the lives and property of the people of Madrid.

The people wanted to continue fighting until they died, but Morla was convinced that prolonging the battle was equivalent to collective suicide. He still spent a few hours touring the city and correcting artillery positions (Fig. 13).

The authorities met and considered that surrender was preferable to avoid useless bloodshed. Morla was detached to go to parley with Napoleon. When he arrived at the French camp, he found the general about to explode with anger. He reproached Morla for the treatment given to the prisoners of the Battle of Bailén and the defeat of the French army in Cádiz, and he threatened to kill all the prisoners if they did not surrender immediately. Morla was convinced that he was very capable of carrying out his threats and returned to Madrid. When he reported the result of the interview to the Defense Junta, they decided to capitulate and send Morla back, who attended the meeting deeply dejected at having to carry out the most painful mission a soldier can

Fig. 13 Picture of the surrender before the emperor, by Carle Vernet

do, surrendering the capital of his country. Later, his enemies took the opportunity to accuse him of being a traitor, a sympathizer of the French, and that he hastened the surrender without waiting for reinforcements. The responsibility for the capitulation was also placed on him when really it was a decision of the majority of the Central Board, whose members immediately fled to Andalusia.

The decline (1808–1812). When Napoleon entered Madrid, Morla's decline began. He returned home to Cádiz, but the people seemed to have forgotten everything he had done for the city and they limited themselves to murmuring that he was a Frenchman. Knowing the exalted character of the people of Cádiz and fearing for his integrity, after a few months he decided to return to Madrid. Also in the capital, he felt that everyone blamed him for having handed over the city to Napoleon. Demoralized and separated from everyone, he began to relate to the new King José I, who treated him with great cordiality. That caused Morla, who felt unfairly treated by his compatriots, to make some statements praising the virtues of the intruding king. When the Central Board found out, they stripped him of all his honors, opened a disciplinary file on him, seized all his assets and auctioned them off, even withholding his salary. That left Morla in a compromised situation since he supported his sisters.

Faced with such a situation, he admitted to being appointed State Counselor to the king. He acted independently, while José I showed him more and more affection. He was never a follower of Napoleon, nor did he take up arms against Spain. Soon after, being sick and depressed, he retired to a small hacienda near Seville. He died in 1812, when he was 65, the victim of a urinary infection, diabetic, and nearly blind.

José I ordered to bury his corpse with military honors (Herrero Fernández –Quesada 1992).

4 On the Circulation of Works

Without a doubt, Morla's most representative work as an author is his artillery treatise titled "Tratado de Artillería para el uso de Caballeros Cadetes del Real Cuerpo de Artillería". The book was written during his time as a professor at the Royal College of Artillery of Segovia, at the express request of the director. At that time, there were no reference textbooks for specific subjects, and teachers dictated their notes during classes. However, the high quality of Morla's work meant that it was used as an essential reference manual for artillery teaching. This was something new and influenced by Enlightenment ideas. The Artillery Treatise was published between 1784 and 1786, becoming the first textbook for artillery college cadets.

The work would have a notable influence on several generations of artillerymen. Due to its content and quality, it earned great praise in Spain and abroad. It was translated into several languages and was used as a text for artillery teaching in the Netherlands, France, and Germany. In reward for this excellent work, he received the rank of lieutenant colonel on April 6, 1784.

Later, in 1803, the work was completed with the edition of a magnificent fourth volume with copper-engraved sheets. The sheets make up a large collection with detailed designs of all kinds of artillery elements, mechanisms and mechanical assemblies. The student cadets themselves made the drawings on the sheets, directed by their Drawing teachers, and the best engravers of the Court manufactured the sheets (Fig. 14).

Because of the War of Independence, all the copies of the Morla treatise disappeared from the Alcazar of Segovia. In 1816, it was decided to print a second edition, which was accompanied by a fourth volume entitled "Colección de las explicaciones de las láminas" (Collection of the explanations of the sheet), as an aid to the interpretation of the plates, which were not printed again due to their high cost (M. D. Herrero Fernández –Quesada 1992).

5 Legacy and Today Interpretation of Contributions

Morla was one of the Spanish historical figures who caused more passion and controversy in his time, a unique character who is still current. Morla was a scholar, a teacher, a scientist, a mechanical engineer, an artilleryman, a strategist, a soldier, and a politician, undoubtedly, he was one of the most notable individuals to come out of the Royal College of Artillery of Segovia. During his life, he had periods of great popularity and recognition among his countrymen, although at the end of his life, his figure was somewhat discredited due to his rapprochement with the French,

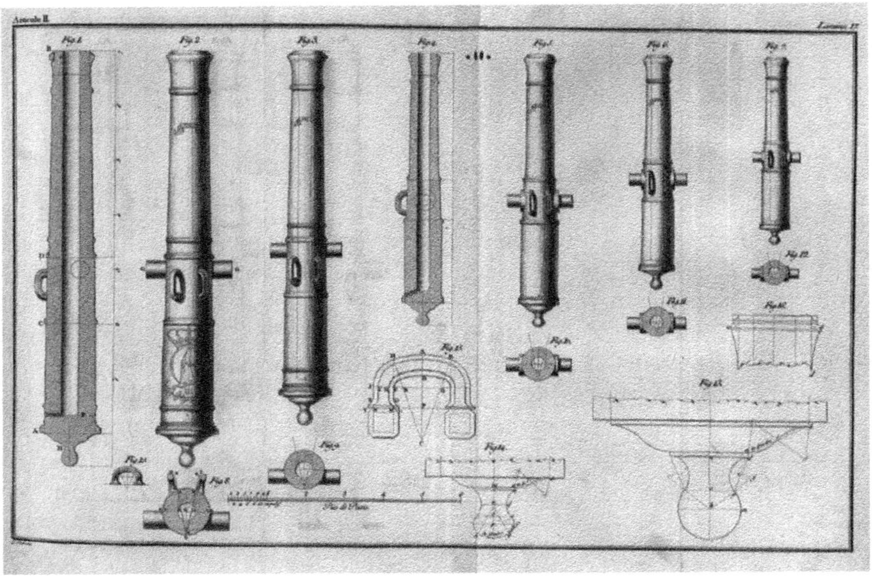

Fig. 14 Sheet with canyon designs from the IV volume of the Treatise (Morla 1803)

and he was accused of being Frenchified and Napoleonic. The French recognized his military rank, and awarded him various honors, some of which, it is documented, he did not welcome (Herrero Fernández–Quesada 1992).

As a result of the 1862 fire in the Alcazar of Segovia, the iconographic gallery of illustrious artillerymen was lost, among which was one of Tomás de Morla. At present, several portraits of the artilleryman are preserved, copies of each other, one in the Artillery Academy and another very similar one in the Army Museum. A bronze bust made from these portraits is preserved in the Military History Museum of Seville. Most of the copper-engraved sheets with the illustrations of his Artillery Treatise are also preserved, which are found among the funds of the Army Museum (Fig. 15).

Fig. 15 Bust of Tomás de Morla in the Historical Military Museum of Seville

References

M. Artola, *Antiguo régimen y revolución liberal* (Ariel, Barcelona, 1983)

M. Artola, *La España de Fernando VII* (Espasa-Calpe, Madrid, 1968)

J. M. Cuenca Toribio: La Guerra de la Independencia: un conflicto decisivo. Encuentro, Madrid (2006)

T. de Morla: Tratado de Artillería para el uso de la Academia de Caballeros Cadetes de Artillería (3 tomos). Imprenta Espinosa, (Segovia) 1784–1786

T. de Morla, *Tratado de Artillería para el uso de la Academia de Caballeros Cadetes de Artillería: Libro de láminas* (Imprenta Real, Madrid, 1803)

T. de Morla: Apuntes autógrafos. (1789–1790)

T. de Morla, *El arte de fabricar pólvora* (Imprenta Real, Madrid, 1800)

J. Fontana, *La crisis del antiguo régimen* (Crítica, Barcelona, 1983)

M. D. Herrero Fernández –Quesada: Ciencia y Milicia en el siglo XVIII. Tomas de Morla, artillero ilustrado. Patronato del Alcázar de Segovia, Segovia (1992)

M. Ruiz Lagos: Biografía de Tomás de Morla. Sevilla (1972)

E. Vega Viguera: La singular vida de Tomás de Morla y Pacheco militar y político jerezano. Boletín de la Real Academia Sevillana de las Buenas Letras, Sevilla (1995)

Guillermo Quintanilla y Fábregas (1867–1929)

J. Tejero Manzanares, F. Mata Cabrera, and F. Montes Tubío

Abstract Guillermo Quintanilla Fábregas was born in San Juan, Puerto Rico, in 1867, and died in Madrid, Spain, in 1929. Quintanilla graduated at the Escuela Especial de Ingenieros Agrónomos of Madrid in 1888, he was appointed the Chair of Agricultural Chemistry at the same School in 1908. For thirty years he was the director of the Estación Agronómica de La Moncloa, where he undertook his outstanding work as both researcher and inventor in the field of industrial machinery, which culminated with Quintanilla, working together with the Marquis of Acapulco, and jointly inventing and patenting a method for the extraction of olive oil under the name of the "Acapulco-Quintanilla System". This patent is considered the first continuous olive oil extraction system, which served as a protype for the development of subsequent versions, and was initially tested in the Granja Agrícola de La Moncloa. Following the successful trials at the Escuela Superior Agraria de Portici (Italy), the patent of the system was adquired by an Italian company for commercial application in the transalpine country. During his career, he managed vast agricultural holdings and key agri-food industries, primarily the sugar and oil industry, and was appointed the Director of Crops at the firm "Azucarera de Madrid, S.A. in 1902. In his homeland, he promoted agricultural research, and for a nine-year tenure he was the director of the Estación Agronómica de Mayagüez (Puerto Rico), which was the precursor to the future Faculties of Agricultural Engineering and Science of the University of Puerto Rico.

Keywords Agriculture · Agricultural engineer · Olive oil · Oil mill · Extraction system · Agri-food industry

---

J. T. Manzanares (✉) · F. M. Cabrera
University of Castilla-La Mancha, Ciudad Real, Spain
e-mail: jose.tejero@uclm.es

F. M. Cabrera
e-mail: francisco.mcabrera@uclm.es

F. M. Tubío
University of Córdoba, Córdoba, Spain
e-mail: ir1motuf@uco.es

R. López-García and M. Ceccarelli (eds.), *Distinguished Figures in Mechanical Engineering in Spain and Ibero-America*, History of Mechanism and Machine Science 43, https://doi.org/10.1007/978-3-031-31075-1_11

1 List of Main Works

- Un aparato denominado «Acapulco de fabricación continua» para la extracción de los aceites, vinos y otros líquidos o jugos (1906).
- Un aparato para la extracción de aceite de oliva, semillas, frutos y residuos de oleaginosos, vino, sidra y demás jugos de frutas, denominado «Acapulco Nº4». (1906).
- Un aparato para la extracción de aceite de oliva, semillas, frutos y residuos de oleaginosos, vino, sidra y demás jugos de frutas, denominado «Acapulco Nº4» (1906).
- Obtención del Aceite por el Sistema Acapulco-Quintanilla (1924).
- Un procedimiento para la extracción del aceite de olivas y frutos oleaginosos, denominado Acapulco Continuo (1908).
- Ministerio de Industria, Turismo y Comercio. Oficina Española de Patentes y Marcas. Archivo Histórico. Exp. nº 44.184.
- Ministerio de Industria, Turismo y Comercio. Oficina Española de Patentes y Marcas. Archivo Histórico. Exp. nº 48.411.
- Actas de "Oleotecnia y aceituna de mesa" (1950).
- Método Acapulco-Quintanilla para la extracción del aceite (1927).
- Extractor-batidor plano que de manera indistinta funciona continua y discontinuamente. Madrid (1925).
- Nuevo Extractor De Aceite Y Batidor De Pasta Oleosa De Funcionamiento Continuo Y Discontinuo—Certificado de adicción por mejoras en la patente núm. 93,072 del 24.03.1925 (1927).
- Un nuevo extractor de aceite (1929).

Quintanilla's career as a researcher and inventor was closely associated to the work of Miguel Antonio del Prado y Lisboa, the VIII Marquis of Acapulco (Figs. 1 and 2).

Miguel del Prado y Lisboa, the VIII Marquis of Acapulco (Madrid 1867—Madrid 1934), was one the most prolific Spanish researchers of his era, who invented and patented devices for the extraction of olive oil owing to his personal financial interests in developispanosh indng the family agricultural holding i.e., an olive plantation located in the province of Jaen in Spain.

Towards the end of the XIX century, in an effort to find an alternative to traditional oil extraction methods, Miguel Antonio del Prado y Lisboa, the engineer Quintanilla, and the chemist Arredondo combined their efforts in order to design a new method for oil extraction using a chemical process. The first prototypes that had been developed were discarded for being unprofitable, and these prototypes were replaced by a device for extracting oil from the pitted and ground paste using a simple cotton filtering cloth. This procedure was termed Acapulco, in honour to its patron.

From 1904 onwards, they registered several inventions at the Spanish Patents and Trademarks office.

Fig. 1 The agricultural engineer Guillermo Quintanilla Fábregas (Mundo del agrónomo 2010)

Fig. 2 Medal of the VIII Marquis of Acapulco (Congreso and internacional de oleicultura 1924)

Thus, under the title «A device called Acapulco continuous production for extracting oils, wines, and other liquids and juices», Miguel del Prado y Lisboa registered a patent with Dossier No: 38.782 in the Spanish Patents and Trademarks Office on 23 July 1906 (PRADO LISBOA 1906a, b, c), which was one of the first patents without a chemical processes, and was the first attempt to eliminate all the rollers and presses from the extraction system. Thereafter, subsequent patents were registered that highlighted the technological innovations of the time in several documents, illustrations, and blueprints.

The machine consisted of three sealed extractors, consisting of semi-cylinders with vertical walls adjacent to each other, with cotton filttering cloths that were placed at the bottom of the cylindrical surface (Quintanilla 1924).

To facilitate the extraction of oil, inverted presure was used by creating a vacuum in the lower chamber, and this chamber or empty tank was connected to the false bottom of the aparatus. The contact surface of the paste needed to be continuously changed in relation to the filtering surface as the olive paste was of low density, very water resistant, and tended to float to the surface. As the device filtered to the bottom part, it was necessary to constantly stir the extracted oil paste, and replace it with the unextracted oil rich paste sitting at the top. As the pores of the cotton cloth were continuously clogged with residue or solid particles, special brushes were used to clean and unclog the filtering surface before introducing a fresh batch of paste. Thus, these devices were equipped with steel brushes to remove unwanted deposits.

A further innovation led to a new patent called *«A procedure for the extraction of olive oil and oleaginous fruits, coined the continuous Acapulco»,* which was registered by D. Miguel del Prado himself in the Spanish Patents and Trademarks Office on 13 October 1908 (Prado Lisboa 1908), Dossier No: 44.184 that stated as follows:

> Patent consisting of a process for the extraction of oil by applying a vaccum, known as the CONTINUOUS ACAPULCO procedure. The olives are placed in the olive pitter to remove the pits, which are then transferred to a washing tank, and then to a tilted cylinder submerged in water with a spiral shaft that draws the pits upwards to separate the pit from the olive paste due to the helicoidal motion of the spiral blade. The olive pulp falls into a tank, and is pumped into a vertical cylinder where it is driven upwards by a pump and the helicoidal motion of the spiralled shaft. The oil is separated from the pulp by a nickel filtering cloth surrounding the cylinder that sits inside a vacuum chamber where the oil is obtained as the pulp rises upwards and the waste is ejected at the top.

Guillermo Quintanilla was a reference for agricultural engineers and a relentless worker who was awarded the Cruz del Mérito Agricultural, and the Orden de Carlos III (Fig. 3).

In addition to the previous joint patents of Quintanilla and the Marquis of Acapulco, another invention was patented by Quintanilla with Dossier number 48.411, under the title «A procedure for producing olive oil referred to as "Acapulco n.º 6"», which was registered in the Spanish Patents and Trademarks Office on 9 July 1910. Other patents were also registered.i.e., Patents No: 44184 with the title: «A procedure for extracting oil from olives and other oleaginosos fruits, called the "Continuous Acapulco"» registered in 1908, and Patent No: 93.072 with the

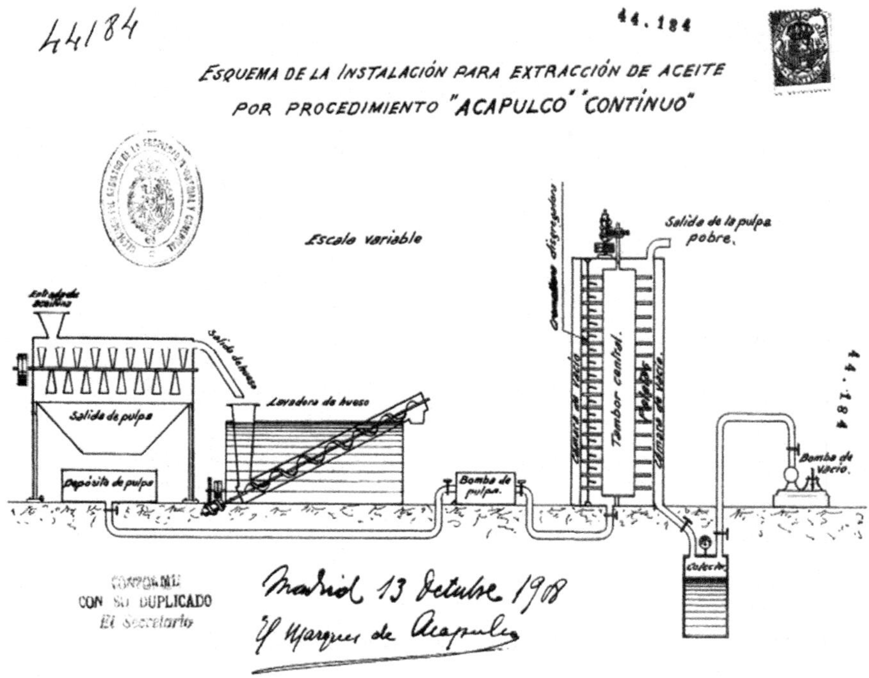

Fig. 3 Diagram of the installation of the "Acapulco Continuo" oil extraction procedure (Industria 2022)

title «Flat extractor-blender that functions either continuously or discontinuously» registered by Miguel de Prado y Lisboa in 1925,

The technical dossier of this patent, registered with the title «A procedure for producing oil from olives called "Acapulco nº6"» Dossier No: ES0048411 A1, Application No: P0048411 (09 July 1910), Applicant: Quintilla Fábregas, Guillermo, Madrid, Spain, stated as follows:

> The purpose of this patent was due to the need to modify certain elements considered to be essential by D. Guillermo Quintanilla, of the procedure called «Acapulco».
>
> Instead of applying a vacuum in the extractor device, it applies an air current produced by the absorption of the vacuum pump. The air passes through the paste and the filtering surface of the extractor device «A» before being transferred to the collectors «D» and «C», and is absorbed by the pump «N».
>
> A shaft runs through the centre of extractor «A», which has a shaft and blades parallel to the shaft that is attached to several supports perpendicular to the shaft. The blades has brushes made of tempered wire coils designed to maintain the nickel filtering cloth clean. The filtering cloth is welded to the mesh of the false bottom.
>
> The device is made from aluminium and the olive pitter continuously removes the pits and ejects them through a lid on the side (Fig. 4).

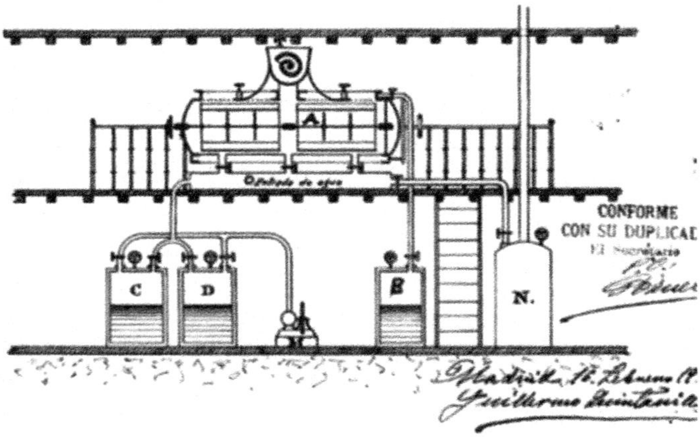

Fig. 4 A procedure for producing olive oil called Acapulco n°6 (Industria 2022)

2 Acapulco-Quintanilla Method

In line with his work, and his determination to innovate the oil extraction system, Guillermo Quintanilla exhibited the Acapulco-Quintanilla 6 system on December 1924 at the VII Congreso Internacional de Oleicultura, prompting the praise of Mallorcan industrialists who were enthusiastic about the successful invention. Quintanilla had modified the system that Miguel de Prado y Lisboa had shown him in Lendínez (Martos), and the audience was enthralled with the successful performance of the revolutionary olive oil extraction system.

The VII Congreso Internacional de Oleicultura, held in Seville in 1924, was one of the most important exhibitions at the time, and highlighted the technological advances designed to simplify the oil extraction processes and raise the quality of the product itself.

As vice-president of the exhibition, Quintanilla played a key role in promoting the innovations of the system and underscored its wider economic impact by highlighting the increased efficiency of the system in increasing oil output, and in reducing the production times and costs of the olive mills. Thus, the introduction of the blenders and thermo-blenders in the oil mills of the 1920s significantly increased profit margins by obtaining greater quantities of the precious oil from the blended paste by exerting less pressure, and by eliminating the second compression of the reground paste used in traditional methods.

Though the Italian, Andrea Acquarone, improved the Acapulco-Quintanilla system in 1940, it failed to be a commercially success.

Likewise, in the mid XX century, the «Société Commerciale d'Usinag pour la Construction d'Outillage» (S.C.U.L.C.O.) of Tours, France, proposed a similar method to the Acapulco-Quintanilla system that consisted only of an olive pitter and a pulp oil extractor that closely resembled the machinery used in the Acapulco-Quintanilla method (Actas de "Oleotecnia y aceituna de mesa" 1950).

3 Development of the Acapulco-Quintanilla Method

The Acapulco-Quintanilla method was considerably different to the initial prototype of the inventors due to several subsequent modifications of D. Miguel de Prado Lisboa and D. Guillermo de Quintanilla; nevertheless, it was based on the same fundamental principles as the initial prototype and all of the subsequent Acapulco systems, where oil was directly extracted from the olive pulp using the surface tension of the liquids and their capillarity.

In 1927, the Spanish Ministerio de Fomento. Dirección General de Agricultura y Montes described the system as follows:

> First, the olives are pitted and ground to a paste. The olives are placed in the oil mills through the olive pitter, which is a two-meter long and twenty-centimetre wide hopper. The walls of the hopper are made of laminated steel with 2 to 3 mm wide and 1.5 to 2.5 cm long slits running perpendicular to the cylinder shaft. The shaft of the drum rotates at a speed of 400–600 r.p.m. and the 8 steel vanes propel the olives against the perforated walls. The blades are not parallel to the cylinder shaft, but are diagonal to enable the movement of the olives from the hopper to the last cylinder.
>
> Through the friction and compression of the olives, the pulp is separated from the pit, and the whole pit is eliminated at the other end of the feed hopper, whilst the paste goes through the slits and falls into a tank and is then sent to the extractor. The extractor unpits at a rate of 800 to 900 kilograms/hour. (Fomento 1927)

In 1925, the ongoing innovations to the continuous oil extraction system led the Marquis of Acapulco and the engineer Quintanilla to jointly register a new patent in the Spanish Patents and Trademarks office, Dossier No: 93.072. Flat continuous and discontinuous extractor-blender (24.03.1925). Publication No: ES0093072 A1, Application No: P0093072 (24.03.1925), Applicant: Miguel del Prado Lisboa, Madrid, Spain.

The technical description of the said dossier (PRADO LISBOA 1925) states as follows:

> The objective is to extract olive oil directly from the olive that has been previously pitted, crushed, and ground [...]. This machine is based on the same principle as the Acapulco procedure, that is, it depends on the different degrees of density, capillarity, and surface tension specific to each of the two liquids composing the paste i.e., the oil and the vegetable water residue. The absorption power of the oil in the vegetable tissue of the olive being much lower than that of vegetable water residue, though it is much greater in the nickel-meshed cloth. Moreover, the surface tension between both liquids is much smaller (less than half) in the oil than in the vegetable water residue. Thus, this device applies a light pressure to control both of these forces in the vegetable water residue that is retained in the cellulose pores, but with sufficient pressure (though small) to control both forces in the oil to separate it from the cellulose and the vegetable water residue so it can pass through the filtering cloth which is in contact with it. As in all Acapulco machines, this new extractor extracts the oil from the paste,

leaving aside most of the vegetable water residue given that it is based on the same scientific principles. Though it is fundamentally the same, the machine is completely different. The main difference is that the filtering cloth affects the flat shape. With slight variations, the flat filtering surfaces of the new extractor can be horizontal, vertical, or oblique. Moreover, the movement of the cogwheel is used to turn the sweeping brushes. Though the most important feature, which is the objective of this patent, is the invention of the flat filtering surfaces. A further novelty of this extractor was that it had both a blender and filter, that is, besides blending it filtered.

The two following new technical dossiers of 1927 (PRADO LISBOA 1927), and 1929 (PRADO LISBOA 1929) improved the design of the extractor patented in 1925.

Dossier No: 102.338. New continuous and discontinuous oil extractor and blender of oily paste – Additional certificate for the improvement of patent No: 93072 of 24 March 1925. Publication number: ES0102338 A1, Application No: P0102338 (20.04.1927), Applicant: Miguel del Prado Lisboa, Madrid, Spain.

Dossier No. - 113.368. A new oil extractor.

Publication No: ES0113368 A1 (01.08.1929), Application No: P0113368 (06.06.1929), Applicant: Miguel del Prado Lisboa, Madrid, Spain, CIP prior to 2006.01: A23N

4 Experimental Trials

The machine was tested in the Moncloa at the Instituto Agrícola Alfonso XII, in the presence of HM and the government of Spain, who observed how the system successfully produced large quantities of oil. The extractor was improved by replacing the cotton cloth with a thick nickel mesh placed on a tin wire at the bottom of the device.

To facilitate the separation of the oil from the paste, the temperature was successively raised so that in the first extraction the oil was completely cold, in the second it was at a temperature of 35 to 40 C, and in the final phase it was at a temperature of 80–100 ° C due to the steam.

This procedure ensured good quality oil was obtained in the first extraction, with the oil being exhaustively extracted from the paste as it moved it from one cylinder to the next.

Finally, the system was simplified by placing two extractors instead of three: in the cold first extraction around 80 to 90% of the oil was obtained, and in the heated second phase, the remainder was extracted using steam.

5 Commercial Applications

Both of the inventors, the Marquis of Acapulco and the engineer Quintanilla, exhibited the system on numerous occasions, and worked closely with the chemist Arredondo and several mechanics to maximise the performance of the system for obtaining high quality oil.

Owing to the enthusiasm of Guillermo Quintanilla, the advantages of the extractor were expounded at the Escuela de Olivicultura de Portici (Italy), and the patent was later adquired by a leading firm of agricultural machinery.

Due to the purchase of the invention, the company was forced to move to Napoles, since a key contractual clause demanded that for the contract to come into effect, it should be applied for an entire campaign in Italy to compare it with an ordinary factory, and a modern factory with the most innovative machinery, which at that time was also another Spanish method known as the «Savatella». After the campaign, the new system proved to be more efficacious than the conventional extraction methods. Thereafter, a committee of experts was appointed, consisting of engineers, technicians, and manufacturers, who issued a favourable report underscoring the advantages of the system.

However, when the system was commercialised in Spain and Italy, it failed to fulfil its initial expectations, and fell short of the success of the system of the Escuela de Olivicultura de Portici (Italy) as the performance of the system was hindered by practical problems that could not be overcome without compromising the main fundamentals of the system.

This failure neither dissuade the Marquis nor Quintanilla, who were convinced of the efficacy of the system for obtaining large quantities of high quality oil in comparison to the conventional methods; thus, in spite of the heavy financial losses, they built their own factories, the former in his olive plantation in Jaen, and the latter in his plantation in Guadalajara, which according to Quintanilla's own words were very successful (Congreso and internacional 1924).

Convinced of the advantages of the Acapulco-Quintanilla system and driven by the prospective financial gains, the Marquis de Acapulco, installed his factory in the family olive plantation in Lendínez (Martos) Jaen, the performance of which was underscored in the proceedings of VII Congress international de la Oleicultura held in Seville in 1924.

The oil mills were modernised by Mariano del Prado y Ruspoli, the X Marquis of Acapulco, who profited from the fiscal reforms of the Spanish Law 152/1963, and a month later, on 25 February 1967, the Cano family purchased the oil mills from Mariano del Prado y Ruspoli.

The current installations of the Acapulco—Quintanilla system, which were modified by the Marquis before his death, were installed in the Lendínez plantation in Jaen. The machinery was still functioning until the decade of the 1990s, and was used by both the Marquis family and the present owners i.e., the Cano family, who eventually deemed the system was no longer profitable for the production of oil as it used excessive amounts of water in comparison to more modern machinery (Fig. 5).

Fig. 5 Current installations of the continuous system in the Lendínez olive plantation, Martos (PÉREZ MARTÍNEZ 2015)

6 The Acapulco System Versus the Hydraulic Presses

The alternative to the Acapulco system that was in use the Spanish oil mills at the beginning of the XX century involved the improved hydraulic press of D. Diego de Alvear y Ward that was introduced in 1834. The initial improvement to the traditional process involved using trollies on rails to transport the loads from the grinders to the presses that were also adapted to the new mechanism.

On 14 January 1907, the Marquis of Cabra, D. Francisco Méndez de San Julián y Belda, registered a patent in the Spanish Patents and Trademarks Office «A new hydraulic press system specific for olives», as described as follows:

"Patent No: 39.849:

> Patent of the Marquis of Cabra, consisting of a hydraulic press and other devices used for unloading the load into the press. The most important features that characterize it is the long distance between the upper and lower bridge that enables very high loads using very low gears, and the use of plates between loads that enhance the stability required to steer the load through the system of rails and wagons to the press. The press does not have a pressing plate, as the actual wagon is the pressing plate itself and has four columns that work as a guide between loads. Prior to pressing, the load is placed in a compressor that slightly compresses it (using a cast iron weight) to make it more compact and easier to transport, as well as a forklift for raising the plates between loads in the wagon.

7 The Outcome of the Competition Between Both Systems

The advent of the Acapulco system beckoned the demise of the hydraulic press used in the oil mills, which was replaced by a filtering extractor with a vacuum chamber that enabled larger quantities of high-grade oil to be obtained at a lower cost.

Nonetheless, the Marquis of Cabra remained an unconditional supporter of the hydraulic press, and a few years later, the engineer José María Soroa claimed:

> In spite of the advantages of the previously mentioned Acapulco-Quintanilla method, the traditional systems cannot be abandoned for more modern ones, at least at present, because the numerous installations of mills and presses cannot be quickly replaced in order to substitute the traditional machinery with the Acapulco-Quintanilla system. Thus, several combinations have been proposed by the said gentlemen themselves, two of which have become the most widespread as follows:

First combination:

(a) The olives are crushed and ground in a stone grinder.
(b) The modernised Acapulco system is applied to the ground paste for an hour and a half in order to extract around 60 to 85% of the oil.
(c) The semi-extracted oil paste is pressed to remove the remaining oil.

Second combination:

(a) The pit is separated from the olive pulp by the olive pitter.
(b) The resulting pulp is placed in the Acapulco extractor, where 60 to 85% of the oil is extracted in under two hours.
(c) The residual pulp of the previous process is mixed with the pits that were separated by the olive pitter, and they are ground again in a stone or other type of grinder.
(d) The resulting ground paste of the previous step is pressed in order to extract the remaining oil." (Congreso and internacional 1924)

8 Conclusion

In 1920, the advertising for the Acapulco system also included the Acapulco mixed system combining presses and grinders (Fig. 6).

The hydraulic presses were still employed in the Spanish oil mills until the end of the XX century, when they were eventually replaced by continuouss oil extraction systems involving the centrifuging of the ground paste.

The partial extraction Acapulco-Quintanilla system was improved by the Alfin system, known in Italy as Sinolea. This system enabled the production of very high quality oil, but was not profitable owing to the need to frequently clean the equipment, and the paste not having the desired temperature for the later centrifugation stage.

The use of these systems that extracted oil by separating the paste during blending was practical for obtaining better quality oil, when the system involved pressure, in

Fig. 6 Early XX century advertising of the Acapulco System, and the Mixed Acapulco System (Congreso and internacional 1924)

comparison to lower quality compressed oil due to the duration of the pressing process, and the difficulties involved in cleaning auxiliary equipment. Thus, the introduction of centrifuge systems unequivocally led to the ultimate demise of this system (Alba et al. 2009).

The olive growing industry is continuously undergoing technological developments to adapt to market changes and worldwide competition. Spain is currently the largest producer of olive oil worldwide, an achievement that has entailed hard work, trial and error, and primarily the impetus of outstanding Spanish entrepreneurs who have invested considerable time, effort, and money to improve the quality and efficiency of oil production, with Guillermo Quintanilla y Fábregas being a leading figure, and an icon to younger generations of entrepreneurs who strive to overcome the challenges in this prestigious Spanish Industry.

References

Mundo del agrónomo. Revista del Colegio Oficial de Ingenieros Agrónomos de Centro y Canarias, N°. 11, 2010.

VII Congreso internacional de oleicultura celebrado en Sevilla bajo el augusto patronato de S.M. el rey Don Alfonso XIII. Madrid: Sucesores de Rivadeneyra (s.a.) 1924.

PRADO LISBOA, M. (1906a). U Un aparato denominado «Acapulco de fabricación continua» para la extracción de los aceites, vinos y otros líquidos o jugos. Madrid.
PRADO LISBOA, M. (1906b). Un aparato para la extracción de aceite de oliva, semillas, frutos y residuos de oleaginosos, vino, sidra y demás jugos de frutas, denominado «Acapulco Nº4». Madrid.
PRADO LISBOA, M. (1906c). Un aparato para la extracción de aceite de oliva, semillas, frutos y residuos de oleaginosos, vino, sidra y demás jugos de frutas, denominado «Acapulco Nº4». Madrid.
QUINTANILLA, G.: «Obtención del Aceite por el Sistema Acapulco-Quintanilla», Actas VII Congreso Internacional de Oleicultura, Sevilla, 1924, pp. 507–514.
PRADO LISBOA, M. (1908). Un procedimiento para la extracción del aceite de olivas y frutos oleaginosos, denominado «Acapulco Continuo». Madrid.
Ministerio de Industria, Turismo y Comercio. Oficina Española de Patentes y Marcas. Archivo Histórico. Exp. nº 44.184.
Ministerio de Industria, Turismo y Comercio. Oficina Española de Patentes y Marcas. Archivo Histórico. Exp. nº 48.411.
Actas de "Oleotecnia y aceituna de mesa",1950.
Ministerio de Fomento. Dirección General de Agricultura y Montes. (1927). Método Acapulco-Quintanilla para la extracción del aceite (pp. 1–11).
PRADO LISBOA, M. (1925). Extractor-batidor plano que de manera indistinta funciona continua y discontinuamente. Madrid.
PRADO LISBOA, M. (1927). Nuevo Extractor De Aceite Y Batidor De Pasta Oleosa De Funcionamiento Continuo Y Discontinuo – Certificado de adicción por mejoras en la patente núm. 93072 del 24.03.1925. Madrid.
PRADO LISBOA, M. (1929). Un nuevo extractor de aceite. Madrid.
PÉREZ MARTÍNEZ, C. Influencias en el proceso de extracción del aceite entre finales del s. XIX y principios del s. XX. Vestigios del Sistema Acapulco. TFM. Universidad de Córdoba. 2015.
Hojas divulgadoras del Ministerio de Fomento. Abril-mayo 1927, año XXI números 7–8–9.
ALBA, J. et ál: «Tecnología de elaboración del aceite de oliva virgen», Ed. GEA, Westfalia S.L. Úbeda, 2009.

José Joaquín Romero de Landa (1735–1807)

J. C. Fortes, A. M. Sarmiento, and J. Castilla-Gutiérrez

Abstract José Joaquín Romero de Landa was a military man, sailor and Naval Engineer of the Spanish Royal Navy. His military life was in the service of the Crown, and he is known today for his professional career as one of the first Spanish naval engineers. He was a naval officer and officer of the Spanish army, and was the first official naval engineer and ship designer of the Spanish Navy. He began to develop this facet at the Guarnizo Shipyard, where he was assigned in November 1765. For four years he devoted himself to the study of shipbuilding. In 1767 he took on his first responsibility with the drawing up of two plans for an 80-gun ship and a 20-gun frigate. In 1769, when he opted to become a construction engineer, he became one of the few officers of the War Officers Corps to join the newly created Corps of Engineers. This article will study, on the one hand, the key formative factors that made José Joaquín Romero de Landa an enlightened engineer and, on the other, how his work was integrated into the shipbuilding systems developed in the eighteenth century.

Keywords Ships · Naval architecture · Military · Marine · Ship designer

1 Biographical Notes

José Joaquín Romero de Landa (Fig. 1), son of Gaspar Romero, captain of horses and commissioner of war, and Mayor de Landa y Muñóz, was born on 27 May 1735 in Galaroza, Huelva (Spain), as recorded in the baptismal certificate of the Parish

J. C. Fortes (✉) · A. M. Sarmiento · J. Castilla-Gutiérrez
University of Huelva, Huelva, Spain
e-mail: jcfortes@uhu.es

A. M. Sarmiento
e-mail: amsarmiento@uhu.es

J. Castilla-Gutiérrez
e-mail: javier.castilla@dimme.uhu.es

R. López-García and M. Ceccarelli (eds.), *Distinguished Figures in Mechanical Engineering in Spain and Ibero-America*, History of Mechanism and Machine Science 43, https://doi.org/10.1007/978-3-031-31075-1_12

Fig. 1 Portrait of José Joaquín Romero de Landa

Archive of Nuestra Señora de la Concepción in that town (https://es.wiipdia.org/wiki/Jos%C3%A9_Joaqu%C3%ADn_Romero_y_Fern%C3%A1ndez_de_Landa).

At the age of seventeen, he began his military career in the Edinburgh Dragoon Regiment, which he joined as a cadet on 27 May 1752. However, his father's preference for the Navy, since as War Commissioner he belonged to the Corps of the Ministry of the Navy, led him to join the Marine Guard Academy in 1754. The great disposition he showed from the outset for the study of mathematics—in 1756 he won three public competitions in this discipline in competition with other marine guardsmen—led Jorge Juan to take notice of him and include him in his group of mathematicians, establishing a solid relationship between disciple and master that lasted until the latter died in 1773.

On 19 November 1756, he embarked for his first sea campaign on the sixty-four guns ship San Fernando, aboard which he was promoted to ensign on 4 December 1757, disembarking on 2 August 1758. On 11 February 1759, he passed from Cádiz to Ferrol on the Conquistador, then embarked on the Glorioso and carried out several privateering operations in the waters off Cape San Vicente during the month of May, later joining Andrés Reggio's squadron; in August he formed part of the squadron which, under the command of the Marquis de la Victoria (Cruz Apestegui 1996), took King Charles III and his family from Naples to Barcelona, passing from there to Cádiz, where he disembarked (Fig. 1).

In 1760 he was promoted to ensign and took part in several operations against privateering in the Mediterranean. In 1761 he was appointed Acting Adjutant Major of the Marine Artillery Brigades, a post he held between Ferrol and Cartagena.

On 1 November 1765, he joined the Guarnizo shipyard under the command of the engineer Francisco Gautier. Between 1766 and 1767 he remained in Guarnizo, working and learning with Gautier and was promoted to frigate lieutenant.

In October 1770, the Corps of Marine Engineers was created. He was part of it from 17 January 1771, as Second Engineer only behind the General Engineer and creator of the corps, Francisco Gautier.

In 1772 he began to collaborate with Pedro González de Castejón, which was an affront to Gautier. When González de Castejón was promoted to Lieutenant General of the Navy in 1774, he appointed Romero Landa as Commander of Engineers, in violation of the Ordinances of the General Corps of Engineers and its Engineer General, Gautier.

In 1775 he was in charge of the preparation of the army's convoys for the first expedition to Algiers and married his cousin, Ana Fernández de Landa y Pérez Rañón. In 1776 he was promoted to captain and, until 1780, he was stationed in Ferrol and replaced the engineer general on his trips to the court.

In 1781 he was promoted to Brigadier of the Navy, maintaining the post of Commander of Engineers. In 1782, Gautier resigned as Engineer General of the Navy and Romero de Landa replaced him as Acting Engineer General. He drafted the Regulations on the timbers necessary for the construction of the King's baxels (1783).

On 28 January 1786, he was promoted to General Engineer of the Navy. In 1789 he was promoted to squadron commander and in 1795 to lieutenant general. He died in Madrid on 5 August 1807.

His would be a military life dedicated to the service of the Crown. However, he will be recognised to this day for his professional career in naval engineering.

"He served H.M. for 55 years from 1752, starting as a cadet in the Edinburgh regiment of dragoons, from where he passed to the royal navy as a marine guard: his military talent, his application to the study of mathematics, his always equal and irreprehensible conduct, together with his exact fulfilment of the commissions entrusted to him in times of peace and war, earned him the promotions corresponding to his career, up to that of general engineer. Applied by order of His Majesty to the theory and practice of naval architecture, he made progress which soon earned him the royal confidence; he was entrusted with the construction of the royal baxels, first in Guarnizo, and later in Ferrol, where he invented for the royal service a mixed system of construction adopted for its evident advantages for the three arsenals of the peninsula and the shipyard of Havana. For these and many other services rendered to the Crown, for his love for the Sovereigns, for his zeal, disinterestedness and unceasing application to the performance of his duties, he has always deserved the confidence of the King and Queen and their ministers and the esteem of the public". This review in the Gaceta de Madrid (Madrid 1807) of 18 September 1807 was a tribute to his figure and recognition of his professional career (Aguado and José Romero Fernández de Landa 1998).

During the eighteenth century, ambitious naval construction plans were drawn up with the aim of maintaining and protecting the overseas possessions and the commercial traffic they generated with the Spanish mainland. Undoubtedly, a powerful navy was needed, both in terms of ships with all their required machines and mechanisms and in terms of capable and trained officers to command it.

However, there was no specific centre for the training of officers. They came from the San Telmo College of Pilots in Seville, the Galleys cadets in Cartagena, or from foreign naval schools. To fill this gap, on 16 June 1717, "Las ordenanzas e instrucciones que se han de observar en el cuerpo de la Marina de España" (Villa and Patiño y Campillo 1882) was published, written by José Patiño, (Milan, 1666—La Granja de San Ildefonso, 1736) (Lafuente and Sellés 1717), intendant general of the Navy and president of the Tribunal de Contratación de Indias. Patiño signed the creation of the first Academy of Marine Guards, founded in Cádiz in 1717, and in 1724 he ordered the construction of the arsenal of La Carraca in the Bay of Cádiz. According to Lafuente and Sellés (Lafuente and Sellés 1717), this was one of the most attractive and influential scientific, educational and institutional experiences of the Spanish Enlightenment.

The requirements for midshipmen candidates were, among others, to belong to the nobility, to be a nobleman, or to be the son of a military officer with a job higher than captain. The aim was to interest the lower nobility, who would see the academy and naval engineering as means of social advancement (Lafuente and Sellés 1988). The son of a Captain of Caballos Cuantiosos and Commissioner of War and of a mother of proven nobility, José Romero Fernández de Landa was able to enter the Academy of Marine Guards in Cadiz in 1754, at the age of 19, not exceeding the age of 20 in fulfilment of another of the requirements for entry. At this centre he would meet many young Spaniards and also some foreigners.

The Academy merged the French model of teaching, which was fundamentally theoretical, and the English model, which was eminently practical. The institution comprised two structures, one teaching (Academy) and the other military (Company), which usually maintained rather tense relations. There was a combination of military and civilian teaching staff, who were called masters and were usually reputed sailors. The teachers at the Academy had great prestige in the life and culture of Cádiz. Proof of this is that they were involved in numerous matters of different kinds. Thus, for example, the mathematics master, Francisco del Orbe (Caravaca de Coca 1717), played a decisive role in the decision of the competition for the design of the new cathedral.

The curriculum, which lasted four academic years, combined military and academic training. It included a practical part on ships, where the pilot and officers were responsible for the instruction of the students. As soldiers, the cadets embarked on the various units, taking part in the campaigns in which they were involved and sharing the same fate as the crew, which in many cases was death and in others imprisonment.

Specifically, of the 129 cadets belonging to the first class of 1718, who embarked on several of the King's ships for practical training, a total of 56 went on to the General Corps of the Navy. In other words, 73 students (more than 60%) did not become War Officers (Plana and La Armada y la Enseñanza Naval 2001).

These data reflect the harshness of the service. As for José Romero de Landa, he began his first sea campaign in 1756 and was promoted to ensign the following year. He thus began his career as an officer in the Navy's War Corps, which was to rise to the position of Lieutenant General of the Royal Navy.

The theoretical part developed at the Academy brought the minds of its officers closer to the Enlightenment ideas that were taking root in Europe at that time. The subjects they studied were Mathematics, Geometry, Trigonometry, Cosmography, Nautics, Fortification and Artillery. The cadets were also taught dance, fencing and languages, as well as artillery, weaponry, shipbuilding and manoeuvres. The method of study consisted of reading the manual and explaining the more complex and doubtful aspects. However, during the first three decades, the academic reality was far removed from the syllabus set out in the Ordinance and Instructions.

In terms of costs, the annual maintenance of the student body at the Cadiz Academy was around seven hundred reales. Added to this was the salary of the teaching staff and the cost of books and scientific instruments. Soon the Academies were considered too costly for the Crown and for the army itself. In 1747, the situation of the Cadiz Academy was in a state of great deterioration, with a lack of teachers, manuals and even rifles for instruction. Rodrigo Pedro Urrutia reported the situation to the secretary of Ensenada who, with the publication of the Ordinances of 1748, initiated a process of improvement that gave greater prominence to academic and teaching life.

In 1752, Jorge Juan y Santacilia (Valverde Pérez 2012) (Novelda, Alicante 1713/ Madrid 1773), appointed Commander of the Royal Company of Marine Guards, initiated the renewal of the teaching staff and recruited highly qualified personnel. He also tried to put into practice the proposals set out in the Ordinances of 1748. To this end, he appointed Luis Godin as director of the Academy and Antonio de Ulloa as lieutenant of the Company. The era of the scientific officer began in the Navy, with great knowledge of Mechanical and Naval engineering. Landa would be an example of this, driven by the knowledge of his professors Rodrigo de Urrutia, Jorge Juan and Antonio de Ulloa (Fig. 2).

In 1754, coinciding with the entry of Romero de Landa, the last important investment in books and instruments was made (Fig. 2) (Lafuente and Peset 1750). This would be a period of splendour for the Academy of Cadiz, where the new textbooks for teaching would stand out. Among the copies that Landa handled was the "Compendio de navegación para el uso de Caballeros Guardias Marinas", by Jorge Juan, published by the Academy of Cadiz in 1757 (Santacilia 1757). It is a textbook that summarised all the knowledge and methodology to be followed for navigation, constituting one of the most significant works on the subject at the time.

The Academy had its own printing press and its editions reflected the cultural and scientific reality of our Enlightenment. Among the works that Landa was able to consult were Jacob Kresa's "Elementos de Geometría de Euclides" (1689) and Tomás

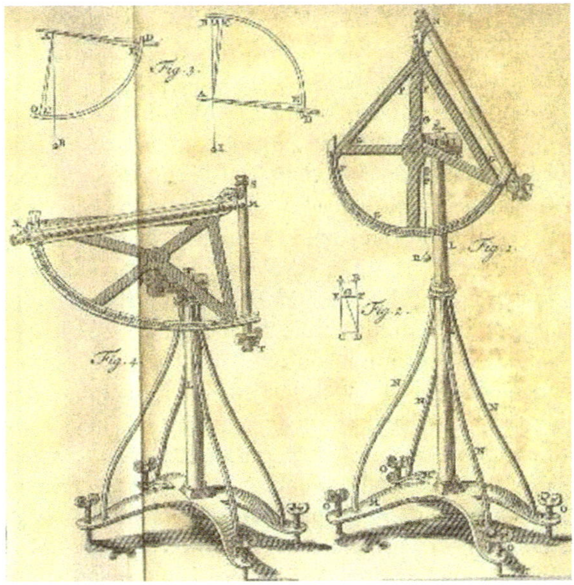

Fig. 2 Meridian degree measuring instrument

Vicente Tosca's "Compendio mathematico" (1707). Also published were manuals by his teachers, such as the "Compendio del arte de la Navegación" (Seville, 1717) and "Trigonometría aplicada a la Navegación" (Seville, 1718) by Pedro Manuel Cedillo y Rujaque; "Aritmética", by Professor Luis Godin; "Geometría y Trigonometría rectilínea", by Vicente Tofiño; "Artillería", by Francisco Javier Rovira, etc. All this bibliography allowed him accedd to a wide knowledge of mechanical and naval engineering, machines and mechanisms.

The curriculum, promoted by Jorge Juan (Santacilia 1771) (Fig. 3) and Godin, strengthened the theoretical preparation of cadets in geometry, analysis, mechanics, and astronomy. Only profitable students, such as Fernández de Landa, were taught the higher subjects: higher mathematics, mechanics, cartography, cannon function and languages. Public competitions were even set up for the most advanced cadets to demonstrate their knowledge, especially in navigation and mathematics.

In his second year, Romero de Landa won three competitions: the first on Analysis, Differential Calculus and Sublime Geometry; the second on Mechanical engineering and the third on Theoretical and Practical Navigation. This gives us a picture of an exemplary pupil.

In order to achieve a high level of teaching, in 1753 the Royal Astronomical Observatory was created, annexed to the Academy of Gentlemen Marine Guards. The aim was to train future naval officers in a science as necessary for navigation as astronomy was at that time. In this way, teaching was complemented by a scientific research institution that provided technical and scientific support to the Enlightenment expeditions of the last third of the eighteenth century (Fig. 4).

Fig. 3 "Tratado de Mechanica" by Jorge Juan 1771

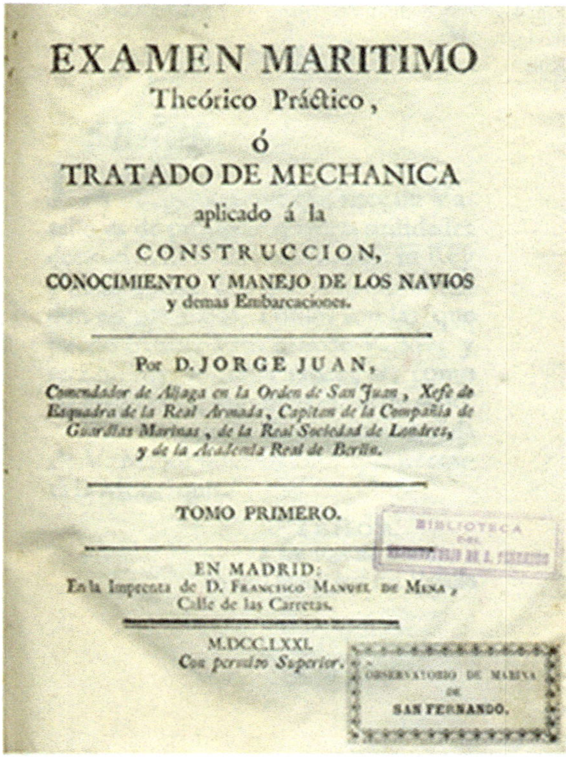

Undoubtedly, this comprehensive training made Landa an enlightened officer and expert in the sciences of navigation, mechanics and shipbuilding. The theoretical foundations and his eagerness for research were deeply rooted. All this, together with the fact that he was assigned to the Guarnizo shipyard in 1765, would mark the development of an important professional career as a ship designer and builder.

2 List of Main Works

His main work was "Reglamento de maderas necesarias para la fábrica de los baxeles del Rey y demás atenciones de sus Arsenales y Departamentos. Formed by Don Joseph Romero y Fernández de Landa, of the Order of Santiago, Brigadier of the Royal Navy and Engineer Director of it". 1782.

The aim of these regulations, as Fernández de Landa himself explains, was to avoid disputes with the shipwrights at the time of receiving the pieces and to put an end to the serious mistakes that were made in the arsenals when measuring the timber in the mountains, which caused a great deal of unnecessary expense. José Romero y

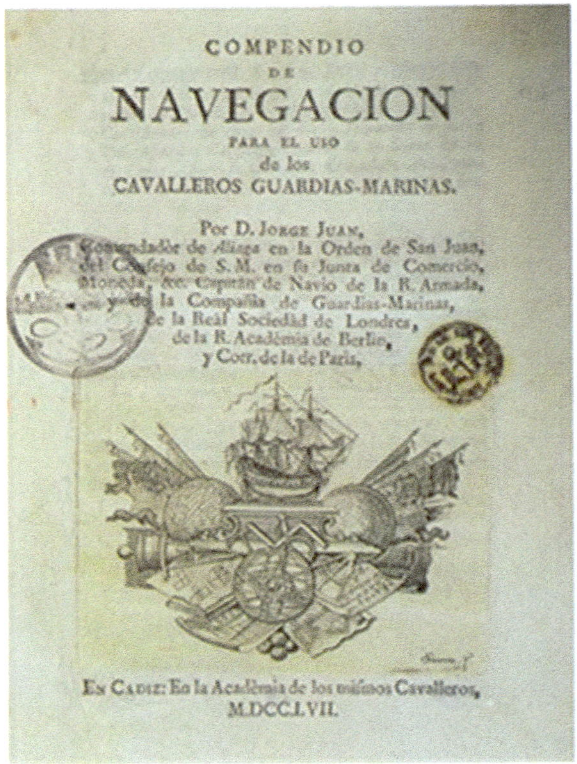
COMPENDIO
DE
NAVEGACION
PARA EL USO
de los
CAVALLEROS GUARDIAS-MARINAS.

Por D. Jorge Juan,
Comendador de Aliaga en la Orden de San Juan,
del Consejo de S.M. en su Junta de Comercio,
Moneda, &c. Capitan de Navio de la R. Armada,
y ... la Compañía de Guardias-Marinas,
de la Real Sociedad de Londres,
de la R. Académia de Berlin,
y Corr. de la de Paris,

En Cadiz: En la Académia de los mismos Cavalleros,
M.DCC.LVII.

Fig. 4 "Compendium of navigation for the use of Knight Sea Guardsmen"

Landa took Gautier's timber regulations as a model in terms of structural layout and dimensions of the pieces, but he introduced a fundamental novelty with respect to the previous timber regulations, the detailed quartering of all classes of ships and the systematic definition of the most important pieces that make them up, establishing a model of the structure of ships and frigates with a unified treatment, that is to say, he established a systematising component.

In the eighteenth century, a profound reform was carried out to place Spain in a prominent position among the European powers and the navy was considered vital as a means of controlling the colonies. New construction techniques were introduced in the Spanish shipyards and from 1782 Romero Landa was its chief engineer and designed the following warships:

8 Ships of 112 Guns:

Santa Ana (1784) (Fig. 5); Mejicano (1786); Conde de Regla (1786); Salvador del Mundo (1787); Real Carlos (1787); San Hermenegildo (1789); Reina María Luisa (1791) and Príncipe de Asturias (1794).

Fig. 5 112-gun ship Santa Ana

San Idelfonso (1785) (Fig. 6); San Telmo (1788); San Francisco de Paula (1788); Europa (1789); Intrépido (1790); Infante don Pelayo (1791); Conquistador (1791) and Monarca (1794).

3 ships of 64 guns:

San Fulgencio (1787); San Leandro (1787) and San Pedro de Alcántara (1788). (Fig. 7).

Fig. 6 Spanish ship San Ildefonso

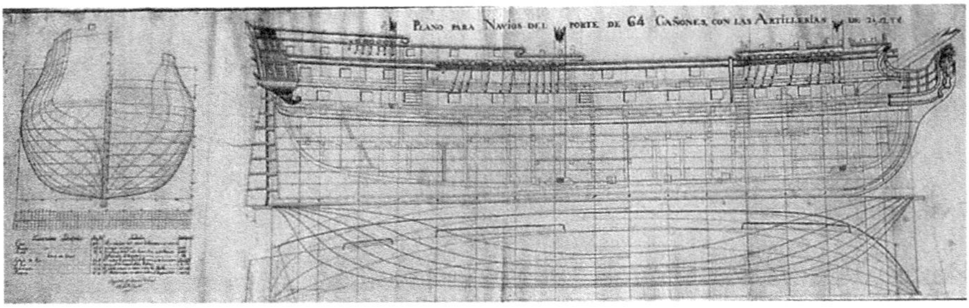

Fig. 7 Drawing of a 64-gun ship

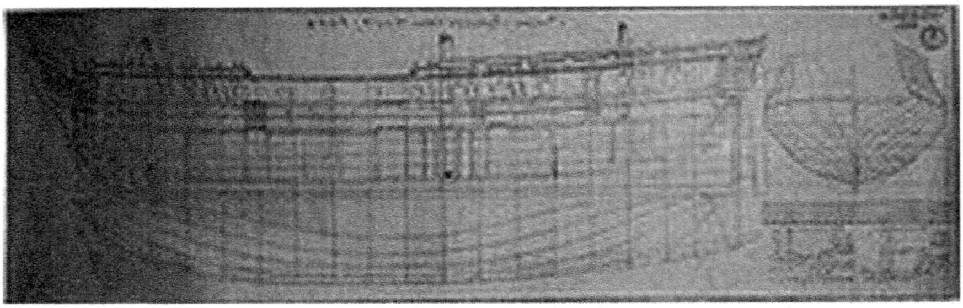

Fig. 8 Drawing for a 34-gun frigate

6 Frigates:

Santa Casilda (1784); Santa Florentina (1786); Nuestra Señora de la Soledad (1788); Mahonesa (1789); Perla (1789) and Preciosa (1791) (Fig. 8).

3 Review of Main Works/Contributions

During the first years of the eighteenth century, shipbuilding continued along the lines of the work carried out at the end of the previous century. However, the appointment of José Antonio de Gaztañeta e Iturribalzaga (Motrico 1656/ Madrid 1728), as general superintendent of the shipyards of Cantabria, would change this situation, marking a clear difference with his predecessors. In the shipbuilding system, José Antonio applied the scientific point of view, as opposed to work based on tradition and custom. Undoubtedly, his contributions would produce variations in the traditional system, beginning with the drawing up of the first modern plans known in Spain, which served as a reference for later authors.

The superintendent reflected his scientific knowledge in several written works. In his document entitled "Proporción entre las medidas arregladas a la construcción de un bagel de guerra de sesenta codos de keel", dated 1712, Gaztañeta introduces criteria, measurements and designs according to the number of cannons that the different ships can carry (Valle and y Gómez 2022).

In 1720 he wrote "Proportions of the most essential measurements for the construction of ships and frigates with the necessary machines and mechanisms of war, which can mount from eighty cannons to one hundred, with the application of the construction of the master barenga, plan and particular profile of a seventy-gun ship, with the lengths, thicknesses and angles of the materials, written by order of the king". It was in this second work that the definitive criteria and the typology to be followed in the construction of Spanish ships, which would be larger and longer than those previously created, would be determined. Its objectives are solidity and durability.

Gaztañeta was therefore the first serious attempt to standardise shipbuilding in Spain and to overcome the old system of private builders, who were responsible for the construction of the Crown's ships without standards or optional inspection of their results. The ships built using Gaztañeta's systems had good seafaring qualities, but they were too costly to build, as the structural elements had to be built in one piece, which required the use of large trees that were becoming increasingly scarce. Moreover, it was essential to work the parts on the ship itself, with the consequent waste of time and resources.

Seeking a solution to this problem, in 1752 the Minister of the Navy, Zenón de Somodevilla y Bengoechea, Marquis of Ensenada, sent the naval captain Jorge Juan y Santacilia to England (Ruiz and Jorge Juan 2013) to spy on English shipbuilding, with the intention of cutting costs and reducing construction time rather than finding technical improvements, and to recruit specialised labour. Ensenada (Abad León 1985), in a letter to the Spanish ambassador in Paris, said: "There are neither builders nor masters of rigging and canvas in France and Spain, and in both kingdoms the economy is very poorly understood… Jorge Juan is already in London and his journey will be very useful to us because in terms of mechanics we are very ignorant, without knowing it, which is the worst thing…".

On his return to Spain, Jorge Juan established the guidelines to be followed in the new shipbuilding. From his meetings with the shipbuilders' boards, one in 1752 and the other in 1754, the new system known as "English style" emerged, which was developed through the Wood Regulations with the rules for the use of wood and to establish the measurements and shapes of the main parts in shipbuilding and ship cutting.

Savings were made in timber because the structural elements were produced by overlapping much smaller elements, even making it possible to use scraps of scrap material. The mass production of similar parts, which were assembled on the floor and mounted on the slipways, allowed the production of series of identical ships in record time, thus optimising the construction systems.

When the Marquis de la Ensenada fell into disgrace, in 1765 the Frenchman Francisco Gautier Audibert (1715/1782) was entrusted with the construction of the ships, and he introduced the French System. Basically, the novelty consisted in the construction of ships of a different design, narrower in beam and with more elongated lines, thus gaining in speed but losing in consistency.

This engineer designed new plans as the basis for his ship models and developed new regulations for wood. It is worth noting Gautier's personal determination to create the Corps of Engineers of the Navy, following the model of its French counterpart established in 1765, and the foundation of the Training Academy. The engineers, whose technical command was located in the Arsenals, were responsible for the construction and care of the Royal Navy's ships. They were also responsible for the construction of hydraulic and civilian buildings and the reconnaissance of the mountains.

In 1774, Gautier resigned from his post, which was refused by the Secretary of State Arriaga. It was in 1782 when his second resignation was accepted and José Joaquín Romero Fernández de Landa, frigate captain and second engineer of the Navy, took up his post as interim, obtaining it in 1786.

Romero Landa consolidated the Corps of Engineers, introduced new criteria for the organisation of the Arsenals and promoted the last attempt to maintain the power of the Spanish Navy, planning and directing the construction of warships, frigates, ships, and others, with all the machines and mechanisms required for their various functions. The defeat at Trafalgar, on 20 October 1805, meant the destruction of the Spanish Navy, which would never recover from this heavy blow.

Fernández de Landa, as a disciple of Jorge Juan at the Academy of Marine Guards in Cadiz, studied the "English-style" system. On the other hand, when he was assigned to Guarnizo in 1765, under the direction of Francisco Gautier, he took part in the implementation of the "French-style" factory in Spain. What path would the new Marine Engineer General follow? Like Jorge Juan or Gautier, he decided to establish his own construction system. His concern was to make the fastest ships, which explains why the ships of this system are characterised by a large sail area. He had new machines for the shipyards, such as saws and fire pumps, and "industrial espionage" continued.

In 1784 he published "El Reglamento de las Maderas Necesarias para la Fábrica de los Baxeles del Rey", (Fig. 9) in which he made a detailed exploded view of various vessels with a scale of Burgos feet. By means of a freehand drawing, the author provides all the details of the dimensions of the wooden pieces that make up various types of ships, frigates and smaller vessels. As it is not addressed to carpenters, the pieces do not include joints, scarps, or any other specifics related to construction carpentry.

In his Regulations, José Joaquín conceived a unique model or type of warship, and his exploded view, depending on a predefined scale, gives us the components of the internal structure of a ship of 100 or more guns, of 74 or 64 guns, of frigates of 34 guns and of other smaller vessels. It is therefore a colossal work, unique in its time.

Its first ship, the San Ildefonso, a 74-gun vessel, was launched in 1785 and underwent an exhaustive programme of sea trials to ascertain its sailing properties and check its advantages over the Navy's best ships. After carrying out comparative sea trials with the gautier "Nepomuceno" on a voyage from Cartagena to Algiers, José de Mazarredo, head of the squadron, declared: "It sailed to windward like the frigates; it steered and tacked like a boat; it had a spacious battery, stable in all positions,

Fig. 9 Romero Landa timber regulations

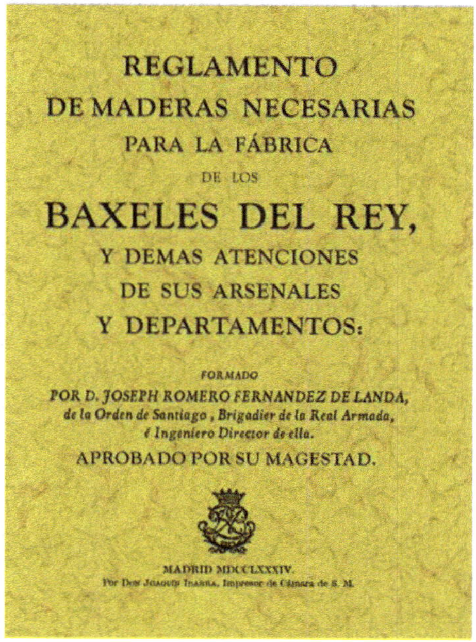

REGLAMENTO
DE MADERAS NECESARIAS
PARA LA FÁBRICA
DE LOS
BAXELES DEL REY,
Y DEMAS ATENCIONES
DE SUS ARSENALES
Y DEPARTAMENTOS:

FORMADO
POR D. JOSEPH ROMERO FERNANDEZ DE LANDA,
de la Orden de Santiago, Brigadier de la Real Armada,
é Ingeniero Director de ella.
APROBADO POR SU MAGESTAD.

MADRID MDCCLXXXIV.
Por Don Joaquin Ibarra, Impresor de Cámara de S. M.

cases and circumstances". Following its good results, the San Idelfonso became the prototype for eight other ships in the series, which were the best 74-gun ships ever built and undoubtedly generated admiration and envy in their English and French opponents.

The ships built according to Romero de Landa's system were rated as excellent. In total they were: the series of 5 ships, whose prototype was the Santa Ana, of three decks and 112 guns; the aforementioned series of 8 ships, whose prototype was the San Ildefonso, of two decks and 74 guns; on the last prototype and applying a reduction of 49.5/52, 3 ships of two decks and 64 guns were built. Finally, the Santa Casilda was the prototype for 6 frigates of 34 guns.

In November 1793, Captain Julián Martín y de Retamosa (Cartagena, 1747, Madrid, 1 February 1827) was appointed First Officer of the Secretary of the Navy, responsible for arsenals and shipyards and shipbuilding. In 1974, as brigadier and second in command to Romero Fernández de Landa, Retamosa designed in Ferrol the ship Montañés, with two decks and 74 guns, under the supervision and preliminary design of Romero Landa. Three ships were to be built following the model of the Montañés and 7 frigates of 34 guns, the prototype of which was the Diana.

Retamosa surpassed Landa's designs in the way he sailed, thanks to the refining of forms that Julián devised and which allowed for good navigation both astern, alongside and bow. In addition, his ships kept their batteries in full bloom in almost any weather, one of the invaluable advantages for a warship, which had been pursued, with greater or lesser success, throughout the century. His would be the last ships of the line of the Spanish Armada.

Romero de Landa continued to work on improving the designs of the mail ships until they became fast brigs in 1802. This is another of the great contributions of the Huelva-born engineer, the considerable reduction in the mass of the general structure of warships, which was very evident in comparison with the English and French construction systems. Many authors claim that this reduction and the elimination of some parts is the basis of the perfection that was achieved.

In short, the differences maintained between the advocates of the different eighteenth century naval construction projects and systems have persisted to the present day. However, despite these differences, the ships of the eighteenth century were able to bring together all the technological know-how of the time. The techniques of brigantine construction were very similar throughout Europe, especially in Spain, France and England. The internationalisation of knowledge occurred because the sea had no borders and ships, which arrived at all ports, became an inexhaustible source of information and exchange of ideas between sailors and engineers.

4 On the Circulation of Works

The methods of design, procurement of materials, construction and maintenance of the Navy's ships evolved throughout the eighteenth century in response to the Fleet's major expansion plans, and were set out in the Ordinances, Regulations and Royal Orders.

From the establishment of trade with America, the naval construction of the ships that made the voyage to the Indies was subject to strict control by the Crown as a fundamental part of this trade.

The most important body of regulations relating to shipbuilding were the following the successive Ordinances that were issued throughout the seventeenth century.

In 1691, a work was published whose purpose was to define a type of ship for the Indies race that would have a shallow draught that would allow it to sail without difficulty up the access road to the city of Seville by going up the Guadalquivir from Sanlúcar de Barrameda, regardless of the time of year and the state of the tide. The author was Antonio Garrote, builder and war captain, and the work was entitled "Compilation for the new factory of Spanish ships, where the proportions and new gauge corresponding to six orders of different sizes are declared, with the usefulness of serving as war ships in the navies of the ocean with all perfection and as merchant ships in the Indian race, dedicated to the Catholic Royal Majesty of our very great Monarch Charles II, may God preserve him".

In his proposal, Garrote unified the project for warships and merchant ships, which had hitherto been treated differently in the Ordinances.

In addition to having a reduced draught, these vessels had to carry more artillery than others of the same size and be endowed with outstanding properties in terms of their handling and steering, which meant that they had to have less hull, masts, rigging and rigging to simplify their handling. In short, they had to be clearly superior to any other ship of the time in order to maintain authoritative control over the traffic in the Indies.

The shipbuilding plan of the Marques de la Ensenada, drawn up at the end of the 1940s, led to the preparation of detailed technical information according to which the different types of ships were to be built. This information was drawn up by a Board of Builders that met in Madrid in 1752.

Francisco Gautier significantly modified the design of ships and frigates, mainly in the structural aspect. In 1769, he published the Reglamento de Maderas de Roble necesarias para fabricar un navío de 70 cañones (Regulations on the oak wood needed to build a 70-gun ship), which follows the outlines of the regulations of the Junta de Constructores of 1752 with some differences that will be analysed later.

José Romero Landa, a disciple of Gautier, whom he succeeded as head of the Corps of Marine Engineers in 1786, published in 1784 the last timber regulations of the eighteenth century: Regulation of timbers necessary for the manufacture of the King's ships and other services of his arsenals and departments, including ships of 100 and upwards, 74 and 64, frigates of 34 guns and upwards, smaller frigates, ocean liners, or similar, and other items of the Departments; with a sixth class of timbers for Boats, Boats, Pikes, Mastels for Planks, Crossbeams, Scantlings, Mirrors, parts for Motorboats, Drums, Anchor stocks, and for buildings. Escoras, Espeques, pieces for Motonería, Tamboretes, Cepos de Anclas and for buildings. Also a regulation of pine spars for spars, timbers, and planking for decks, dead works, oars, and other service purposes.

The process of arming ships was regulated from the beginning of the eighteenth century due to the large number of elements involved and the need to standardise their characteristics in order to standardise and speed up supplies. Throughout the century, successive Armament Regulations were published that incorporated the changes in the equipment and operation of ships.

In the last third of the eighteenth century, the technical information that was prepared during the design stage of a new ship formed a body of documentation covering all aspects of the construction.

The Arsenals Ordinance of 1776, in Title XXIII, establishes the information to be prepared by the Engineer General when the order for the construction of a ship is given. This information shall consist of three equal plans "... calculated and with all the profiles detailing the distribution of its holds, decks and accommodation; showing the timber and iron necessary for its construction, its dimensions and those of its rigging...". This is the compulsory rule and all the projects to be drawn up from then on must comply with it.

The calculations of hull geometry were carried out by integrating the trapezoidal method, as Jorge Juan explains in detail. To calculate the displacement, the value of the specific weight determined by Jorge Juan was used.

Pierre Bouguer and Jorge Juan had clearly established that the parameter that determines the angle of heel produced by the action of the wind with a crosswind component, i.e. the hold of the sail, was the value of the distance between the ship's centre of gravity and the transverse metacentre.

Despite the recognised importance of this value, it was not specifically calculated nor was it used as a determining criterion for setting the value and stowage of the ballast, which was used as a corrective element for the hold of the sail, which was set on the basis of the corresponding values in similar ships and adjusted, if necessary, during the Sea Trials according to the hold of the sail and the trim necessary to achieve good steering observed during these Trials.

In 1771 Jorge Juan had established a procedure to determine the height of the ship's centre of gravity by carrying out an experiment that consisted of moving all the crew's artillery, bullets, boxes and chests to one side and placing water-filled pipes hanging from the ends of the spars where crew members were also located, with the aim of producing a heeling moment.

By measuring the angle of heel produced (Δ), and the heeling moment caused by the displaced weights ($\int p - \pi$), the desired distance was obtained:

$$\frac{\int p \cdot \pi}{P.sen\Delta} \tag{1}$$

p: weight moved

π: transverse distance

P: displacement of the vessel

This procedure is essentially the same as the one currently used for all newly built vessels.

However, no record has been found of this experimental measurement of the distance between the centre of gravity and the transverse metacentre being carried out systematically on ships of the late eighteenth century, which on the other hand is carried out today on all newly built ships.

The waterline or sailing line defined by the fore and aft draughts was established on the plan and normally had a positive trim, i.e. aft, of one or two feet. The suitability of this waterline was checked during Sea Trials and if necessary its position was altered in the light of the ship's performance during the Sea Trials.

It was common for the original plan to bear the date of approval of the plan signed by the Secretary of the Navy. Several copies of these plans were made, which were also signed by the person who drew them, and which were used in the Arsenals for the entry into the careening docks and the arming and disarming processes.

5 Legacy and Today Interpretation of Contributions

Romero Landa drafted and published, in 1782, a Regulation on Timbers concerning the typologies of the frigates and ships of the time. It is considered that the main objective of the regulation was to try to reflect in the supply of timber the reduction in thickness that was considered necessary to reduce the weight of new ships. And, on the other hand, with the aforementioned regulation, the discussions that appeared in wood supplies could be reduced by establishing criteria for substituting oak for pine.

The regulation states that its purpose was to reduce the confrontations between wood suppliers at the time of receiving the material. The regulation also included prices for each of the five classes of ships, as well as the sixth class for miscellaneous purposes and for spars and planking. It also indicated that in the event that a piece was smaller than the size specified in the Regulation, it would be considered to be of the class to which this size corresponded.

These regulations defined the oak pieces necessary for the construction of three types of ships (100, 74 and 64 guns) and two types of frigates (over 36 and under 36 guns), establishing the measurements for each type of ship.

The Reglamento de maderas consisted of a volume printed on 35 × 25 cm. sheets, with a total of 77 numbered pages and 12 unnumbered pages. It lacks an index and the title page is simple, without drawings or ornamentation, as shown in Fig. 9. Within the regulations there is a classification by type, and it is the number of guns that is the determining factor in distinguishing a ship from a frigate. The Mercedes, for example (Fig. 10), because of its number of guns, belonged to the frigate type, with fewer guns than the ships.

For the time, this regulation was highly innovative, and brought the regulatory environment of eighteenth century shipbuilding to a close. Romero Landa's first ship, the San Ildefonso, a 74-gun vessel, was launched in 1785 and underwent an exhaustive programme of sea trials to ascertain its sailing properties and to test its

Fig. 10 Nuestra señora de las Mercedes frigate

advantages over the best ships in the Navy. After comparative sea trials with Gautier's Nepomuceno on a voyage from Cartagena to Algiers, José de Mazarredo, head of the squadron, declared: "It sailed to windward like the frigates; it steered and tacked like a boat; it had a spacious battery, stable in all positions, cases and circumstances".

To finish talking about Romero Landa, it should be pointed out that, as General Engineer, to whom the Ordinance of Engineers granted the power to direct the construction of ships, he gave the order for the construction of the frigate "Nuestra Señora de la Mercedes" in Havana. This frigate was based on what was proposed in the "Reglamento de Maderas" and the plans correspond to designs by Don José Romero de Landa.

This was the famous frigate of the "Odissey" case. Although there is no documentary record of the plan from which this frigate was built, it can be conjectured that it was the plan of the frigate Santa Florentina, built in Cartagena, as the dimensions were similar and Romero Landa had carried out the project tests.

The Reglamento de las Maderas Necesarias para la Fábrica de los Baxeles del Rey, better known as the Reglamento de Romero Landa, hereinafter (RRL), is the most relevant work for understanding the evolution and perfection of warships in Spain at the end of the eighteenth century.

This work, which almost borders on perfection in what it represents, is unparalleled by any other in the naval world. All of Romero de Landa's developments have subsequently had a great impact on Mechanical engineering applied to naval architecture.

It may be thought, and rightly so, that there have been many who, with their works of architecture and naval construction, formed similar descriptions and/or regulations, from Gaztañeta, Jorge Juan, La Junta de Constructores, Bryant, and Gautier, to highlight the most relevant of the eighteenth century in Spain.

But in no work in Spain, England or France, have they had such a brilliant foundation.

The RRL is the culmination of his work, but it is considered that his genius lies in conceiving a unique model or type of warship, and that his exploded view, depending on a predefined scale, gives us the components of the internal structure of a ship of 100 or more guns, or 74 guns, or 64, or frigates of 34 or more or fewer guns, or smaller vessels.

This is the colossal nature of his work, which many authors do not dissociate from the economic context of eighteenth century Spain, but his genius is undeniable and his objectives are well defined at the head of the regulations.

References

https://es.wiipdia.org/wiki/Jos%C3%A9_Joaqu%C3%ADn_Romero_y_Fern%C3%A1ndez_de_Landa, last accessed 2022/11/30

C. Cruz Apestegui, El Marqués De La Victoria Constructor Naval. Jornadas de Historia Marítima, XV. Madrid. Cuadernos Monográficos del Instituto de Historia y Cultura Naval, 59–75 (1996)

Gaceta de Madrid, Volumen 2, Imprenta Real (1807)
J.M. De Juan-García Aguado, José Romero Fernández de Landa. Un ingeniero de marina en el siglo XVIII. Universidad de Da Coruña. Monografías Nº 59. Julio (1998)
A. Villa, Patiño y Campillo. Reseña histórico-biográfica de estos ministros de Felipe V formada con documentos inéditos y desconocidos en su mayor parte, Madrid, Sucesores de Rivadeneyra (1882). En Biblioteca Virtual del Principado de Asturias
A. Lafuente, M. Sellés, El proceso de institucionalización de las Academias de Guardiamarinas de Cádiz 1717–1748. Instituto Arnau de Vilanova. CSIC (1991)
A. Lafuente, M. Sellés, *El Observatorio de Cádiz (1753–1831)* (Ministerio de Defensa. Instituto de Historia y Cultura Naval, Madrid, 1988)
J.M. Caravaca de Coca, Apuntes para la historia. La figura de don Franscico Antonio de Orbe, profesor de matemáticas y primer director de la academia en 1717 (2015)
M. Alía Plana, La Armada y la Enseñanza Naval (1700–1840) en sus Documentos. Aproximación a las Reales Ordenanzas reguladoras, desde una perspectiva jurídico-administrativa y pedagógica. Tesis, UNED (2001)
N. Valverde Pérez, Un mundo en equilibrio. Jorge Juan (1713–1773). Editorial: Marcial Pons, Ediciones de Historia. Colección: Ambos Mundos, Madrid (2012)
L. Antonio, J.L. Peset, J.L.: Las Academias Militares y la inversión en ciencia en la España Ilustrada (1750– 1760). In VI Congreso de Historia de la Medicina. Barcelona. 29 de septiembre de 1979
J.J. Santacilia, Compendio de navegación para el uso de los caballeros Guardias Marinas. Imprenta de Marina, Cádiz (1757)
Santacilia, J.J.: Examen marítimo teórico práctico o tratado de mecánica aplicado a la construcción, conocimiento y manejo de los navíos y demás embarcaciones. 1771. Madrid
García del Valle y Gómez, Jesús: Rol de la Metrología en la construcción naval y navegación en la armada española de los siglos XVII y XVIII. Revista Española de Metrología, Nº 7. http://www.e-medida.es/documentos/Numero-7/naval#1. last accessed 2022/11/30
Martínez Ruiz, E.: Jorge Juan: su misión en Londres y la construcción naval española. Mariano Juan y Ferragut. En Jorge Juan y la ciencia ilustrada en España. Instituto de Historia y Cultura Naval. Cuaderno monográfico N.º 68. Madrid, (2013)
F. Abad León, El Marqués de la Ensenada, su vida y su obra. Editorial Naval, 1985, vol. I, Ensenada al embajador en París, 24–3–1749, p. 227

José Ruiz-Castizo (1857–1929)

P. Zulueta Pérez

Abstract This paper tries to analyse the figure of the mathematician, physicist, and inventor José Ruiz-Castizo y Ariza (1857–1929), Professor of Rational Mechanics at the University of Zaragoza and Professor of Exact Sciences at the Central University of Madrid. Throughout a life dedicated to science, he carried out numerous jobs in his role as a teacher and researcher. Many of them, including his "Treatise on Rational Mechanics", were published in monographs and scientific journals of the time. However, the main objective of this article is to present José Ruíz-Castizo not only through his academic activity but also as an inventor of scientific instruments. For this, we will focus on one of his main inventions, specifically the "Tangential evaluation Cartesian planimeter" designed by him from other existing planimeters. From this instrument designed to measure the area of any plane figure, rectilinear, curvilinear or mixtilinear, as well as other applications derived from the previous one, he patented and materialised a new proposal based on his previous work on the theory of curves.

Keywords History of MMS · History of IFToMM · José Ruiz-Castizo y Ariza · Rational mechanics · Topographic instruments · Measurement of irregular areas · Planimeter

1 Introduction

The Enlightenment century in Spain led to the establishment of foundations for the development of scientific activity in the eighteenth century. Due to the European influence and the principles of the Enlightenment, the two fundamental characteristics of that period were implanted: utility and rigor. At that time, the "experimental natural philosophers"—the enlightened scientists—who were the inventors or professional engineers, were greatly influenced by the ideas of Isaac Newton (1643–1727) on the

P. Zulueta Pérez (✉)
University of Valladolid, Valladolid, Spain
e-mail: patriciabeatriz.zulueta@uva.es

R. López-García and M. Ceccarelli (eds.), *Distinguished Figures in Mechanical Engineering in Spain and Ibero-America*, History of Mechanism and Machine Science 43, https://doi.org/10.1007/978-3-031-31075-1_13

laws of motion, his mathematical model and his experimental method, and they had a great interest in applied mechanics established long ago by Galileo Galilei (1564–1642) and Guidobaldo del Monte (1545–1607). Scientists and engineers turned out to be technology professionals during the industrialization process that spanned the eighteenth and nineteenth centuries, first with the culture of the Enlightenment and later with the culture of the nineteenth century.

During the seventeenth century in Spain, other paths for the dissemination of culture were proposed in parallel as an alternative to extreme university immobility. It is well known that the enlightened monarch Charles III (1716–1788) promoted teaching and research in newly created institutions. Among these, the creation of the new Academies—originated from meetings and private cultural gatherings—and the promotion of the existing ones that had been founded in the last period of the Habsburgs should be highlighted in a fundamental way. Likewise, the birth of the Economic Societies of Friends of the Country was of great relevance, which is an initiative started—as in the case of the Academies—from a group of individuals who talked about the social issues of the moment, in addition to the discussion on cultural and scientific matters.

During the second half of the eighteenth century, the creation of various scientific institutions took place in Spain. Among them were the Royal Botanical Garden (1781), the Astronomical Observatory (1790), the Royal College of Surgery of Saint Charles (1771)—poster Faculty of Medicine—and the Royal Academy of the Three Noble Arts of Saint Ferdinand (1752)—later called Royal Academy of Fine Arts of Saint Ferdinand- which, in addition to the Chair of Mathematics, housed a Museum of Natural History. All of them, located close to each other, acquired the rank of university during the enlightened period.

All these institutions collaborated, to a great extent, to introduce the postulates of modern thought in Spain. The Enlightenment brought with it important novelties in Spanish classrooms because, although there were serious problems that opposed any change within its stale structures, it managed, to a certain extent, to introduce them alongside the creation of new chairs and the publication of new books and educational programs destined to the implantation of the new knowledge of modern science.

During those years and given the influence of French engineering and its textbooks in our country, a great effort was made by the authorities and intellectuals to start promoting and valuing the works written in our language. Towards the middle of the eighteenth century, there was growing concern about the lack of textbooks in the Academies. It was in this context and in view of the scarce scientific activity in Illustrated Spain that translations of foreign works were carried out. Approximately in 1770, the publication of scientific texts and manuals began to flourish, leading to higher levels of teaching, commissioned by an original work that included advances from the rest of Europe (García 2015). The great demand for technical works occurred both in civil organizations, as in the case of the Royal Academy of Saint Ferdinand, where texts and primers were prepared for the studies of mathematics and perspective, as well as in the military.

Later, the Napoleonic invasion and the absolutist monarchy of Ferdinand VII led to a crisis that made the aforementioned scientific institutions disappear with the exile, in many cases, of the men of science in our country.

Throughout the nineteenth century, the different scientific disciplines became independent, experimental and mathematical methods were imposed, and with this, a growing role of the physical-natural sciences was achieved. In 1847, the Royal Academy of Exact, Physical and Natural Sciences was founded. The scientific disciplines of the universities began to have their own identity when the Faculties of Sciences were implanted in Spain in the second half of the eight hundred. The society of the moment recognized engineers as the scientific authority in the country: Civil Engineers, the Industrialists, those of Mines or the Military.

The Exact and Physical Sciences of the Faculties became, as a result of the Law of Public Instruction of 1857, better known as the Moyano Law, in pre-paratory subjects for entry into Special Schools (González 2002). In those years, the closeness between architects and engineers began; these groups, for years, shared studies in a common Preparatory School created by Royal Decree on November 6, 1848, for the degrees of Civil Engineers, Channels and Ports, Mines, and Architecture. The idea was supported by Juan Bravo Murillo (1803–1873), Minister of Commerce, Education and Public Works, who recovered a project in 1822 from a Polytechnic School for the joint preparation of all these studies. With this new proposal, the Special School of Architecture seemed to understand the importance of mathematics, the analysis of materials, mechanics, and everything related to engineering in the training of its students. The fundamental premise was to create affinity with engineers through the acquisition of scientific knowledge and the updating of technical advances appropriate to construction (Sebastián 2016).

Through the Moyano Law of 1857, which was in force in our country until 1970, it was established that the highest education—which could be received in the Faculties, in the Superior Schools and in the Professional Schools with different access requirements—qualified for the exercise of certain professions. It was in those years that higher engineering studies were also created. The opening that had occurred with the creation of the Royal Academy of Exact, Physical and Natural Sciences in 1847 and the revitalization of higher education, led to the resurgence of the study of mathematics, which was considered then a tool of practical use. The definitive modernization of mathematics came with José de Echegaray (1832–1916), politician, engineer, mathematician, academic, and Nobel Prize winner for literature, who, along with other authors, raised the study of this science to the level of the most advanced countries. The work of mathematicians was that of wise importers of science.

On the other hand, the eighteenth and nineteenth centuries were also times of intense development for physics in all its branches (mechanics, thermodynamics, optics and electromagnetism) from the principles established in previous centuries by scientists such as Newton and Galileo. This evolution was due, first, to the illustrated ideas and, later, to the spectacular development of the technique that led to the industrial revolution. Despite the successive impulses of the enlightened, it was not until the middle of the nineteenth century that the teaching and study of physics began to be effectively promoted, as well as the development of experimental laboratories and cabinets (García 2015).

2 Biographical Notes and Academic Activity

On the occasion of the death of José Ruiz-Castizo in 1929, the Spanish mathematician born in Valladolid, José Barinaga Mata (1890–1965) wrote a 12-page handwritten obituary in his honour in which he takes a tour of his life and work (Barinaga, 1929). Barinaga had been his disciple at the Faculty of Sciences of the Central University of Madrid and assistant to his chair. In this testimony and in the monographic work on the author written by Claudia Vela Urrego (Vela 2013), we will position ourselves to expose the most relevant data of the biography of our character.

José Ruiz-Castizo y Ariza was born in Fuentes de Andalucía (Seville) on December 13, 1857, into a humble family whose father worked in the fields. His independent character and his strong personality can already be seen in the years in which he attended secondary education at the Provincial Institute of Seville, which ended in 1876, during some of which he was an instructor at a school in Andalusia.

He began his university studies at the Faculty of Sciences of Seville in 1877. He completed two courses at that institution, after which he continued studies at the Central University of Madrid until obtaining a degree. In Madrid, he studied subjects such as Descriptive Geometry, Line Drawing, Mechanics and Geodesy, showing an inclination from the beginning towards Mechanics in which he obtained the qualification of Honours.

After completing his degree, he left Spain in 1881 with the intention of working as a teacher in Latin America. As we can read in the work of Oyuela dedicated to Professor Ramón Rosa (1848–1893) (Oyuela 2007), ideologist of the Liberal Reform of Honduras in 1876 framed in the positivism of Augusto Comte (1798–1857), in the reopening speech of The Central University, Dr. Rosa stated the necessary measures for the execution of the new university program. Among them, and for the management of the syllabus, three Spanish teachers were hired: Antonio Abad Ramírez Fontecha, Santiago Guerrero López and José Ruiz-Castizo y Ariza. The trip to Honduras began on October 26, 1881, passing through Amapala and finally arriving in Tegucigalpa on December 14 of that same year. Due to illness, José Ruíz-Castizo had to return to Spain a few months after his arrival, while Abad and Guerrero began working on February 4, 1882, the first as Professor of the Faculty of Medicine and Surgery and as director from the National College, and the second as Professor of Roman Law at the Faculty of Jurisprudence and Political Sciences at the Central University, and as a professor of Rhetoric and Poetics at the National College.

Back in Madrid, José Ruiz obtained a doctorate in physical and mathematical sciences in 1883 (Vela 2013).

As for his academic career, it began in Madrid with the position of Assistant in the General Preparatory School of Engineering and Architecture that he occupied from 1888 to 1892. This school, established as such since 1885 and suppressed in 1892, was

Fig. 1 Photograph of José Ruiz-Castizo y Ariza (Barinaga 1929)

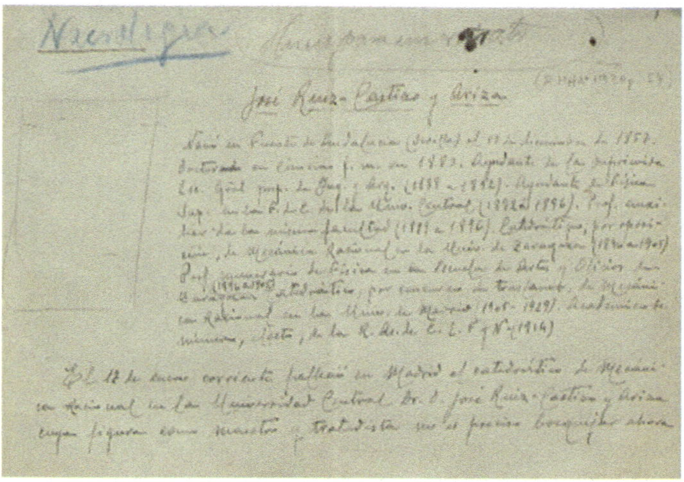

José Ruiz-Castizo y Ariza

Fig. 2 Home page of the manuscript "Necrología", by J. Barinaga (Barinaga 1929)

born, as explained above, to impart common knowledge to those who wanted to enter the schools of Civil Engineers, Mines, Forestry, Agronomists and Industrialists and the school of Architecture, which created a great controversy at that time (Montero 1886). Apparently, the work of Ruiz-Castizo as Assistant in the Preparatory School consisted of conserving the cabinet of machines and its facilities and studying the instruments; all this in addition to teaching in the subjects of Infinitesimal Calculus, Topography, Drawing, Descriptive Geometry and Physics (Vela 2013).

Then, between 1892 and 1896, he was also a Senior Physics Assistant in the Faculty of Sciences of the Central University in Madrid. In those years, from 1889 to 1896, he also held the position of Assistant Professor of the same faculty.

In 1896, he obtained the Chair by opposition of Rational Mechanics in the Faculty of Sciences of the University of Zaragoza, which he held until 1905. His chair examination was based on the following main axes: kinematics, statics and dynamics, fluid mechanics and principles of mechanics. After different study plans, teachers and various treatises used in the subject of Rational Mechanics, after obtaining the chair, our character took over this subject in 1902.

During the years of his stay in Zaragoza, José Ruiz-Castizo was also a full professor of Physics at the School of Arts and Crafts in that city (Barinaga 1929).

Starting in 1905, Ruiz-Castizo returned to Madrid when he obtained, by means of a translation competition, the Chair of Rational Mechanics at the Central University, which he held until 1929.

Among other positions, he was also a Corresponding Academician of the Seville Royal Academy of Literature, member of the Royal Spanish Society of Mathematics and of the Spanish Association for the Progress of Sciences, and Academic of number, elected, for the Royal Academy of Exact, Physical and Natural Sciences, since its appointment in 1914.

Our character died in Madrid on January 17, 1929.

3 Main Published Works and Contributions

José Ruiz-Castizo wrote several literary essays throughout his academic life that he published in the magazine *La Naturaleza,* which began in 1890, and other periodical publications of the time.

During the first teaching period in Madrid, he completed various publications and began to work on his inventions, which we will discuss later.

Below is a list of the most notable works written by our character—articles, essays, brochures, opuscules and treatises—arranged chronologically.

- *Analytical study of a fourth-order locus*: 93-page opuscule published in Madrid in 1889 when he was Assistant of Topography and Drawing at the General Preparatory School of Engineers and Architects. The text was divided into four parts: Definition, Construction and discussion, Tangent and Area, highlighting in this

Fig. 3 Book cover "Analytical study of a fourth-order locus" (Ruiz-Castizo 1889)

last part the devices used to calculate the squares. This work was registered by its author as intellectual property (Ruiz-Castizo 1889) (Fig. 3).

- *The laws of balance of the wedge.* Several articles published in the magazine La Naturaleza in 1894, which were dedicated to the analysis of this simple machine.
- *The electro-optics and the ideas of Maxwell.* Article published in the journal La Naturaleza in 1895.
- *The centre of gravity.* Article published in the magazine La Naturaleza 1895.
- *Theory of the new mechanical integrator for the three orders $\int ydx$; $\int y2dx$; $\int y3dx$, called tangential valuation cartesian integrometer*: 32-page brochure published in 1898 (Ruiz-Castizo 1898). On this instrument, he wrote six articles in the Revista de Obras Públicas (1896–1898). As we will see later, this instrument is one of his best-known inventions, reported by the Royal Academy of Sciences and its construction was subsidized by the Ministry of Development. It was acquired by various scientific offices and used by the Geographic and Statistical Institute (Fig. 4).

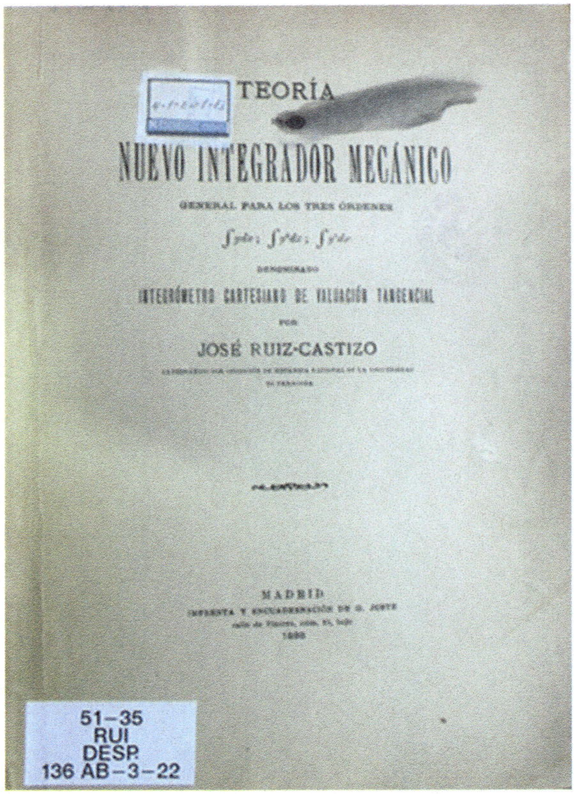

Fig.4 Book cover "Theory of the new mechanical integrator ..." (Ruiz-Castizo 1898)

- *On the hypotheses that serve as the foundation of rational mechanics.* At the solemn opening of the 1903 to 1904 academic year at the University of Zaragoza, he read the opening speech written by him, which was 98 pages long (Ruiz-Castizo 1903) (Fig. 5).

  Ruiz-Castizo defined Mechanics in this Discourse as

  > Mixed science of observation and mathematical analysis, today and always subject to revision as far as its foundations are concerned.

In this text, he also presented his opinions about the concepts of space, time and movement, absolute and relative, explaining that

> I must not limit my speech to a mere exposition of alien ways of thinking, or to a negative criticism, passed out and sterile. I will be forced, I say, to establish affirmations, oppose arguments and even try the refutation of some doctrine that is not in accordance with my scientific beliefs, even within the narrow limits that a work of this nature requires ... (Ruiz-Castizo 1903)

He also explained that

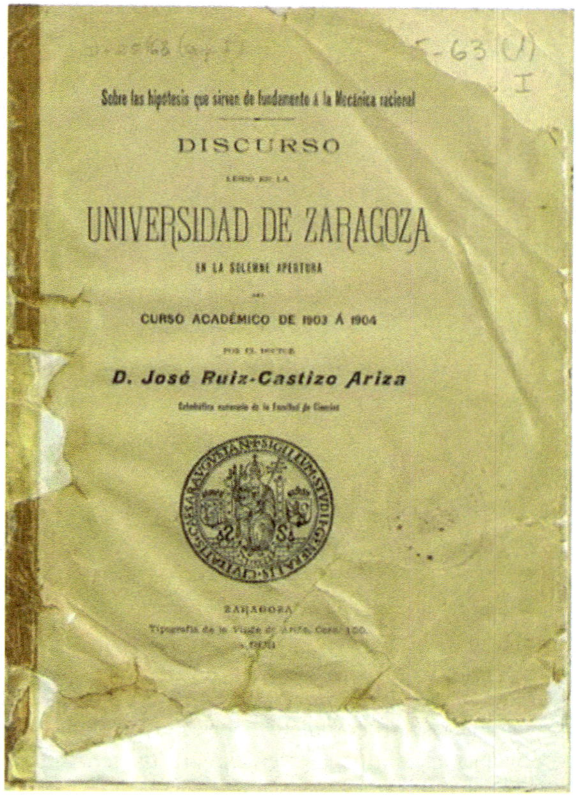

Fig. 5 Book cover: "On the hypotheses that serve as the foundation of rational mechanics" (Ruiz-Castizo 1903)

> The teachings of experience were necessary: it was necessary for the genius of the Galileo and the Kepler and the Newton to apply his powerful sagacity to the contemplation and analysis of the phenomena that reality offers, to see the emergence of mechanical science with a positive and solid base (Ruiz-Castizo 1903).

Mechanics was then going through a moment in which the basic hypotheses were discussed. Faced with the diffuse theories of Ernst Mach (1838–1916), the concrete theories of Henri Poincaré (1854–1912) were in vogue, precursors of the great events that would occur immediately afterwards and that produced a profound transformation in the foundations of Rational Mechanics:

> Doubting everything and believing everything, are two identically comfortable solutions, which equally dispense us from reflecting (Ruiz-Castizo 1903).[1]

- *The total eclipse of the Sun. Description of the phenomenon and summary exposition of its causes and circumstances of greatest interest...*, from 1905 in collaboration with D. Gabriel Galán (Fig. 6).

[1] The author is citing Henry Poincaré.

Fig. 6 "The total eclipse of the Sun": Book cover and First Chart, drawn by J.R. Castizo (Ruiz-Castizo and Galán 1905)

- *Integrating wattmeters*: He devised three different models of integrating wattmeters. His complete theory and the plans of the three models were published in an article entitled *Integrating wattmeters* in the Journal of the Royal Academy of Exact, Physical and Natural Sciences in 1909.
- *The fundamental principles of Rational Mechanics.* A first chapter of Dynamics of 51 pages published in 1909 in the Journal of the Royal Academy of Exact, Physical and Natural Sciences.
- *Demonstration of the Gaussian principle.* Published in the Hispanic/American Mathematical Magazine, 1922.
- *Compendium of the cubic plane curves*: it was another notable work that he published in the Revista matemática Hispana/Americana, 1925.
- *A theorem on the Theory of forms.*
- *Oblique Cartesian Axes Applications to Analytical Mechanics.*
- *Original experiment on properties of cloudy liquids,* published by Victoriano García de la Cruz.

However, among the works written by the author, his *Treatise on Rational Mechanics, which is appropriate for teaching in the Faculties of Sciences and in Special Schools,* deserves special mention (Ruiz-Castizo 1907). The work was prepared during his stay as a professor in Zaragoza, but the publication of the first three parts of this treatise in two volumes (Fig. 7) occurred when he arrived in Madrid to occupy the chair at the Central University. The first volume published in 1907 included Part One: *General Theory of Vector Systems* and Part Two: *Kinematics*. The third part, *Statics*, was published in a second volume in Madrid in 1910. The treaty was left unfinished because the continuation of this work, dedicated to dynamics, only published *The fundamental principles of Rational Mechanics* and *Demonstration of the principle of Gauss*, mentioned above, leaving the rest, as Barinaga explains, in "*sheets that the author kept careful*" (Barinaga 1929).

In the prologue of the Treaty, we can read the definition of Mechanics made by the author (Ruiz-Castizo 1907):

> Mechanics is a physical-natural science that tries to establish the laws that regulate the succession of positions of natural beings; affirm and clarify how each one passes from one position to another, in how long and for what causes, or what changes will necessarily occur due to causes fixed in advance... Mechanics proposes to deduce the laws of the movements of physical beings and the dependencies between these movements and their causes... Mechanics would be the entire Science of the Universe... That Mechanics is a science of quantity, a mathematical science, is of course evident.

He also carried out numerous applied mechanics tests devising various integrating wattmeters. All the graphic and written information on the wattmeters is collected in the journal of the Royal Academy of Exact Physical and Natural Sciences (1909). Another of Ruiz Castizo's inventions was a rotary distributor for steam engines, of which he published nothing.

Other published works were those referring to his most relevant invention, the tangential planimeter. In these publications, he described how to calculate the area limited by a curve and clarified that the use of his planimeter was only conceived with the construction of a machine. The articles published by Ruiz-Castizo on this invention, the publications made by other authors and the patents requested by the inventor will be explained in detail in the following section of this work.

4 Dissemination of His Works: Inventions and Patents

During the four years that his served as Assistant of the General Preparatory School of Engineering and Architecture of Madrid lasted, José Ruiz-Castizo was in charge of the conservation of the machine cabinet where he became familiar with the instruments used by the engineers. Among these was the planimeter, a surveying device on which he began to think about a new model that would improve its design and applications.

At that time, he also worked on other inventions, such as a system of hollow rubber with an elastic steel core applicable to the wheels of velocipedes, for which he filed a patent application in 1893. He also devised improvements in other instruments and mechanisms such as the case of a system of thermal and hydraulic rotary motors; a rotary distributor, regulator and inverter mechanism for reciprocating motion steam engines; and some integrating wattmeters. For all these instruments, he filed patent applications between 1893 and 1919 (Vela 2013).

During the period in which he was occupying the chair of Rational Mechanics in Zaragoza, he continued working on his new version of the planimeter, of which he

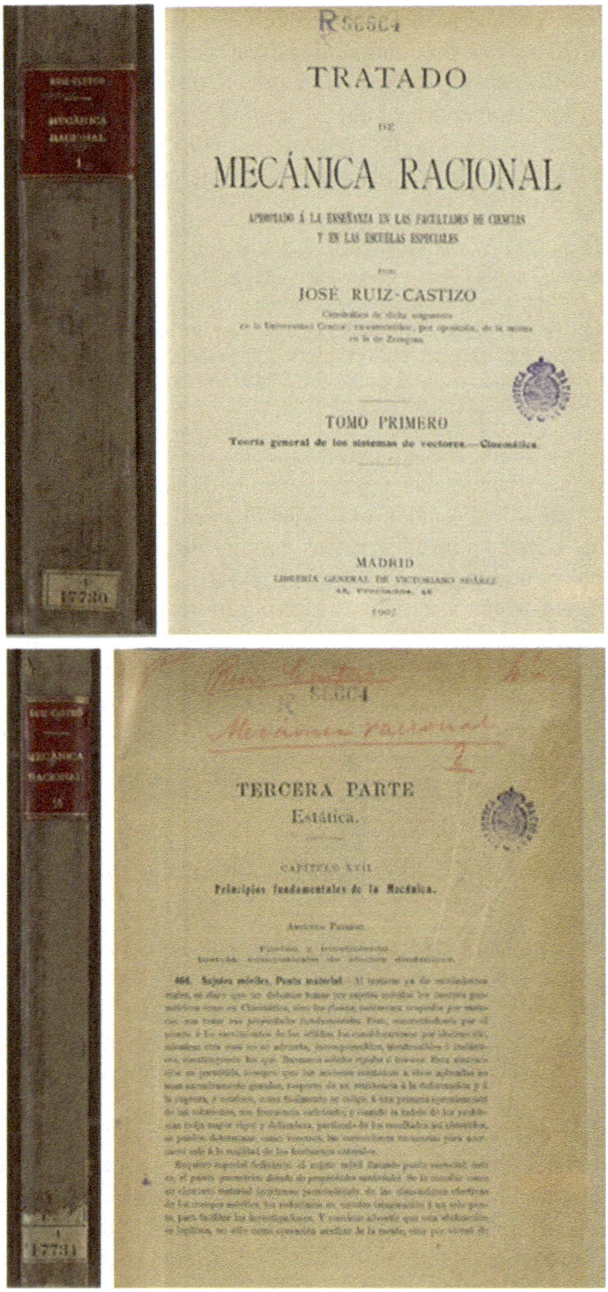

Fig. 7 "Treatise on Rational Mechanics". Volume I: Part I and II and Volume II Part III (Ruiz-Castizo 1907)

published six articles in the Revista de Obras Públicas between October 1896 and January 1898 (Ruiz-Castizo 1896), and applied for two patents of invention, the first dated in Madrid in July 1895 (Ruiz-Castizo 1895) and the second in Zaragoza on December 9, 1897 (Ruiz-Castizo 1897).

From 1878, with the promulgation of the Law of July 30 under the reign of Alphonse XII, the term "privilege of invention" was definitely replaced by "invention patent". Article 3 of this law clarifies the scope of patents (Sáiz 1995):

> ... The machines, apparatus, instruments, procedures or mechanical or chemical operations that in whole or in part are of their own invention and new or that without these conditions are not established or practised in the same way in the Spanish domains. New industrial products or results, obtained by new or known means, provided that their exploitation establishes a branch of industry in the country.

Comparing the dates on which Ruiz-Castizo applied for his patents and on which he published the advances of his invention, it seems that the way of proceeding in the world of invention, at the end of the nineteenth century, was similar to the one that currently governs based on the condition of *novelty* as a requirement of patentability of the current Spanish legislation on patents: first patent and then publish.

When referring to the planimeter, in general, it is obligatory to mention the uses of this instrument as a surveyor's apparatus. Geodesy, topography and surveying are sciences that teach to delimit the relative position of points on the Earth's surface and to represent them on a plane. The precision of the measurements and the extension of the surface to be represented is what distinguishes one from the other. Thus, geodesy considers the surface of the earth as a whole as it truly is, taking into account its sphericity. The topography represents more limited surfaces, not considering the sphericity of the earth, and, finally, the surveying descends to particular details that fix property limits, crop separation, and measurement of areas. Surveying can therefore be considered a part of practical geometry, which deals with the measurement of land. Formerly, land surveying was considered a branch of topography for the delimitation of surfaces, areas and rectification of limits and was based on geometric principles, engineering, trigonometry, mathematics and physics. Surveying instruments have always been exact and precise and were used to obtain information necessary for engineering (Castella 1939).

Among the historically existing methods to measure agricultural areas are measurements obtained "in situ", in areas of little extension, and measurements derived from mechanical procedures on the representation in a plane. In the latter case, for the measurement by mechanical means of areas of irregular plane figures, one of the most commonly used instruments has been the planimeter.

> For the mechanical measurement of flat surfaces nothing more ingenious and perfect could be devised than the planimeter. It is truly astonishing to see how such a simple instrument, which anyone can handle, solves quickly and without any mental effort on the part of the operator, a problem as complicated as the determination of an area limited by any contour (Navarro 1896).

Apparently, until 1855, there was no news about the invention of the planimeter. It was Carl Maximilian von Bauernfeind who, in a note in Dingler's Polytechnisches

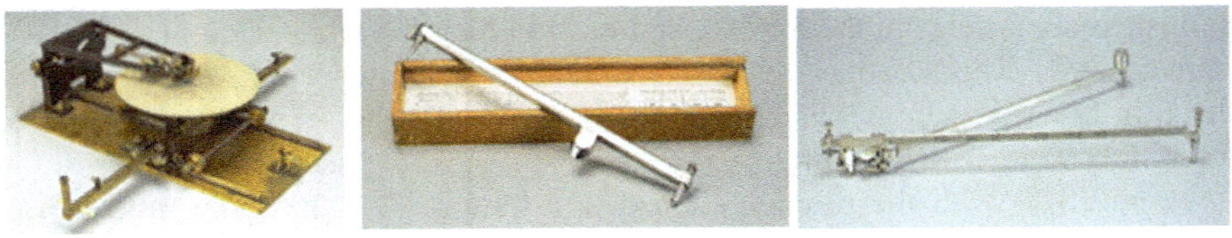

Fig. 8 Cartesian planimeter, Starke and Kemmerer, 1880. Planimeter, Prytz, 1886. Polar planimeter, Amsler, 1900. National Geographic Institute (Catálogo de instrumentos del Instituto Geográfico Nacional 2011)

Journal, first named Johann Martin Hermann, an employee of the cadastre in Bavaria, as its inventor in 1814. However, it is also true that a professor at the Academy of Fine Arts of Florence, Tito Gonella, showed a planimeter that he had devised to academics in 1824, and in 1825 to Leopold II of Tuscany, who incorporated it into his Cabinet of Mechanics. Gonella's instrument underwent modifications by the Swiss engineer Oppikofer and by Ernst, Wetli and Hansen, and with their names, it was introduced to the market years later. Other well-known planimeters were the polar planimeter from Amsler, a Swiss manufacturer of scientific instruments; the Lang compensation planimeter; the Coradi disk precision planimeter; the Starke Cartesian; the Danish captain Prytz, and the Bushnell planimeter (Pasini, 1940) (Fig. 8).

In his text entitled *Notes on the use of planimeters*, of 1896, the Spanish Agronomist Leandro Navarro (Navarro 1896) (Fig. 9) explained the following:

> There are two systems of planimeters that we have to deal with; the so-called Wetli and Starke and the one known by the name of J. Amsler-Laffon and more commonly called the Amsler planimeter. We intend to say nothing about the curious theory of these marvellous instruments which, with that of integrometers and integraphs, is the most surprising and original that human ingenuity has been able to produce.

Another of the publications of the time in which the characteristics and operation of various planimeters and their effectiveness as surveying instruments were explained was the work of the Forest Engineer Eugenio Guallart Elías of 1898, *Monograph of the counter planimeter and mainly of the models of Amsler and its derivatives* (Guallart 1898) (Fig. 10). According to the author, the Amsler planimeter is:

> one of the most useful and beautiful instruments of applied mechanics and the most precious auxiliary of the engineer, and thus it is explained that in numerous publications it has repeatedly been the subject of works of different importance, developing the theory of its operation.

The basic operation of the planimeter is based on the sliding of a punch along the contour of a figure, being able to directly obtain the area enclosed by the same from the difference of readings in a drum.

When dealing with other planimeters, Guallart refers in his text to the Cartesian Planimeter of tangential valuation, invented by José Ruiz-Castizo, making a brief description of the device and explaining very basically the improvements incorporated by its inventor with respect to the existing models.

Fig. 9 Book cover. *Notes on the use...* (Navarro 1896)

4.1 *The Tangential Planimeter of José Ruiz-Castizo Y Ariza*

As previously mentioned, it was in the Revista de Obras Públicas between 1896 and 1898, where our author published in six instalments the analytical foundation and the exhaustive description of his best-known invention, the mechanical integrator for the three orders. He also provided a detailed description of this instrument in the brochure published in 1898 (Ruiz-Castizo 1898) (Fig. 4). The instrument was informed by the Royal Academy of Sciences and subsidized by the Ministry of Development in 1894 for the construction of the prototype, which was carried out in the Talleres Bastos y Laguna of Zaragoza. The definitive version of the planimeter was made in a Swiss house, and today, it is kept in the National Geographic Institute of Madrid. It was acquired by various scientific offices and was widely used in the Geographic and Statistical Institute.

According to its inventor, the Cartesian tangential valuation integrometer has three uses: that of a simple planimeter, that of an integrator of static moments and that of a third-order integrator. All of them have been extensively explained and justified mathematically and mechanically (Ruiz-Castizo 1898).

Fig. 10 Book cover. *Monograph of the...* (Guallart 1898)

In the archives of the Historical Library of the School of Mining Engineers of Madrid, is currently available a very interesting book entitled *Pantograph-Planimeter,* also by Eugenio Guallart professor at the Special School of Forestry Engineers, published in 1895 (Guallart 1895) (Fig. 11). When this specimen was examined in the aforementioned library, it was observed that inside it was stapled an independent document entitled *Tangential Planimeter* (Planímetro Tangencial... 1897) (Fig. 12). The author of this booklet and the date of publication are not specified, although the text refers to the year 1897. The first page of this document shows an engraving showing the planimeter as it was built later. A broad description of this invention is given in its pages. Along with the engraving, we can read:

> A device of great precision, easy and comfortable to use, and of remarkable accuracy and security in measurements. Its originality is that the integration is carried out by drawing a special curve, the integratrix, whose arc is proportional to the area sought. The fundamental movement is a simple rolling, without any sliding, so that the plane of the integrating wheel is at each moment tangent to the said curve, which is described at the same time as valued by that in said movement, and hence the name tangential with which the apparatus has been proposed by its author" (Planímetro Tangencial... 1897).

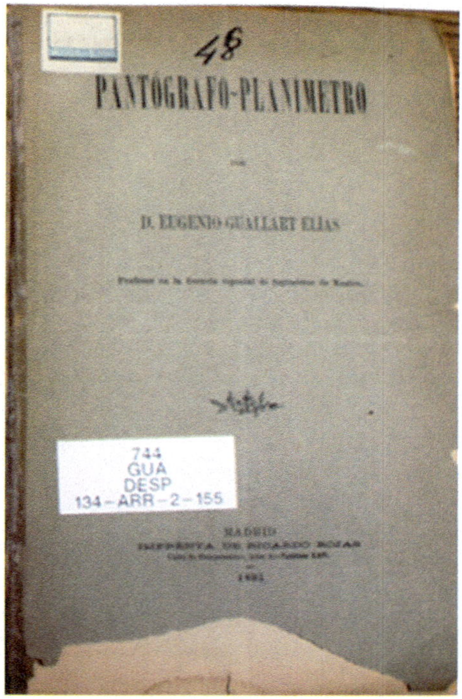

Fig. 11 Book cover. *Pantograph...* (Guallart 1895)

Additionally, Guallart's book (Guallart 1895) also contains inside a manuscript on Wetli's planimeter with sketches and mathematical explanations (Fig. 13).

Both the book and the two documents contained therein were loaned to the library in 1916 by Eusebio del Busto, Mining Engineer and professor of Topography at the Madrid School of Mining Engineers, as can be read in an annotation signed by hand and hand letter in the book itself.

As previously noted, the planimeter is an instrument used to measure areas of irregular shape represented on a plane surface, which by means of geometry or trigonometry would not be possible to obtain. Both the analytical principle of the invention and its kinematic combination, the advantages it entails, and the detailed description of the built prototype are extensively explained by the author in the aforementioned reference works and exhaustively in the six articles published by him in the Revista de Obras Públicas (Ruiz-Castizo 1896).

In the first of these articles, dated October 1896, Ruiz-Castizo explains that the analytical principle is different from that of other planimeters, reducing the problem of squaring a given area to that of rectifying a new perfectly defined curve. The inventor's intention was to accurately assess a flat area of curvilinear contour by means of mechanical integrators. It is a mechanical method of obtaining surfaces.

Analytical principle: Below is a synthesis of the author's explanations (Fig. 14).

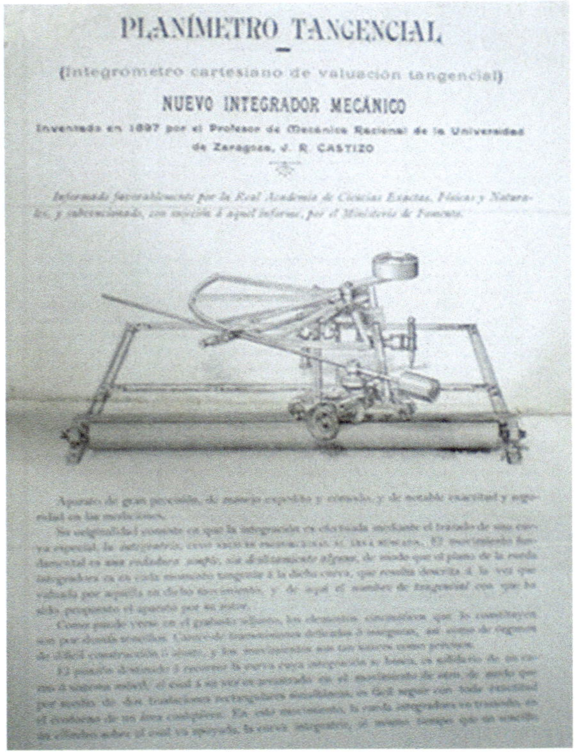

Fig. 12 Home page. *Tangential Planimeter...* (Planímetro Tangencial 1897)

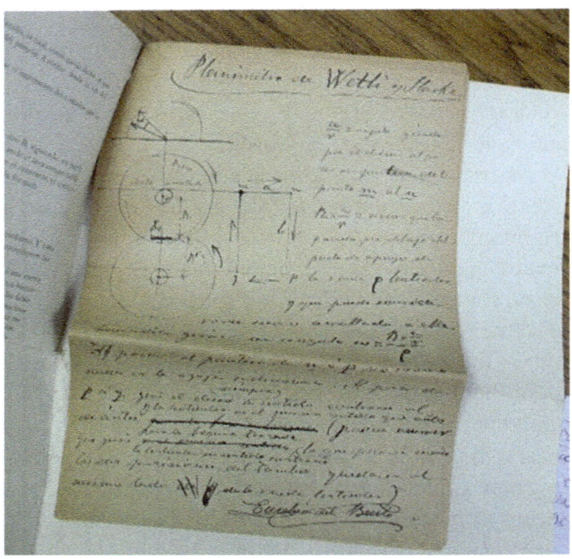

Fig. 13 Interior manuscript of Guallart's book (Guallart 1895)

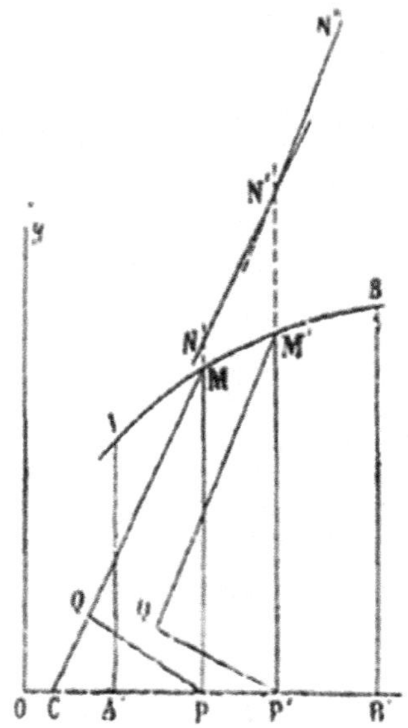

Fig. 14 1st paper Revista de Obras Públicas (Ruiz-Castizo 1896)

- An AB curve is considered from which it is intended to calculate the area between points A and B and their corresponding ordinates A 'and B'.
- The ordinate MP of any point M on the curve becomes the hypotenuse of a right triangle MQP at Q.
- The point M goes along the curve AB and dragging with it the triangle described MQP so that it always maintains its condition of rectangle at Q with the corresponding ordinate hypotenuse and always keeps the leg QP = **a** constant, of arbitrary length, and that will be called *a constant of integration.*
- The second leg MQ, apart from changing position and magnitude, will vary in direction continuously as a guideline of the tangent to a new curve called *quadratrix,* defined by the following conditions:
- Through an arbitrary point N of the ordinate MP, a line NN 'is drawn parallel to the variable leg MQ.
- The intersection of this parallel with the new ordinate M'Q 'will be the point N' through which a parallel N'N" to M'Q 'is drawn.
- The previous steps are repeated for other points, N" and successive ones, obtaining new lines parallel to the legs M" Q", and so on repeatedly.
- The envelope of this system of lines NN', N'N" … will be the new curve sought, the *quadratrix.*

After mathematically relating the two curves considered, the author deduces that the area AA'BB 'is equal to the product of the constant "**a**" (base of integration) by the length of the arc of the new curve between the two ordinate extremes that make up that area.

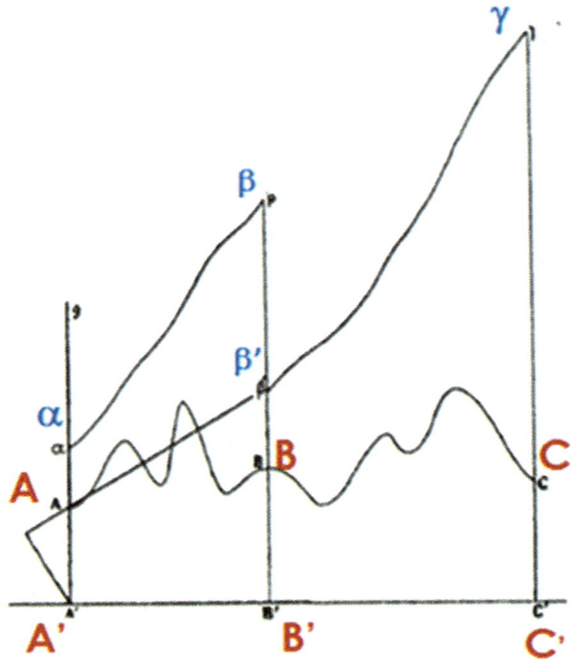

Fig. 15 2nd paper Revista de Obras Públicas (Ruiz-Castizo 1896) (Coloured letters are added)

Therefore, as we will see below in the explanation of its mechanical operation, if a punch coinciding with point M of the initial curve is applied in the planimeter, it would be enough to traverse its contour to obtain the new corresponding curve and consequently the total area.

In the second article published, in July 1897 (Ruiz-Castizo 1896) (Fig. 15), the author presents a new image in which he represents not only an initial curve ABC from which the area is to be calculated but also the *quadratrix*.αββ 'γ obtained by the above explanation.

Mechanical operation: Next, we will expose the mechanical operation of the planimeter by means of a kinematic combination that Ruiz-Castizo also presents in the articles of the Revista de Obras Públicas. Particularly, in the first one, using an exclusively mechanical means, he explains how by means of this combination it will be possible to trace and measure the new curve, the *quadratrix*, and thereby obtain the searched surface.

The components will be the following (Fig. 16):

- A fixed rule XX' and another mobile YY' on the previous perpendicular to each other.
- A third rule YDD', bent at a right angle in D and articulated in the rule YY', that will adopt positions as a guideline of the tangent to the curve that we want to draw as did the leg MQ of the triangle in the mathematical explanation.

Fig. 16 1st paper Revista de Obras Públicas (Ruiz-Castizo 1896)

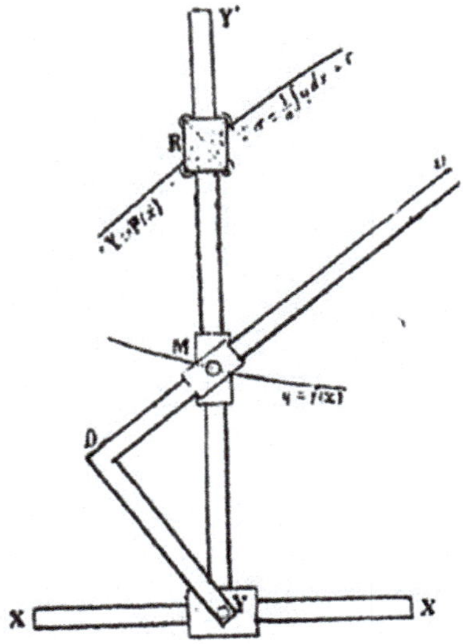

- A double M slide, whose two parts are also articulated with each other, with a housed punch that runs through the initial curve from which we must calculate the area it comprises.
- A slide R, related to M by some means established for it, with an arbitrary distance MR, that slides by YY'.
- A rotating armature positioned in R, with rotation axis perpendicular to the drawing plane and a V wheel with axis parallel to the drawing plane and plane parallel to DD', equipped with a lap counter. Wheel V will move tangentially to the curve described by its point of contact with the drawing or *quadratrix* curve.

If the wheel rests on the paper or plane of the drawing, when the punch housed in M travels the initial curve, R must participate in the translational movement parallel to XX' that the YY' rule has.

Wheel V will move rolling with a simple cycloidal movement (Fig. 17) without any slippage of its edge with the paper, and the place of its contact points with it will be the *quadratrix* curve since its tangents are parallel to the corresponding positions of the guideline.

On its way, the integrating wheel V will trace the quadratrix curve, so it will be enough to know the number of turns made during the operation to obtain the length of

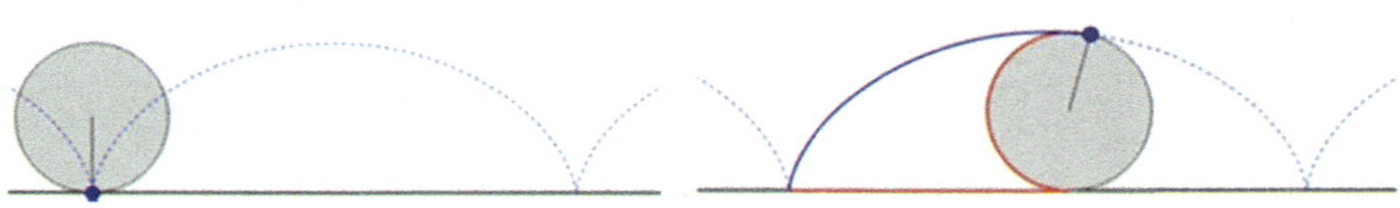

Fig. 17 Simple cycloidal movement

Fig. 18 Cabinet instrument. Author: José Ruiz-Castizo y Ariza. Planimetre tangentielle, *Société Genevoise pour la construction d'Instruments de Physique et de Mathématiques*, núm. 6. Circa 1910s (Catálogo de instrumentos del Instituto Geográfico Nacional 2011) (Heras 2011)

the quadratrix arc described, knowing the radius r of the wheel V, and consequently the searched area -or the integral that was intended to be valued- taking the DY dimension of the rule as a constant of integration.

In the following articles published in the Revista de Obras Públicas the author presents, apart from its mathematical foundation and mechanical functioning described, the advantages over other existing models of other authors, the specific applications of the instrument, and the improvements in the new versions of this one. In the fourth article, dated December 1897, Ruiz-Castizo explains his agreement with the Swiss house *Société Genevoise pour la construction d'Instruments de Physique et de Mathématiques* for the construction of the apparatus (Catálogo de instrumentos del Instituto Geográfico Nacional 2011) (Fig. 18) and exposes the modifications made to the prototype. In the fifth article, of January 1898, he exhaustively describes the final model, defined in the previous article, explaining the changes contemplated, as well as each of the pieces that compose it. The main modifications are as follows:

- Substitution of the square and its attachments by the set of a new wheel, whose radius will be the integration constant mentioned above, and a rod resting on the rim of the wheel.
- Elimination of the non-useful part of said wheel.
- A fundamental change was the placement of a free-turning boxwood cylinder to draw the envelope curve. Wheel V, due to its cutting rim, creates an imperceptible but sufficient groove on the cylinder for them to be dragged by contact. Wheel and cylinder mesh by friction.
- Location of the meter within reach of the operator without the need for an additional lens.

The fifth article also defines the effective dimensions of the instrument and its scope:

- X travel of the punch: 25 cm
- Y travel of the punch: 16 cm
- Maximum angle: 65°
- Wheel radius = 12 cm (integration constant)
- Integrating wheel radius V = 1,326 cm
- 1 full turn of V accuses: $12 \times 2\pi \times 1{,}326 = 100\ \text{cm}^2$

Hence, the thousandth of a turn, which is appreciated by the counter, corresponds to 10 mm^2.

As a curiosity, in the magazine Memorial de Ingenieros del Ejército, number 10 of the year 1898 (N. de U 1898), we can read an article dedicated to Ruiz-Castizo and his invention, signed by N. de U., in which he defines the planimeter as.

> A new device devised by him, to determine, by a simple tour of the contours of the figures, their areas, static moments and moments of inertia, which indicates the article that is of great use to engineers. After explaining its operation, in a similar way to that carried out by the inventor in his publications, he makes a final simplified explanation different of the calculation of the area enclosed by the curve.

It exposes in its pages the existing link between the development of the arc of the integrating wheel (sheave) given by the counter, and that of the area covered by the closed curve travelled by the punch (stylet). Calling this area, **A**, **s** the development of the arc of the integrating wheel (sheave) and **r** the radius of the circle (constant of integration), there is the relationship $\mathbf{A} = \mathbf{s} * \mathbf{r}$.

From this moment on, the author of the article states the following theorem:

> *The area A enclosed by the contour covered by the stylus, is that of a base rectangle equal to the development s of the sheave arc run by a point of its own and of height equal to the constant r.* He explains below: *Although the author does not state it in this way...* referring to the inventor.

From the different versions of his tangential planimeter, José Ruiz-Castizo obtained two invention patents, the complete documents of which can be consulted at the Spanish Patent and Trademark Office in Madrid.

In the first patent applied for on July 5, 1895 (Ruiz-Castizo 1895): Planimeter and Cartesian integrator for tangential evaluation. First design of the Planimeter. Royal Conservatory of Arts. File number 17695 (Fig. 19), the inventor, only described the mechanical operation of the planimeter.

The second patent is dated December 9, 1897 (Ruiz-Castizo 1897): Planimeter and tangential integrator. Second planimeter design. Royal Conservatory of the Arts file 21,880 (Fig. 20).

One year after the application for the first patent, in June 1896, Ruiz-Castizo submitted a report to the Royal Academy of Exact Physical and Natural Sciences with the intention of requesting a grant for the construction of the instrument in the Swiss house (Vela 2013). The Academy appointed the Spanish mathematician Eduardo Torroja y Caballé (1847–1918) as evaluator, who issued a favourable report to the

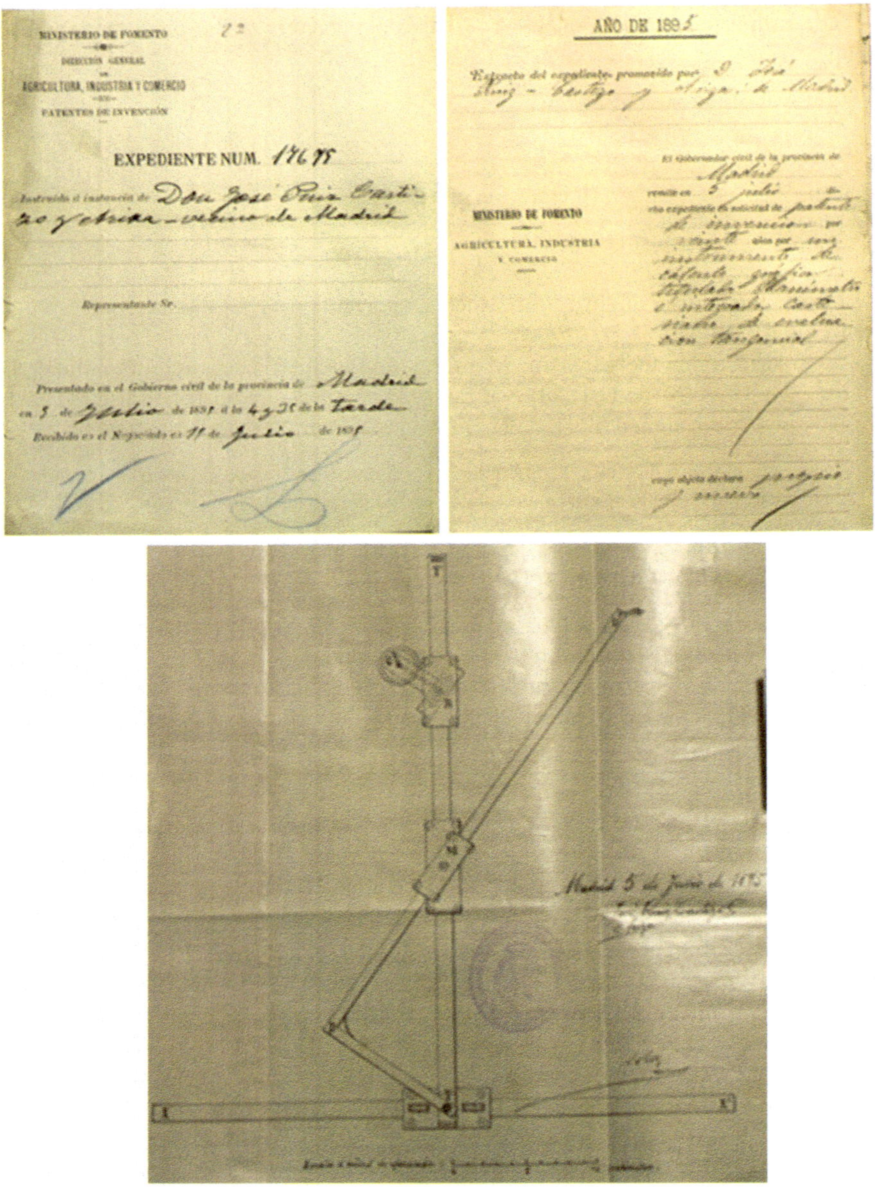

Fig. 19 First patent of the planimeter (Ruiz-Castizo 1895)

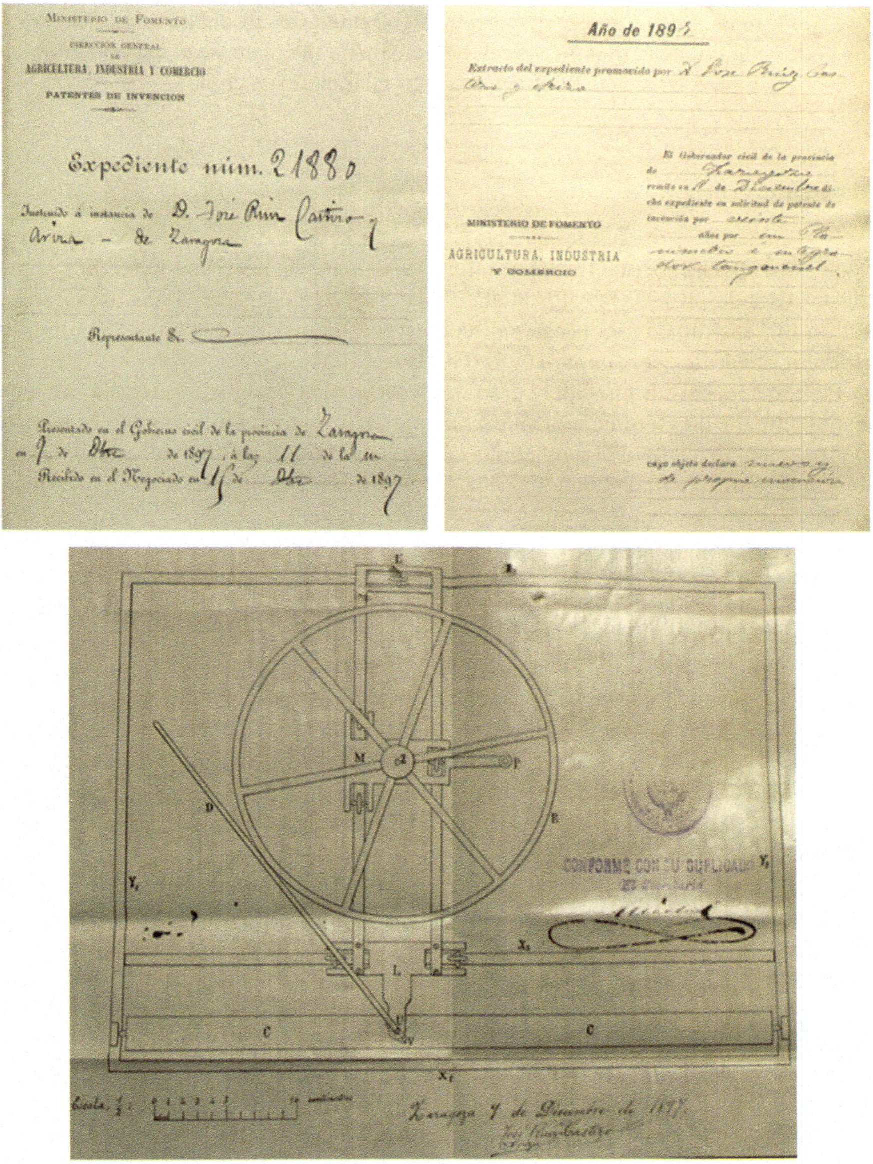

Fig. 20 Second patent of the planimeter (Ruiz-Castizo 1897)

inventor's request. In his report, Torroja highlighted, as an outstanding advantage over other planimeters, that the wheel did not slip on the plane as in the other models and that the quadratrix was left traced, which guaranteed accuracy.

5 Final Considerations

Since their invention, planimeters in their various forms have had numerous applications (Randolph 1958) (Allos et al. 1984). Subsequent research has been carried out on this instrument, its uses, and its derived and evolved versions, detailing its different scientific uses (Foote 1998) (Gradstein 1934).

Planimeters have been widely used for decades in all technical-scientific subjects that require the measurement of an irregular area, from topography to engineering or biomedicine. In the field of mathematics, it was used for the numerical calculation of solutions of differential equations because the measurement of an area is equivalent to the calculation of an integral.

Likewise, planimeters have been materialized in which the measurement mechanism was both mechanical and electronic; however, their use gradually decreased as they were displaced by computational graphic techniques applied to digitized images (McGinn 1970).

Throughout history, it is essential to highlight, here and now, the importance that analogue instruments have had, which was neglected in many cases. As has been noted, some of them, such as planimeters, first invented at the beginning of the nineteenth century, were developed essentially after the theoretical legitimation of Jakob Amsler's polar planimeter after 1856. Originally designed to measure surfaces, they materialized theoretical integral calculus and gave results even when mathematical calculations did not offer them.

These small and practical instruments quickly spread among engineers in European countries. Specifically, in England, planimeters gained importance with the particular involvement of some engineer-physicists. Thus, the initial system of a roller rotating on a cone, or a disk was integrated into more complex devices. Examples of this are Lord Kelvin's harmonic analyser (1824–1907) applied to the periodic movement of the tides, or the differential analyser of Douglas R. Hartree (1897–1958) designed several decades later following Vannevar Bush (1890–1974) (Durand-Richard 2010).

Returning to our character of knowledge, the Spanish scientist José Ruiz-Castizo y Ariza entered the field of Mechanical Engineering, which was so present in the nineteenth century, contributing his designs: a new design of rubber bands for velocipedes, a meter for electrical energy, a system of thermal and hydraulic rotary motors, a rotary distributor mechanism, or a regulator and inverter of the movement in steam engines. He applied for invention patents for all of them. In addition, he not only devised and patented his inventions but also materialized a relevant surveying instrument: the Cartesian planimeter and integrator for tangential evaluation, whose construction is preserved in the Museum of the National Geographic Institute, and which turned out to be of great use for engineering since its invention in the last decade of the nineteenth century.

Finally, who knows if the natural inclination that physicist and mathematician José Ruiz-Castizo showed towards mechanics from his beginnings as a student—evidenced in the excellent grades he obtained in this matter—together with his stay as an Assistant in the Preparatory School of Engineers and Architects—in which he was in charge of the instrument cabinet—or, perhaps, his childhood among agricultural machines and tools—living in a family whose father worked in the fields—all this together with his exceptional mind and scientific training, he could become in his fruitful facet an inventor of instruments.

References

E. García, La regeneración científica en la España del cambio de siglo, Biblioteca Virtual Miguel de Cervantes, Alicante (2015)

F.A. González, La Matemática en el panorama de la Ciencia española, 1852–1945, (En el 150 Aniversario del nacimiento de Santiago Ramón y Cajal y Leonardo Torres Quevedo) La Gaceta de la Real Sociedad Matemática Española, 5(3), 779–809, (2002)

J.A. Sebastián, Arte, ciencia e industria en la arquitectura madrileña, 1870–1936: (Hierro, acero y hormigón armado como agentes renovadores), PhD thesis, Autonomous University of Madrid, Madrid, (2016)

J. Barinaga, Necrología. Manuscript, Madrid (1929), http://simurg.bibliotecas.csic.es/viewer/image/CSICAR000137643/1/last accessed 2022/12/07

C.T. Vela, Estudio sobre el físico-matemático e inventor José Ruiz-Castizo y Ariza (1857–1929), PhD thesis, University of La Rioja. Logroño (2013)

L. Oyuela, Ramón Rosa. Plenitudes y desengaños, Ed. Guaymuras, 2nd edition, Tegucigalpa (2007)

E. Montero, La Escuela Preparatoria de Ingenieros y Arquitectos, Revista de Obras Públicas, suplemento al número del 15 de febrero de, 4ª serie, Tomo 4. Número **3**, 49–54 (1886)

J. Ruiz-Castizo, *Estudio analítico de un lugar geométrico de cuarto de orden* (Imprenta de La Guirnalda, Madrid, 1889)

J. Ruiz-Castizo, Teoría de un nuevo integrador mecánico general para los tres órdenes órdenes ∫ydx; y2dx; ∫y3dx, denominado integrómetro cartesiano de valuación tangencial. Madrid Imprenta y encuadernación de G. Juste (1898)

J. Ruiz-Castizo, *Sobre las hipótesis que sirven de fundamento a la mecánica racional: discurso leído en la Universidad de Zaragoza en la solemne apertura del curso académico de 1903 a 1904* (Discursos de apertura de curso, Universidad de Zaragoza, Tip. de la Viuda de Ariño, Zaragoza, 1903)

J. Ruiz-Castizo, G. Galán, *El Eclipse Total de Sol, de 1905* (Librería Internacional de Adrián Romo, Madrid, 1905)

J. Ruiz-Castizo, *Tratado de Mecánica Racional* (Librería General de Victoriano Suárez, Madrid, 1907)

J. Ruiz-Castizo, J.: Un nuevo planímetro. Revista de Obras Públicas, 43, tomo II (19), 374–376, (1896) 44, tomo II (1149), 362–365 / (1153): 471–473 / (1161): 669–673, (1897) 45, tomo I (1166), 55–59 / (1170), 130–136, (1898)

J. Ruiz-Castizo, Planímetro é integrador cartesiano de evaluación tangencial. Solicitud de patente de invención, Expediente No. 17695, Madrid (1895)

J. Ruiz-Castizo, Planímetro é integrador tangencial. Solicitud de patente de invención, Expediente No. 21880. Zaragoza (1897)

P. Sáiz, Propiedad Industrial y Revolución Liberal. Historia del sistema español de patentes (1759–1929). Spanish Patent and Trademark Office, Madrid, (1995)
L.G. Castella, *Lecciones de Topografía* (Bosch Casa Editorial, Barcelona, 1939)
L. Navarro, Notas relativas al empleo de los planímetros. Celestino Apaolaza, Ed., Madrid, (1896)
Pasini, C.: Topografía. Ed. Gustavo Gili, 2nd edition revised according to italian 5th edition, Barcelona, (1940)
Catálogo de instrumentos del Instituto Geográfico Nacional, Madrid (2011) https://www.ign.es/web/ign/portal/ic-coleccion-instrumentos/-/coleccion-instrumentos/CTCinstrumentos, last accessed 2022/12/15
E. Guallart, *Monografía del planímetro de contador y principalmente de los modelos de Amsler y sus derivados* (Est. Tip. de la viuda e hijos de Tello, Madrid, 1898)
E. Guallart, *Pantógrafo-Planímetro* (Imprenta de Ricardo Rojas, Madrid, 1895)
Planímetro Tangencial (Integrómetro cartesiano de valuación tangencial) Nuevo Integrador Mecánico, inventado en 1897 por el Profesor de Mecánica Racional de La universidad de Zaragoza J. R. Castizo, (1897)
Heras, A.E de las.: Instrumentos topográficos de la E.T.S. de Minas de Madrid. School of Mining Engineering of Madrid, Madrid, (2011). http://oa.upm.es/32391/1/instrumentos_topograficos.pdf, last accessed 2022/12/15
N. de U. "Otro integrador mecánico". Memorial de Ingenieros de Ejército, Revista mensual, cuarta época, tomo XV, número X, (1898). http://www.bibliotecavirtualdefensa.es/BVMDefensa/i18n/catalogo_imagenes/grupo.cmd?path=28870, last accessed 2022/12/15
M.L. Randolph, Modifications of a Planimeter for the Construction of Integral Curves. Rev. Sci. Instrum. **29**, 796 (1958). https://doi.org/10.1063/1.1716344
J.E. Allos, S. Saadallah, E.A. Hasso, A Digital Planimeter for Industrial Applications, in IEEE Transactions on Industrial Electronics, IE-31 (2), pp. 152–156, (1984). https://doi.org/10.1109/TIE.1984.350060
R.L. Foote, Geometry of the Prytz planimeter. Rep. Math. Phys. (1998). https://doi.org/10.1016/S0034-4877(98)80013-X
S. Gradstein, A Simple Photo-Planimeter. Rev. Sci. Instrum. 5(9) (1934). https://doi.org/10.1063/1.1751859
J.H. McGinn, A New Type Planimeter. Rev. Sci. Instrum. 41(3) (1970), https://doi.org/10.1063/1.1684524
M.J. Durand-Richard, Planimeters and integraphs in the 19th century. Before the differential analyser. Nuncius, 25(1), 101–124 (2010). https://doi.org/10.1163/182539110X00064

Leonardo Torres Quevedo (1952–1936)

H. Rubio, J. C. Garcia-Prada, C. Castejon, and J. Meneses

Abstract This chapter presents a compilation of the works of Leonardo Torres Quevedo, as a prolific engineer and scientist. The document starts with a brief biography of his life and the rest of the manuscript is focused on describing the main contributions of his written work and his machines. It will describe his work in the development of air shuttles; his designs and patents in the field of aerostats; his theories and designs of analogue, entirely mechanical calculating machines, known as algebraic machines; his postulates on Automatics and his main proposals for automatic electromechanical machines: the electromechanical arithmometer, the telekino or the chess players.

Keywords Leonardo Torres Quevedo · Algebraic machine · Self-rigid airships · Cable car · Automatics and digital machine

1 Introduction

Maurice D'Ocagne (eminent French engineer and mathematician, President of the French Mathematical Society) in 1930, in the pages of Le Figaro, defined Leonardo Torres Quevedo as "*the most prodigious inventor of our time*" (Torres-Quevedo Polanco 1951). And the professor of the École Polytechnique was not wrong, because

H. Rubio (✉) · C. Castejon · J. Meneses
University of Carlos III of Madrid, Madrid, Spain
e-mail: hrubio@ing.uc3m.es

C. Castejon
e-mail: castejon@ing.uc3m.es

J. Meneses
e-mail: meneses@ing.uc3m.es

J. C. Garcia-Prada
National University of Distance Education Madrid, Madrid, Spain
e-mail: jcgprada@ind.uned.es

R. López-García and M. Ceccarelli (eds.), *Distinguished Figures in Mechanical Engineering in Spain and Ibero-America*, History of Mechanism and Machine Science 43, https://doi.org/10.1007/978-3-031-31075-1_14

between 1887 and 1930, Torres Quevedo registered a total of 23 patents in Spain, and in other countries, mainly in France.

The Spanish Property Register has a total of 20 patents registered in the name of Torres Quevedo, as well as 3 certificates of addendum for improvements introduced (Industrial Property Register 1988). His first patent (no. 7348) was granted on 12 October 1887 for an aerial funicular railway system with multiple cables, followed by the remote control or telekino (1903) and his revolutionary semi-rigid airship system (1906). The last invention dates from 1930. all of them can be classified in eleven sections or subjects, corresponding to funicular railways, signalling, the telekino, dirigible balloons, shorthand machines, binave, railway interlockings, typewriters, book paging, projectable pointer and didactic projector. No patent exists for the chess automaton.

His essay '*Ensayos sobre Automática*' (*Essays on Automatics*), published in the *Revista de la Real Academia de Ciencias* (Journal of the Royal Academy of Sciences) in 1914, is perhaps the most important milestone in Torres Quevedo's scientific production. As García Santesmases pointed out in García Santesmases (1980), Leonardo laid the foundations for the designs of automata by means of commutation functions, that is why he is considered a precursor of cybernetics. In the aforementioned essay, Leonardo systematises the principles already used to build two fundamental machines, the telekino and the chess player, and briefly describes a third, the electromechanical arithmometer, which was presented in Paris in 1920. In 2006, Leonardo Torres Quevedo's telekino was awarded an IEEE Milestone.

However, the main reason why Mr. Leonardo Torres Quevedo appears in this book is for his contribution to machine science in its most mechanical conception. His ingenuity and original mechanical contributions in funiculars, the mechanical devices used in airships, his original teaching mechanisms, etc., are unquestionable. But, in the authors' opinion, his fundamental contribution to mechanical engineering is his research into calculating machines, as well as the original mechanisms invented for them.

Torres Quevedo revolutionises the concept of 'kinematics'; he designs and builds analogue calculating machines, purely mechanical, conceptually brilliant and implemented with very complex mechanisms; and he was the first to consider the limit of analogue calculating machines, and the need to evolve towards digital (discrete) machine models, no longer totally mechanical but electromechanical. Torres Quevedo arrives at today's automatics by evolving from mechanics to cybernetics.

It is also worth mentioning that Torres Quevedo was also an eminent philologist: one of the intellectuals of his time who was one of the staunchest defenders of Esperanto (Barrio Unquera 1852); the main promotor of the Spanish-American Technological Dictionary (Torres-Quevedo Torres-Quevedo 2007); and a member of the Royal Spanish Academy of Language, occupying the seat previously held by Mr. Benito Pérez Galdós (García Merayo 2013).

This document is structured in the following sections: after this introduction, it continues with a brief biography of Torres Quevedo and, then, his main inventions are presented, as well as the description of the scientific manuscripts on which they are based: aerial shuttles, analogue machines (algebraic machines and integrator), digital machines (eletromechanical arithmometer, telekino and chess machines) and airships. Finally, other works and inventions by Torres Quevedo are briefly mentioned (system of symbols for describing machines, automatic scales, projectable pointer, co-ordinate indicator, etc.) and conclusions are drawn.

2 Biography of Leonardo Torres Quevedo

Leonardo Torres Quevedo (Fig. 1) (García Santesmases 1980; García Merayo 2013; González de Posada 1992; Rodríguez Alcalde 1974; Fuente Merás 2003; II and III Symposium 'Leonardo Torres Quevedo: su vida, su tiempo, su obra' 1987; Symposium 'Ciencia y Técnica en España de 1898 a 1945: Cabrera, Cajal y Torres Quevedo' 1898; Proceedings and lectures and papers in digital format, 1987) was born on 28 December 1852, in Santa Cruz de Iguña (municipality of Molledo, Cantabria, Spain), his parents were Luis Torres Vildósola y Urquijo and Valentina Quevedo de la Maza. The Torres Quevedo family lived most of Leonardo's childhood and youth in Bilbao (Spain), where his father worked as a railway engineer, although they used to spend long periods of time at his mother's family home in Santander (Spain).

Fig. 1 Portrait of Leonardo Torres Quevedo

Torres Quevedo completed his secondary studies at the *Instituto de Enseñanza Media* in Bilbao. In 1868, he went to Paris for two years to complete his studies. Then, he returned to Spain and settled in Madrid, where his family had moved because of his father's work.

In 1871 he entered the Official School of the Corps of Civil Engineers, finishing his degree in 1876 and graduating at number four of his year. During this period, the siege of Bilbao (1874) by the Carlist troops took place and the young Torres Quevedo enrolled as a volunteer and took part in the defence of the city.

He began his professional career working for the same company as his father, the *Compañía de Ferrocarriles del Noroeste* (Spanish Railway Operator on XIX century). He soon resigned from the Corps of Civil Engineers (*'to focus on thinking about my things'*, he says) (González de Posada 1992). He began a period of ten years devoted to intensifying his studies and researching and developing all the ideas in his mind. He started this period with a long journey through Italy, France and Switzerland, to see first-hand the scientific and technical advances of the time. Fortunately he was well off thanks to the inheritance from relatives of his father.

After returning from his trip to Europe, he settled in Valle de Iguña (Cantabria), where he married Luz Polanco y Navarro on 16th April 1885, in Pontolín (Molledo, Spain). In 1887 their first child (Gonzalo) was born. He would eventually have eight children.

In the surroundings of the '*Casa de Doña Jimena*', his first family residence in Portolín, he conceived and tested his new aerial ferry system (cableway), to later develop a longer one over the river León. In 1887 he patented the aerial ferry in Spain, France, Germany, the United Kingdom, Switzerland, the USA, etc.

In 1889 he moved to live in Madrid. The engineer combined his mathematical, physical and technical studies and research with cultural and social gatherings. In 1890 he presented his aerial ferry project in Switzerland, but it was rejected, and he was even mocked by the press.

In 1893 he presented the'*Memoir on algebraic machines*' to the Royal Academy of Exact, Physical and Natural Sciences of Madrid. During the following years he focused on the study and resolution of algebraic equations by means of mechanical principles, building several analogue calculating machines. As a reward for the work of these years, in 1901 he was appointed Director of the recently created Laboratory of Applied Mechanics (later, of Automatics). That same year he joined the Royal Academy of Exact, Physical and Natural Sciences of Madrid, becoming its president in 1928.

The following years were frenetic, Torres Quevedo stand out as a brilliant inventor:

- In 1902, he patented the Telekino, the first remote control system (using Hertzian waves). In 1906, he tested it in the '*Abra de Bilbao*' (commertial port of the Cantabrian Sea), in the presence of Alfonso XIII, King of Spain at that time.
- He was involved in the design, construction and operation of self-rigid airships: they were developed industrially, first in Spain and then by the French company ASTRA. From the researching point of view, he managed to develop many

different systems inside the airship, protected in successive patents, presented in 1902, 1906, 1911, 1914 and 1919.

- He continued to develop cableway projects: in Spain, in 1907, a funicular was built in San Sebastian on Mount Ulía, the first cableway for passenger transport in history. However, the aerial ferry that most contributed to its fame was the the "Spanish Aerocar", built in 1916 on the 'Whirlpool' of the Niagara River (Canada), which is still in operation today.
- In 1912 Leonardo built his first'Chess Player', the first ever manifestation of artificial intelligence. The 'Chess player automaton' was the first chess player system built with electrical relays where, playing with the king and a rook, he gave checkmate to a player who had only the king. A second prototype was built in 1920 and presented in Paris in 1922 by his son Gonzalo.
- In 1914 he published 'Essays on Automatics. Its definition. Theoretical extension of its applications', a fundamental work of world science where the principles stated, years later, would be known as automatics, cybernetics and artificial intelligence.
- In 1920, in Paris, he presented his 'Electromechanical Arithmometer', a fully digital calculator connected to a typewriter. According to today's philosophy, it would be the first computer in history.
- In 1930 he obtained the patent for the last of his inventions to be registered in Spain: the didactic projector.

Leonardo spent the last twenty years of his life developing and promoting his inventions. It is interesting to note that, during this time, he was one of the most active defenders of Esperanto as a universal language, although this brought him a great deal of criticism.

At the end of his life, Torres Quevedo was recognized,with numerous national and international awards, among which we highlight the following:

- In 1916, King Alfonso XIII awarded him the Echegaray Medal of the royal Academy of Sciences of Madrid. Also in the same year, he was awarded the Parville Prize by the Academy of Sciences in Paris.
- He was proposed, in 1918, as Minister of Public Works of the Government of Spain, but declined the post.
- In 1919 he was appointed President of the Spanish Society of Physics and Chemistry. That same year he was awarded the Grand Cross of Carlos III.
- In 1920 he was appointed Corresponding Academician of the Mechanics Section of the Paris Academy of Sciences. Also, in that year, he became President of the Spanish Mathematical Society. On 31 October 1920, he joined the Real Academia Española de la Lengua (Royal Spanish Academy of Language), in seat with letter N, previously occupied by Benito Pérez Galdós.
- In 1921 he was appointed President of the International Spanish-American Union of Scientific Bibliography and Technology. He is also awarded the Grand Cross of S. Tiago da Espada in Portugal.
- In 1922 he was appointed Commander of the Legion of Honour of the French Republic.

- In 1923, the University of Paris named him *Doctor Honoris Causa*. Also this year he became an Honorary Academician of the Société de Physique et D'Histoire Naturelle, Geneva (Switzerland).
- In 1924 he was appointed Honorary President of the Spanish Mathematical Society.
- In 1925 he was awarded a Doctorate *Doctor Honoris Causa* by the University of Coimbra (Portugal) and promoted to Corresponding Member of the Hispanic Society of America.
- In 1926. In Spain, he became Honorary Inspector General of the *Corps of Civil Engineers*.
- In 1927 he was appointed as a foreign associate academician of the Paris Academy of Sciences (there were only twelve associate academicians).
- In 1928 he became president of the Real Academia de Ciencias Exactas, Físicas y Naturales de Madrid (Royal Academy of Exact, Physical and Natural Sciences of Madrid). In 1934 he resigned from the presidency and was appointed Honorary President.
- In 1929 he was elected Honorary Member of the *International Committee of Weights and Measures* in Paris.
- In 1934 he was awarded the Order of the Republic Band (Spanish).
- In 1935 he was appointed Honorary President of the Spanish-American National Board of Scientific Technology and Bibliography.

Leonardo Torres Quevedo died in Madrid, in the middle of the Spanish Civil War, on 18 December 1936, ten days before his eighty-fourth birthday. The event went unnoticed.

3 Aerial Ferries (Cableways)

Cableways (García Santesmases 1980; González de Posada 1992, 1990; II and III Symposium 'Leonardo Torres Quevedo: su vida, su tiempo, su obra' 1987; Symposium 'Ciencia y Técnica en España de 1898 a 1945: Cabrera, Cajal y Torres Quevedo' 1898; Rui-Wamba and Sáenz Ridruejo 1995; los transbordadores de Torres Quevedo 2012) occupy an important place among the works developed by Torres Quevedo. He devoted his first research to these devices and the first system he patented and built was a cable car. A cableway devised and built by Torres Quevedo in 1916, the Niagara Spanish Aerocar, the famous funicular near Niagara Falls, is still in operation today.

Torres Quevedo's experimentation with ferries or funiculars began with the construction and testing of a pair of funicular models near his house in Portolín (Cantabria, Spain). The first aerial ferry was an animal-driven machine, with 200 m of span, stretched over the meadow of Los Venerales, and the second was a cable car of about 2 km, over the Iguña valley.

The first aerial ferry, built in 1887, was about 200 m long and covered a distance of 40 m where the seat or basket of the device was a bench made of trunks, with a capacity for one person, which was drived by animal power: two cows were the traction motor. This experiment was the basis for the application of his first patent: a 'multiple wire aerial funicular track system', which was registered in Spain the same year. The following year, Leonardo extended the patent to the USA, France, Italy, Great Britain, Prussia, Austria-Hungary and Switzerland.

This first patent (no. 7348) was granted on 12 October 1887 and described an aerial cableway system of multiple cables tensioned /strained by the action of counterweights, in a controllable and uniform way, so that the posible breakage of any cable would not be catastrophic, giving it a great safety value. This system would be applied to the following funiculars developed by Torres Quevedo.

The second project of aerial ferry was to cross the valley of Iguña, crossed by the river Leon, between the peak of Pando and the Picones. He built the cableway over the river Leon and it had great repercussions due to the high speed provided by a mechanical engine. It was only used to transport goods, not people.

In 1890 Mr Torres Quevedo travelled to Switzerland to present his funicular model. Switzerland had already begun to use ferries for the transport of goods and showed great interest in these devices, due to the complex morphology of the terrain, but Torres Quevedo's project was rejected, even received with irony and sarcasm by the Swiss press.

Seventeen years had to pass before Torres Quevedo, sponsored by his friends from Bilbao, returned to work on aerial ferries, executing the third project of cableway: 'the funicular of Monte Ulía' (Fig. 2), in San Sebastián (Spain), inaugurated on September 30, 1907. The route was about 280 m long, running over a watercourse. Although the difference in height between the start and the end was 24 m (start at 200 m., finish at 224 m.), the distance to the ground was greater, making the route more spectacular.

The funicular of Monte Ulía started at the top of the mountain and reached the Peña del Águila with an almost horizontal route. The departure and arrival stations were connected by means of six steel cables forming two series of three, which served as rails on which a platform slid, fitted with wheels, from which hung the basket. The basket could accommodate a maximum of eighteen passengers. An endless cable pulled the platform and the nacelle/basket, driven by a 12 HP electric motor.

Safety issues were solved by the fact that it had six rails, which gave it maximum safety even in the event of one of the cables breaking. Furthermore, the tension supported by the cables of the rails did not depend on the load being transported, since it had a system of counterweights that allowed the tension on the cables to be uniform and independent of the overload.

The work was carried out by the '*Sociedad de Estudios y Proyectos de Ingeniería de Bilbao*', built in 4 months, at a cost of just over 50,000 pesetas (Spanish coin at the time). In the first seven summer seasons it transported more than 60,000 people, '*without any accident*'.

Leonardo Torres Quevedo's aerial ferry became one of the icons of the city of San Sebastián in those years, widely reproduced in photographs and postcards of

Fig. 2 Postcard showing the Funicular of Monte Ulía arrival

the time (Fig. 2). In 1917 the funicular ceased to operate. However, the Monte Ulía funicular was the first passenger aerial ferry in the world and would open the way for other cable cars in many countries.

There is no doubt that the most famous Torres Quevedo's cablecar is the Niagara Falls funicular in Canada (Fig. 3). It crosses a distance of 580 m to go over to the whirlpools of the Niagara gorge on the Canadian side. It was built between 1914 and 1916 and was an entirely Spanish project from start to finish, designed by a Spaniard and built by a Spanish company. A bronze plaque marks this fact at the entrance to the cable car station.

The success of the Monte Ulía funicular encouraged him to install a much larger one in North America. In 1911, together with Valentín Gorbeña, Leonardo travelled to the United States and Canada. The licence was obtained/awarded by Antonio Balzola, in 1913, in the name of the company '*Sociedad de estudios y obras de ingeniería*' and, for its execution, a Spanish company was set up in 1914, '*The Niagara Spanish Aerocar Company Limited*', with a capital of 110,000 dollars, and a planned concession of 20 years. The headof the company was Gorbeña and the executive board was formed by Torres Quevedo and prominent Basque industrialists and promoters.

The project was delayed and construction took place between 1915 and 1916, under the direction of Gonzalo Torres-Quevedo, a civil engineer and son of the inventor. The tests were carried out on 15 February 1916 and the official inauguration took place on 8 August 1916. The following day it was opened to the public.

Fig. 3 Images of the inauguration and the current situation of the "Niagara Spanish Aero Car"

Initially, the ferry was proposed to have a route close to Niagara Falls but it disturbed the view of the falls. It also had a station in Canada and another in the USA, with the resulting border control problems. For these reasons, finally, the ferry was built about 4.5 km from Niagara Falls downstream at a place where the river forms a whirlpool, with its ends in Ontario, Canada, between Colt's Point and Thomson Point stations. The journey is 580 m long, with a height of 60 m, and the nacelle, built in Spain and then assembled in Canada, with a capacity for 45 passengers (24 seated and 21 standing, plus the driver).

The inauguration announcements emphasised the safety of the funicular and the fact that it was in operation during all year. Thus, the aerial ferry, with minor modifications, has operated uninterruptedly for more than 100 years without major accidents from its inauguration day to the present. Today, it is still a major tourist attraction.

The transport system consists of six carrying cables, which are supported by two towers over the Niagara River, some 550 m apart. The cables on one side are anchored in a concrete block, embedded in the rock that serves as a foundation. At the other end of the cables are counterweights, free to move vertically, which ensure uniform tension on the cables.

The transport trolley consists of a set of 12 wheels linked, six by six, by two horizontal axles, one on each side of the trolley. The two axles, with six wheels each, are anchored to a structure made up of two semi-circular profiles and a radial system of cables that coincide in a node (and axis) from which radial cables supporting the trolley basket hang.

The nacelle is a rectangular-shaped enclosure measuring 6 × 3 m, with a capacity for 24 seated passengers and a central corridor suitable for a further 21 passengers. Transversely, the two compressed arches are braced with a system of cables in Saint Andrew's cross configuration. The transversal stability of the carriage is also ensured by inclined cables that connect the four corners of the basket with the four edges of the axles.

The aerial ferry is completed with the trolley traction and retaining systems and the braking and safety devices. The access and landing platforms are integrated into the cable support towers, which also house the counterweights and motors that ensure the movement of the basket/nacelle.

The deflection of the six carrying cables, three on each side of the basket, logically depends on the span of the constructed counterweights, varying from about 14 m, without the basket, to a maximum value of 30 m, with the basket loaded with 45 passengers.

One of the end of each cable is attached to a 10 Tm counterweight, consisting of a steel cell in which cast iron bars are placed. Each cable is thus subjected to a constant tension of 10 Tm, irrespective of the actual load of the ferry, which will only influence the deflection of the cables. The constant tensioning of the cables significantly improves their behaviour. Moreover, the eventual breakage of one cable does not lead to the breakage of the other two, as they were oversized to cover this possibility.

The idea of safety is the essence of the design of the aerial ferry system, for this reason, the funicular was greatly oversized in many of its parameters, but even so, there are aspects that it does not solve: in the case of the breakage of one of the three cables located on one side of the basket, the bending stresses on the axles would intensify and the deflection of the two remaining cables would increase significantly, with the result that the carriage would tilt sideways and, with it, the basket, with a certain risk for the users.

These comments are not intended to detract from the merit of the aerocar. The safety of the device does not only depend on the reliability of the system of supporting cables, even though these are essential elements, but on its general design and the quality of the details. Torres Quevedo was undoubtedly aware of this, but also of the technological limitations at the time and of the simplistic calculation procedures available for the project dimensioning and verification. For this reason, he used experimental methods to validate the safety of the system. The structure of the carriage and the basket were subjected to prior experimental tests and, before putting the system into service, a load test was carried out by placing a weight on the basket equivalent to three times the maximum number of passengers that could occupy it.

In addition, the shuttle has the right qualities of a good mechanical design: It is simple (no place for ornamentation), all its elements are essential, easily inspectable, easily replaceable, low and easy maintenance and, as a whole, very efficient. This ensures its continuous correct operation and a long service life.

As a result of the success of the inauguration of the Spanish Aerocar, other cableway projects appeared, such as: a funicular for Havana (Cuba); and several in Spain: another funicular for Bilbao, over the Nervión; one that would connect Monte Igueldo with Monte Urgull, in San Sebastián, crossing a span of more than 1,600 m over the bay of La Concha; the Ebro ferry for Zaragoza; and even a simple proposal for a funicular for Madrid that would cross the Parque del Oeste; there was even a simple proposal for a funicular to the Alhambra, in Granada. For various reasons, none of these projects accomplished.

4 Analogic Machines

4.1 Background to Torres Quevedo's Algebraic Machines

Calculation machine can be classified, according to how their actions are regulated, as analogue machines (use continuous variables, are commanded by continuous forces) and digital machines (use discrete variables, are driven by intermittent forces) (Quevedo and y la conquista del aire. Centenario de la botadura del dirigible 'Torres Quevedo' 2007; González de Posada and González Redondo 2004).

In analogue calculation machines or analogue computers, numbers are defined by the measurements of certain physical quantities: lengths, displacements, angles, etc. In analogue calculators, mathematical (algebraic) equations are transformed into analogue physical systems and the mathematical problem is solved by means of a mechanical physical model. In other words, the analogue calculation machine is a physical analogue system that behaves according to the physical model of the mathematical equation.

The phases in the evolution of calculation machines, according to the type of construction, can be summarised in four historical periods: the geometric (or pre-physical) type; the mechanical type; the electromechanical type; and the current electronic type. Torres Quevedo made very important contributions to the research on mechanical and electromechanical calculation machines.

The first calculus systems (Puig Adam 1953; Huélamo Martínez 1995; Randell 1982a) appeared in the seventeenth century with Napier's abacus or the Pascaline, but it was really in the nineteenth century when the great evolution of these systems took place, with the invention of integrators and multipliers (mechanical devices that performed such operations). J. Rowning is considered the inventor of the first machine to solve equations efficiently (in 1770). This machine plotted algebraic curves by finding the solutions graphically. He was followed in 1822 by Charles Babbage, presenting his design of the Difference Engine and his extraordinary but failed invention in 1834, with his Analytical (discrete) Engine.

As the nineteenth century progressed, analogue machines appeared which worked by finding the balance between weights, such as Lalanne's balance (1840) and Grant's balance (1896) or with hydrostatic balances such as A. Desmanet's balance (1898) or G. Meslin's balance (1900).

In the second half of the nineteenth century, many authors studied continuous analogue devices: Wetli (1849), Amsler (1854), Maxwell (1855), Thomson (1860), Coradi (1875) and Kelvin (1876). In 1863, E. Stamm presented an invention which dealt with the subject in its generality and which would be improved by F. Carducci in 1892, but both presented profound difficulties of a mechanical nature.

The calculating instruments available at the end of the nineteenth century were arithmetical (tables), geometrical (abacuses), mechanical (systems based on statics) and kinematic (which can be considered as the 'calculation machines' of the time).

Torres Quevedo's research and designs of machines in the field of analogue machines were initially of the mechanical type. His theoretical concepts and technological inventions are based on the kinematics of mechanisms: in the machines, the relationships between certain displacements or angles rotated by the moving elements of the machines are established, these relationships being marked by mathematical formulae. Torres Quevedo called them algebraic machines, the name by which they have gone down in history.

The stage of the analogue machines of Torres Quevedo begins with his first scientific report, where his theories on algebraic machines are methodically exposed, which he presents, in 1893, to the General Directorate of Public Works of Spain, asking for help. The General manager of Public Works requested a report from the Royal Academy of Sciences of Madrid, which returned a very favourable opinion (issued on 15 January 1894 and drawn up by the academician Eduardo Saavedra). In view of the report, the General manager decided (on 22 December 1894) to grant financial support for research. This scientific report, entitled Memoria sobre las Máquinas Algébricas, de Torres Quevedo' (report aobut algebraic machines, by Torres Quevedo) was published in Bilbao in June 1895.

In 1895 he travelled to France where he presented the report 'Sur les machines algébraiques' (a more advanced version of the Spanish report) and a machine as a demonstration model, at the Academy of Sciences in Paris and at the Bordeaux Congress of the'Asociation pour l'Avancement des Sciences'.

4.2 Report of 1893

In the 1893 report (Torres-Quevedo Polanco 1951; II and III Symposium 'Leonardo Torres Quevedo: su vida, su tiempo, su obra' 1987; Quevedo and y la conquista del aire. Centenario de la botadura del dirigible 'Torres Quevedo'. 2007; Randell 1982b; Torres Quevedo 1895a, 1895b, 1895c, 1895d, 1895e, 1895f, 1895g) the engineer examines the mathematical and physical analogies on which analogue calculus is based and how to establish mechanically the relations between them, expressed in mathematical formulae. The manuscript proposes a machine capable of solving equations of up to eight terms, equations of real or complex coefficients, with real and complex solutions.

Torres Quevedo points out that the machine he proposed was not related to those that had preceded it: the machines of W. Schickard, B. Pascal, Leibnitz or Ch. Babbage. Regarding the latter, Babbage's famous differential machine, he states that the mechanical technology used by its author was not precise enaugh to be viable.

In the initial theoretical study of this report, Torres Quevedo makes important considerations on the definition of kinematics:

In undertaking this study, I believe it is necessary to highlight the close analogy that exists between a machine and an algebraic formula, rectifying the most generally accepted concept of machine, the one formulated by Ampère when defining kinematics.

Could Ampère, in defining the machine, not have dispensed with the idea of time, for the same reasons that led him to dispense with the idea of force?

It is irrefutable that he could have done without it, and if he did not do so, it was because under the name of Kinematics he confused two completely different theories: Pure Kinematics, which studies movement in itself and needs to use to the concept of time to define velocity, accelerations of various orders, etc., and the Geometrical Theory of Mechanisms, which does not need to take into account the idea of time at allon the contrary, it is necessary to eliminate it, when studying a mechanism, in order to determine the relationship between the movements under consideration'.

He even defines the concept of a mechanical machine, in accordance with the above assertions:

A machine is an instrument that links several moving parts and mechanically imposes certain relations between the simultaneous values of their displacements.

These relations will usually be formulated in one or more equations, and it will be said with complete propriety that, in constructing the machine, the equations established between the values of the displacements considered have been built. It is sufficient to bear this analogy in mind to understand the possibility of obtaining machines which perform certain algebraic calculations.

And then the fundamental question is asked about the use of mechanical machines as calculators:

Can any formula be constructed? The question has not been solved or even set out. Many inventors have designed apparatus applicable to the resolution of equations, but all have simplified the kinematic problem, resorting to an artifice, to employ a machine that fits the usual definition, that changes the direction and speed of a given movement.

And Torres Quevedo answers the question.

'Can any formula be constructed' with his 'algebraic machines'.

4.3 *Report of 1900*

Later, in 1900, Torres Quevedo presented the report "Machines à calculer" to the Paris Academy of Sciences (Torres-Quevedo Polanco 1951; II and III Symposium 'Leonardo Torres Quevedo: su vida, su tiempo, su obra' 1987; Quevedo and y la conquista del aire. Centenario de la botadura del dirigible 'Torres Quevedo'. 2007; Torres Quevedo 1895h; Torres Quevedo and Máquinas algébricas. R. O. 1339). A new, more detailed project of the machine presented in the evolved manuscript of 1895 and, at the same time, another model of machine for calculating the solutions of second degree equations.

This document is divided into two parts: the first part is theoretical and considers artefacts which, according to the memoir itself, are pure abstractions, built with infinitely rigid materials, without friction, etc.; the second part provides practical solutions and recommends the use of logarithmic scales on drums (called arithmophores) and endless mechanisms. Thus, Torres Quevedo insists on his previous arguments:

> It is shown in my report that by following this method, it is possible, in pure theory, to develop any system of analytically defined relations.

The aim of the report is to present the physiology of his "Machine for solving eight-term equations", which he describes in brief, as well as other calculation machines. Torres Quevedo describes the proposed machines as follows:

> I place at the disposal of the Academy the detailed project of my machine and several models that I have built to test some new mechanisms. My machine contains nothing more than ordinary mechanisms or mechanisms already tested by me; it is to be believed, therefore, that the results I have just announced will be obtained.
>
> Among the models I present, there is one that serves to calculate the real roots of trinomial equations. It allows these roots to be obtained very quickly and with a high enough accuracy to be usefully applied.

This manuscript was examined by a commission of three members: M. Deprez,. H. Poincaré and P. Appell, in 1900, and was published in the collection "Mémoires présentés par divers savants à l'Académie des Sciences de l'Institut National de France". The report is very positive and the opinion concludes that:

> 'In short, Mr. Torres has given a general and complete theoretical solution of the problem of the construction of algebraic and transcendental relations by means of machines; moreover, he has effectively constructed machines that are easy to use for the solution of certain types of algebraic equations that are frequently encountered in applications'. 'The Commission asks the Academy to order the insertion of Mr. Torres' report in the collection of foreign scholars'.

Furthermore, the commision state that:

> 'The new machine is intended to produce in a continuous and automatic manner the successive values through which a rational and integer polynomial passes, as the variable increases or decreases; and in this concept it could be called a polynomial generator'.

In this memory, simple mechanisms were described which, theoretically, allow the mechanical execution of: the four arithmetic operations; the construction of functions of one and several variables; and that of $y' = dy/dx$. Symbols are proposed to represent each of these mechanisms and the way of linking them together, and it establishes how, by combining them, the mechanical construction of any function or system of functions, however complicated they may be, can be represented symbolically.

With this work, Torres Quevedo was the first to approach and solve the problem, in a complete way, of constructing mechanically, by kinematic procedures, any algebraic or transcendental relation, including imaginary and differential functions, establishing general principles that allow, theoretically, to achieve it. They also apply to different cases or machines.

Torres Quevedo considers that the conditions that an algebraic machine must fullfil to overcome the practical difficulties are three:

- Use logarithmic arithmophores to obtain better precision between large outliers and homogeneous relative errors.
- No contact transmissions and always use geometrical links to avoid slippage errors.
- Have only endless mechanisms so as not to limit the range of variable values.

Afterwards, Torres Quevedo published in the Revue de Mécanique the work "Sur la construction des Machines Algébriques" where he explains in detail the "Calculator that builds the function that calculates the roots of the polynomial of eight terms", where he shows diagrams of the composition of the machine, the project of the logarithmic arithmophores, the exponential trains, the endless spindles and the general structure. This machine was not built at that time, but began to be built in 1910 and was completed in 1920. He concludes the report by describing his demonstration prototypes that mechanically implement how to permute roots.

Torres Quevedo, with these ideas of algebraic machines, built a machine to solve algebraic equations of eight terms, obtaining their roots (even complex ones) with thousandths precision; a machine to solve second-degree equations with complex coefficients and an integrator.

4.4 *Machine for Calculating the Roots of an Eight-Term Polynomial*

Torres Quevedo's calculation machine (see Figs. 4a, b, and 5a and b) was a theoretical breakthrough and a practical genius (Torres-Quevedo Polanco 1951; Torres-Quevedo Torres-Quevedo 2007; García Merayo 2013; II and III Symposium 'Leonardo Torres Quevedo: su vida, su tiempo, su obra' 1987; Quevedo and y la conquista del aire. Centenario de la botadura del dirigible 'Torres Quevedo'. 2007).

(a) (b)

Fig. 4 Algébrica machine: **a** Demonstration model, 1895. **b** Commercially available model

(a)

(b)

Fig. 5 **a** External view of the algebraic machine. **b** Evolved model of a worm screw

Torres Quevedo, in his "Memoria sobre máquinas algébricas" of 1893, specified a machine for solving algebraic equations which solves in a general way an equation of eight terms, of the form (Eq. (1)):

$$A_1\,x^{n1} + A_2\,x^{n2} + \cdots + A_8\,x^{n8} = 0 \tag{1}$$

where the positive and negative terms are separated so that we arrive at an expression of the type (Eq. (2)):

$$\alpha = \frac{A_1\,x^{n1} + A_2\,x^{n2} + \cdots + A_5\,x^{n5}}{A_6\,x^{n6} + \ldots + A_8\,x^{n8}} \tag{2}$$

The values of the coefficients are inscribed in the corresponding arithmophores (8 pairs of circles at the top front, with front crank, see Fig. 5a) and tuning the variable x (the pair of circles at the right end of the machine, with the side lever, see Fig. 5a) until it is verified that $\alpha = 1$, solution of the equation. By continuing to vary the value of x, the different solutions (roots of the equation) are obtained. This machine is capable of calculating the roots of an algebraic equation, up to polynomials of degree 8, with an accuracy of 1/1000.

Related to the machine's function and constitution, Saavedra (his evaluator for the Spanish Academy of Sciences) said that the purpose of this machine was.

> *...to solve numerical equations of all degrees with continuous magnitudes. Its purpose is to produce in a continuous and automatic way the successive values through which a rational and integer polynomial passes, as the variable increases or decreases, and in this concept it could be called a polynomial generator. The new machine is composed of three essential components: the quantity generator, the monomial generator and the addition generator.*

The three generators are mechanical systems which establish the appropriate relationships between their movements in order to represent the corresponding algebraic operations.

The machine consists of an "arithmophore" (see Fig. 6b) to convert the numbers to logarithmic scale, exponential trains (see Fig. 6a) and an "endless spindle" (see Fig. 5b and 6c).

The quantity generator or arithmophore (see Fig. 6b) is an original mechanism by Torres Quevedo, composed of two wheels connected to each other, to convert the numbers to logarithmic scale: by means of a logarithmic scale on a wheel graduated from 1 to 10 (to obtain all possible mantissae of the logarithms) and another wheel counting turns (to represent the increasing or decreasing characteristics of the logarithms).

The purpose of the arithmophore is to represent numbers, using the logarithmic scale ('Arithmophore' means just that, bringing numbers to the scale). By the use of logarithms products are transformed into sums, in order to construct polynomials that are sums of monomials.

The main disk (V) of two discs that make up the arithmophore invented by Torres Quevedo, is logarithmically graduated from 1 to 10, whereas the auxiliary one (V′)

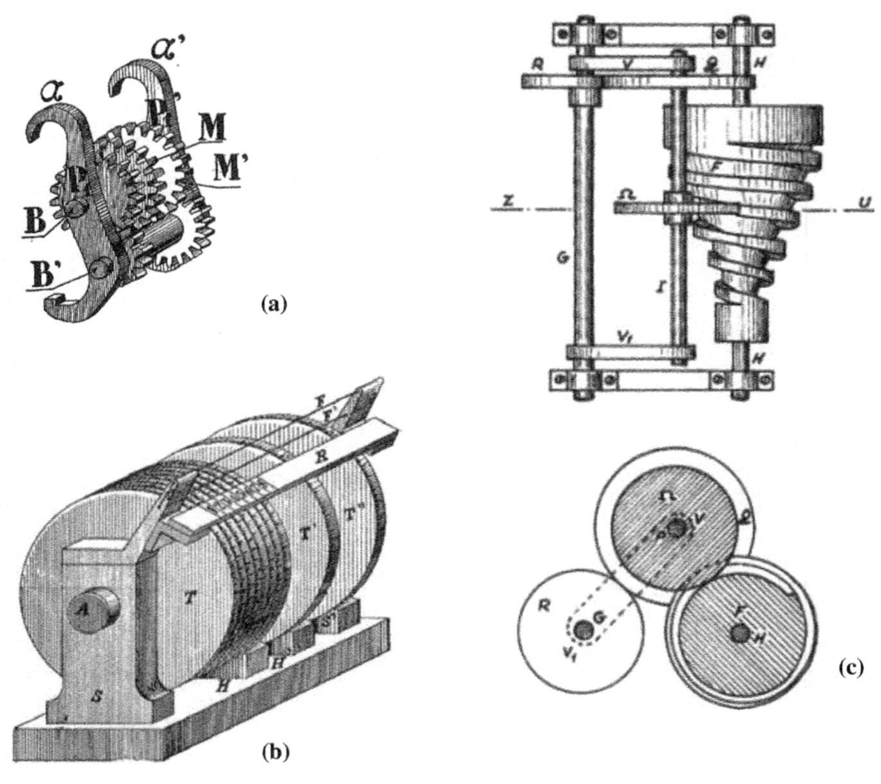

Fig. 6 Main mechanisms of algebraic machines: **a** Monomial generator; **b** Drum arithmophore; and **c** Diagram of the worm screw, with elevation and plan

is divided into equal parts and moves driven by the main one. In this way, each time the disc V makes a complete turn, disk V' advances one division. Disc V' has 16 positive and 16 negative divisions, so that the numbers between 10^{-16} and 10^{16} can be inscribed by means of both discs. In other arithmophore models, its composition was given by several groups of rotating wheels or drums T, T' and T'', as shown in Fig. 6b.

The monomial generator (see Fig. 6a) is an epicyclic train which, by using logarithms, converts the operation necessary to construct a monomial, such as the one in Eq. (3), into a sum:

$$\log(A \cdot x^n) = \log(A) + n \cdot \log(x) \tag{3}$$

This device is not absolutely original to Torres Quevedo, a similar mechanism had already been used by Stamm in 1863.

The monomial generator was a mechanical system that allowed establishing a relationship of the proportional type, $p = n\text{–}x$, or exponential $p = x^n$, between the speeds n of two gears. One such mechanism used by the inventor, in its simplest form, is shown in Fig. 6a.

The summation generator or polynomial generator is an 'endless spindle' (see Figs. 5b and 6c) and represents the core of the machine. According to its reviewers '*it is the most curious and original part of the invention*' and is a '*mechanism for which there is not the slightest precedent*'. It allows the logarithm of a sum to be obtained as a sum of logarithms, solving the complex and difficult problem of its mechanical representation.

The endless spindle is a mechanical system that allows the ratio of Eq. (4) can be calculated, expressing the logarithm of a sum as a sum of logarithms, so that if the value of x verifies the ratio of Eq. (5), combining the endless spindle with an adder gives Eq. (6).

$$y = \log(10^x + 1) \tag{4}$$

$$\frac{A}{B} = 10^x \tag{5}$$

$$y + \log(B) = \log\left(\frac{A}{B} + 1\right) + \log(B) = \log(A + B) \tag{6}$$

Each spindle consists of a series of helical teeth arranged on a quasi-conical surface ending in two gearwheels. This system is absolutely original to Torres Quevedo.

The machine in Fig. 5a has six spindles, one for each + sign, the four in the numerator and the two in the denominator. It is a fully mechanical, analogue calculating machine in which the angular displacements of its arithmophore are proportional to the logarithms of the quantities represented on it. The endless spindle expresses the logarithmic functional relationship of Eq. (4) where the logarithm of a sum is converted to a sum of logarithms. The machine can be used to solve eight-term algebraic equations with real or complex solutions and with results approximate to the nearest thousandth.

The total algebraic machine works as follows: the value is calculated according to Eq. (7), for different values of x and the various coefficients.

$$\alpha = \frac{A_1\, x^a + A_2\, x^b + A_3\, x^c + A_4\, x^d + A_5\, x^e}{A_6\, x^f + A_7\, x^g + A_8\, x^h} \tag{7}$$

If the numbers are expressed in logarithmic scale, whenever the value of the parameter α is 1, the corresponding value of x is a real root of Eq. (8):

$$A_1\, x^a + A_2\, x^b + A_3\, x^c + A_4\, x^d + A_5\, x^e - A_6\, x^f - A_7\, x^g - A_8\, x^h = 0 \tag{8}$$

The purpose of the worm screw is to establish the relationship between the displacements of two wheels r and q (Eq. (9)):

$$r = \log(10^q + 1) \tag{9}$$

If, for example, we consider a sum of two terms u and v (Eq. (10) and (11)), we try to mechanically link two arithmophores where these variables u and v are represented with another one in which $u + v$ is inscribed.

$$u = A_m \cdot x^m \tag{10}$$

$$\mathrm{v} = A_p \cdot x^p \tag{11}$$

The result would be Eq. (12):

$$\log(u + \mathrm{v}) = \log\left[\mathrm{v} \cdot \left(\frac{u}{\mathrm{v}} + 1\right)\right] = \log(\mathrm{v}) + \log\left(\frac{u}{\mathrm{v}} + 1\right) \tag{12}$$

Then the problem consists in the mechanisation of *log(u/v + 1)*. This is easy to implement, considering Eqs. (13) and (14), Eq. (12) would look like the expression of Eq. (15), which relates the variables V and V'.

$$\log\left(\frac{u}{\mathrm{v}}\right) = V \tag{13}$$

$$10^V = \frac{u}{\mathrm{v}} \tag{14}$$

$$V' = \log\left(\frac{u}{\mathrm{v}} + 1\right) = \log(10^V + 1) \tag{15}$$

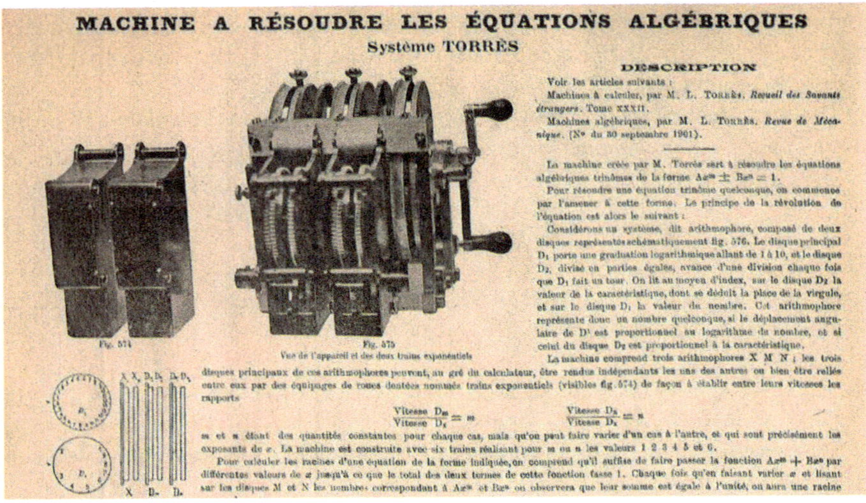

MACHINE A RÉSOUDRE LES ÉQUATIONS ALGÉBRIQUES

Système TORRÈS

DESCRIPTION

Voir les articles suivants :

Machines à calculer, par M. L. Torrès. *Recueil des Savants étrangers*. Tome XXXII.

Machines algébriques, par M. L. Torrès. *Revue de Mécanique*. (N° du 30 septembre 1901).

La machine créée par M. Torrès sert à résoudre les équations algébriques trinômes de la forme $Ax^m \pm Bx^n = 1$.

Pour résoudre une équation trinôme quelconque, on commence par l'amener à cette forme. Le principe de la révolution de l'équation est alors le suivant :

Considérons un système, dit arithmophore, composé de deux disques représentés schématiquement fig. 576. Le disque principal D_1 porte une graduation logarithmique allant de 1 à 10, et le disque D_2, divisé en parties égales, avance d'une division chaque fois que D_1 fait un tour. On lit au moyen d'index, sur le disque D_2 la valeur de la caractéristique, dont se déduit la place de la virgule, et sur le disque D_1 la valeur du nombre. Cet arithmophore représente donc un nombre quelconque, si le déplacement angulaire de D^1 est proportionnel au logarithme du nombre, et si celui du disque D_2 est proportionnel à la caractéristique.

Fig. 574

Fig. 575

Vue de l'appareil et des deux trains exponentiels

La machine comprend trois arithmophores X M N ; les trois disques principaux de ces arithmophores peuvent, au gré du calculateur, être rendus indépendants les uns des autres ou bien être reliés entre eux par des équipages de roues dentées nommés trains exponentiels (visibles fig. 574) de façon à établir entre leurs vitesses les rapports

$$\frac{\text{Vitesse } D_m}{\text{Vitesse } D_x} = m \qquad \frac{\text{Vitesse } D_n}{\text{Vitesse } D_x} = n$$

m et n étant des quantités constantes pour chaque cas, mais qu'on peut faire varier d'un cas à l'autre, et qui sont précisément les exposants de x. La machine est construite avec six trains réalisant pour m ou n les valeurs 1 2 3 4 5 et 6.

Pour calculer les racines d'une équation de la forme indiquée, on comprend qu'il suffise de faire passer la fonction $Ax^m + Bx^n$ par différentes valeurs de x jusqu'à ce que le total des deux termes de cette fonction fasse 1. Chaque fois qu'en faisant varier x et lisant sur les disques M et N les nombres correspondant à Ax^m et Bx^n on observera que leur somme est égale à l'unité, on aura une racine

Fig. 7 Algebraic machine for sale by the company "Château Père et Fils"

If you have an arithmophore that represents *u/v*, i.e., whose angular displacements are the values of *V*, and you can link it with another arithmophore that gives *V'*, you have solved the problem (Torres Quevedo and Máquinas algébricas. R. O. P., 1341). The worm screw that performs this operation has a bottle shape with an elongated neck (see Fig. 5b).

An early model of this machine was marketed in France (see Figs. 4b and 7).

4.5 *Machine that Mechanically Performs the Equation X2 – Px + q = 0, with Coefficients and Imaginary Roots*

Another of Torres Quevedo's analogue machines was designed to calculate the roots of a trinomial equation (Torres-Quevedo Polanco 1951; Torres-Quevedo Torres-Quevedo 2007; García Merayo 2013; II and III Symposium 'Leonardo Torres Quevedo: su vida, su tiempo, su obra' 1987; Quevedo and y la conquista del aire. Centenario de la botadura del dirigible 'Torres Quevedo'. 2007), that is, a second degree equation with real or complex coefficients. Figure 8 shows the inner scheme of its mechanism as well as an illustration of a real prototype of it.

It is a small machine, built before 1900, which establishes, between four points which can be moved in a plane and represent the values of the quantities p, q, x_1 and x_2. Calling x_1 and x_2 the two roots and establishing the mechanical links corresponding to Eqs. (16) and (17), the second degree equation is solved for all the values of p and q.

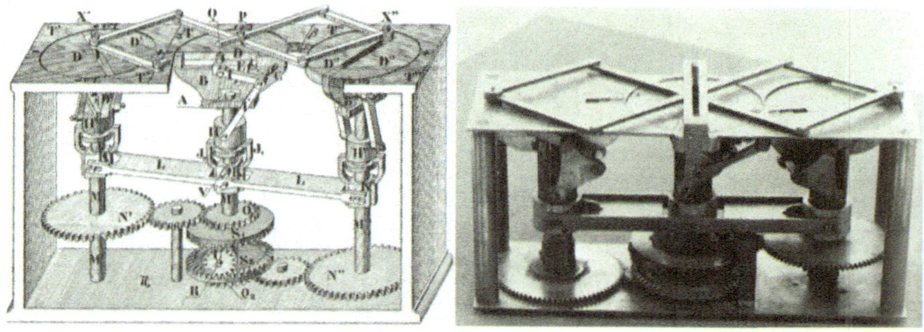

Fig. 8 Mechanical diagram and prototype of the machine that solves trinomial equations

$$x_1 + x_2 = p \tag{16}$$

$$x_1 \cdot x_2 = q \tag{17}$$

This machine (see Fig. 8) is significant because, theoretically, with the addition of other similar ones and considering that in algebraic equations of any degree the coefficients are symmetrical functions of the roots (sums, products or sums of products), it would be possible to build imaginary functions and solve equations of any degree with imaginary coefficients and roots.

In the developed model, the permutation of the two roots is done by making a coefficient describe a closed contour enclosing one of the two critical points. The device does not have an endless mechanism so the range of coefficients and variables is limited.

4.6 Machine for Integrating First Order Differential Equations

The copper sphere integrator (see Figs. 9 and 10) was presented at the Paris Academy of Sciences under the title "*Construction mécanique de la liaison exprimée par la formule y' = dy/dx*" and is a machine for integrating first order differential equations (García Santesmases 1980; Torres-Quevedo Torres-Quevedo 2007; García Merayo 2013; II and III Symposium 'Leonardo Torres Quevedo: su vida, su tiempo, su obra' 1987). There is no relation between the sphere of this apparatus and that of the Thomson integrator, the idea of this device being entirely original to Torres Quevedo.

Equation (18) is constructed mechanically on this machine and solved.

$$y' = \frac{dy}{dx} \tag{18}$$

Fig. 9 Prototype of the machine for integrating first-order differential equations

Fig. 10 Detail of the copper wheel of the machine for integrating first-order differential equations

It consists of a copper sphere (see Fig. 10) which can rotate freely around its centre. The centre of the sphere remains fixed by the action of five wheels, tangent to the sphere. Three of the wheels are placed so that their points of tangency with the sphere are in the same meridian plane; and the other two wheels have as their points of tangency with the sphere the poles of the maximum circle mentioned above. The motion of these wheels and the rotation of the sphere are related in such a way that the differential equation is solved, with the corresponding initial conditions: two of the wheels have axes that do not change position, whose rotations are the variables x and y; the third wheel can rotate, always remaining normal to the surface of the sphere, and the direction of its axis determines the value of y'.

The combination of movements of the wheels results in an action that can be broken down into two other combined movements, perpendicular to each other: one on the ordinate axis and the other on the abscissa axis (Cartesian coordinate system), represented on a board (see Fig. 9). The system is reciprocal, so if the indicator point on the board describes a curve $y = f(x)$, the coordinate axes are displaced and this movement of the coordinate axes is transmitted to the wheels and, in turn, the wheels generate the rotation of the copper sphere. To solve the problem, it is only necessary to place a gear train that forces the movements of the three mobiles (x, y, y') *to* vary according to the relation of Eq. (18).

The links that establish this condition impose no limit on the variations of the movements of the three mobiles (x, y, y') and, moreover, in this apparatus, slippage is practically negligible.

This integrator machine makes it possible to reduce the construction of a first-order differential equation between two variables to that of an equation in finite terms, establishing new links between the three mobiles under consideration that impose the relation $f(x, y, y') = 0$.

5 Digital Machines

5.1 Annual Report

Another fundamental work by Torres Quevedo is the manuscript entitled*'La Automática. Su definición. Extensión teórica de sus aplicaciones'* (Automatics. Its definition. Theoretical extension and its application), published in the magazine of the Real Academia de Ciencias de Madrid, in January 1914, and in the Revue Générale des Sciences Pures et Appliqués, on 15 November 1915, under the title *'Essais sur l'Automatique. Sa définition. Etendu théorique de ses applications'*.

In this technical report (Torres-Quevedo Polanco 1951; II and III Symposium 'Leonardo Torres Quevedo: su vida, su tiempo, su obra' 1987; González de Posada and González Redondo 2004; Randell 1982a; Randell 1982a; Thomas 2008; Rodríguez Alcalde 1966) Torres Quevedo studies the properties that automata must possess: they must react considering the conditions (orders) they are receiving and also the orders received previously, that is to say, they must have memory. It is shown that, from a theoretical point of view:

> it is always possible to build an automaton whose acts, all of them, depend on certain more or less numerous circumstances, obeying rules that can be arbitrarily imposed at the moment of construction.

He had already begun his publications on automata, with an essay on electromechanical calculation machines, in a congress in Buenos Aires, in 1910 (González de Posada and González Redondo 2005; Torres Quevedo 1857).

The report on Automatics is a fundamental and pioneering work in the history of automatic science, computer science, cybernetics and artificial intelligence. Already the author himself wrote about his electromechanical machines:

> ...these machines belong to a new chapter of machine science which might be called Automatics.

To get an idea of the transcendence of this work, let us take a look at García Santesmases' assessment of it:

> 'one of the most important aspects of Torres Quevedo's work, which alone would have made him universally renowned, is that of a precursor of present-day Automatics, whose name he introduced...', and '(in the Essays on Automatics) the foundations of Automatics were established, presenting prophetic points of view which, even today, can be considered valid'.

Very briefly, in this 1914 report:

- The theoretical foundations of Automatics are set out.
- Systems for performing arithmetic operations by digital processes are presented, introducing the idea of switching circuits by means of relays, the only possibility at that time.
- An original procedure for comparing two quantities is developed.
- A study is made of the work of Ch. Babbage and his famous analytical machine, stressing that the cause of its failure was its exclusive use of mechanical procedures: Torres Quevedo established the maximum threshold for purely mechanical calculation machines.
- As a consequence of the previous analysis, Torres Quevedo put face in the use of electromechanical systems for machine operation, highligthing that this type of system will be the future.
- The design of a simple automaton is proposed.

In fact, the aforementioned report includes the complete scheme of a machine capable of calculating automatically: a detailed description is given, as a specific invention, of an*'automaton prepared to calculate the value of the formula:* $\alpha = a\text{–}x(y - z)^2$ *without help'*, for series of values of *x*, *y*, and *z*. In such a way that*'the automaton must execute all the calculations, print the results and warn that the operation is finished'*. This machine is described in detail, but a physical prototype was never made.

Finally, it should be noted that there were mainly three devices or machines resulting from Torres Quevedo's conceptions of Automatics (prior, simultaneous and subsequent to the report on Automatics): the telekino, the chess player and the electromechanical arithmometer.

5.2 Electromechanical Arithmometer

In 1920, Torres Quevedo's electromechanical arithmometer (see Fig. 11-left and 12), which he considered to be an automaton, was presented at the Academy of Sciences of the 'Institut de France-Paris', accompanying his report 'Arithmomètre électromécanique', a text that uses the 1914 report as a reference.

The electromechanical arithmometer (II and III Symposium 'Leonardo Torres Quevedo: su vida, su tiempo, su obra' 1987; Huélamo Martínez 1995; Randell 1982a;

Fig. 11 Left, Torres Quevedo's electromechanical arithmometer from 1920. Right, an arithmometer by Thomas, from 1822

Fig. 12 An electromechanical arithmometer, currently on display in the 'Torres Quevedo Museum'

Randell 1982a; Thomas 2008; Rodríguez Alcalde 1966) consists of an electromechanical calculator machine (with relays) connected to a typewriter on which the numbers and operations (the four arithmetic operations) are typed in the order in which they are to be performed. The result is automatically printed out on the typewriter and then the typewriter is put into operator mode so that another operation can be performed.

The electromechanical arithmometer of Torres Quevedo is, fundamentally, the automation of Thomas' arithmometer (see Fig. 11-right):

> '... my apparatus is based on the same principles as that of Thomas of Colmar but differs completely in its operation. In mine all the movements are automatic and it is the way of automating them that I am going to talk to you about today...' and he continues '... they have no connection with algebraic machines', i.e. the operations are carried out with discrete magnitudes.

With this device, for the first time, a calculation machine that had an electromagnetic memory and was capable of comparing numbers (in the division operation) appeared. This electromechanical machine can be considered the first digital calculator.

Figure 11-left and 12 show photographs of one of these arithmometers with the arrangement of its mechanical parts, the typewriter and the wiring and electrical parts.

Torres Quevedo's electromechanical arithmometer had the following main features:

- Much higher calculating capacity than all previous calculation machines (as an adding machine it can operate with five-digit addends up to a total sum of 9,999,999).
- As an arithmometer, it is capable of performing the four arithmetic operations: addition, subtraction, multiplication and division, with numbers from three to five digits, depending on the operations.
- It can be remotely controlled by means of an ordinary typewriter equipped with electrical contacts on its number keys and a device for automatically writing the results of the operations transmitted by the automaton (remote control capability which allowed the centralisation of a series of automatic calculators serving different departments).
- The input of the data for the programmed operation and the result obtained were carried out on the typewriter, the only component manipulated by the user: the calculations were automatic and the arithmometer electrically commanded the typewriter to display the results. To operate, all you had to do was type the corresponding symbol between the data.
- It has an electromechanical memory by means of an inscriber where the automaton registers the numbers until the user writes the sign of the operation.
- For the different operations, it is equipped with an automatic regulator (brain) that controls the way it works, depending on the type of operation.

The first arithmometer was built in the Laboratory of Automatics, which was directed by Torres Quevedo himself, and with this device, for the first time in history, an electromechanical calculation machine was develop, incorporating the concept of remote control and the use of a memory. Another electromechanical arithmometer and another mechanical one were almost completely built, with improvements on the first one.

5.3 The Chess Automaton

The chess player (Torres-Quevedo Polanco 1951; II and III Symposium 'Leonardo Torres Quevedo: su vida, su tiempo, su obra' 1987; Rodríguez Alcalde 1966) is a practical example, pioneer of the theoretical conception of discrete machine, exposed by Torres Quevedo in the manuscript *'Automatics. Its definition. Theoretical extension and its application'* and considered a precursor of Artificial Intelligence. It is an electromechanical system that is programmed following the rules of the game of chess, where any endgame of white rook and king against the black king, which is controlled by a human opponent, is resolved in a deterministic way and could achieve checkmate from any initial position in a few moves.

Although the function of the chess player was limited to the endgames of chess games, Torres Quevedo proved that advancement in automatics was possible with the execution of automata.

Two versions of the chess automaton were built, one in 1912 (see Fig. 13) and the other around 1920 (see Fig. 14). The first version or first chessmaker was developed at the Sorbonne's Laboratoire de Mécanique and was presented at the Paris Fair in 1914. Torres Quevedo and his son Gonzalo built the second chess player and it was presented by Gonzalo in Paris, in 1922. In terms of programming, the second chessplayer did not differ much from the first, but operational improvements were introduced.

In general terms, the second chess player works considering that the movement of the white pieces is a function of the movement of the black king. The 64 squares of the chessboard (8 rows by 8 columns) are made up of three metal pieces, separated by insulating material; the central part is circular (connected to the positive terminal) and the lateral ones are triangular, connected respectively to a horizontal and a vertical electrical conductor (see Fig. 14, right). The black king has a silver mesh at its base that brings the central part of the square into contact with the triangular ones, which closes two electrical circuits that move two slides, one horizontal and the other vertical, until they reach two positions that define the position of the king on the board.

Fig. 13 The first chess automaton by Torres Quevedo. Left, together with other instruments of the Laboratory of Automation, exhibited at the Congress of the Spanish Association for the Progress of Sciences (Seville 1917). Right, detail of the first chess player, today exhibited in the 'Torres Quevedo Museum'

Fig. 14 Second chess automaton, exhibited in the 'Torres Quevedo Museum'. Left, the complete piece of furniture. Right, detail of the composition

Similarly, the positions of the white king and rook are defined by four sliders, two for the white king and two for the white rook. When the black king moves one position, the corresponding sliders move and, by means of the appropriate contacts, the circuits are closed which, in turn, act on the white pieces, placing them on the appropriate squares, in accordance with the strategy of the game. The white pieces carry hidden steel balls, which are moved by movable electromagnets located under the board and activated for each position of the black king.

When a check situation occurs, a phonograph record plays the phrase 'check on the king'. When 'checkmate' is reached, the disc also sounds, but in addition a luminous sign indicating'check' appears. Under these conditions, an electromagnet removes the voltage applied to the board so that the game cannot continue. The automaton has won.

5.4 The Telekino

The telekino (see Fig. 15 and 16) was the first radio control device invented in the world and it was the pioneer in the field of remote control (Torres-Quevedo Polanco 1951; II and III Symposium 'Leonardo Torres Quevedo: su vida, su tiempo, su obra' 1987; Rodríguez Alcalde 1966; Torres Quevedo 1860, 1653; Andrés Hernández 2006).

The name Telekino was chosen by Torres Quevedo as a combination of the Greek words: "tele" (at a distance) and "kino" (movement), so that both words together mean "movement at a distance". In Torres Quevedo's own words, the invention*'comprises in principle a telegraphic transmission with or without wires determining the position of a needle which governs a servomotor reacting any mechanism'*.

In the final paragraphs of the 1914 report, the author stated:

> 'It was the study of the telekino that led me in this new direction. The telekino is, in short, an automaton which executes the orders sent to it by means of wireless telegraphy. Moreover,

Fig. 15 The telekino at the Fronton Beti-Jai (Madrid), in 1905 (Photograph by La Ilustración Española)

Fig. 16 Telekino model exhibited in the 'Torres Quevedo Museum'

to interpret the orders and to act in the desired way at any given moment, it must take into consideration different circumstances'. In other words, it could be said that he takes his invention of Automatics from 1914 back to 1903.

Torres Quevedo began to work on the Telekino in 1901, in order to remotely test the handling of airships without risking human lives. His aim was to remotely control a ship, balloon, torpedo, etc.

On 10 December 1902, Leonardo Torres Quevedo presented a patent application in France for *'Système dit Télékine pour commander à distance un mouvement mécanique'* (Telekine system for remotely controlling a mechanical movement), registered under number 327,218. It is the world's first remote control system (in this case, using Hertzian waves).

On 10 June 1903, Torres Quevedo presented a patent application with the Spanish Industrial Property Registry under the title: *'A system called 'Telekine' for remote control of a mechanical movement'*, registered under number 31,918. The Spanish patent was issued on 19 September 1903.

On 9 December 1903, Torres Quevedo filed another application for an addendum to the Spanish patent under the title *'Improvements made to a system called Telekine for remote control of a mechanical movement'*. At the same time, he also applied in France for a 'Première addition' to his French patent. In the certificate of addition to the patent, Torres Quevedo improved the telekine by means of what he called 'delayed contact'. On 12 February 1904, the certificates of addition to both his Spanish patent and his French patent were issued.

In 1903, Torres Quevedo presented the Telekino to the Paris Academy of Sciences, giving a brief experimental demonstration (Comptes Rendus de l'Academie des Sciences, 3 August 1903).

In the 'Bulletin de la Société Internationale des Electriciens' of June 1906, a note by M. Devaux was published, in which he presented, as his own, *'a new application of the discoveries relating to Hertzian waves'*. M. Devaux had directed the manoeuvres of a torpedo at a distance in the port of Antibes. Torres Quevedo asked the Director of the Bulletin to recognise his priority in the invention from 1903 (Pérez Yuste and Salzar Palma 2004), but the Société's regulations did not allow the claim to be published in the Bulletin, so the controversy remained open. However, in 1907, the Paris Academy of Sciences recognised Torres Quevedo's scientific priority, as far as the invention of the'delayed contact' was concerned, by publishing in the Comptes Rendus the communication presented in June on 'Le Télékine et la Télémécanique'.

The Telekino has three distinct parts (see Fig. 16): a radio receiver, a multi-position rotary switch and, finally, two servo motors that can be used to move a mechanical system.

The radio signal is picked up by an antenna and converted into electrical pulses by a cohesor. Each impulse triggers an electromagnet which closes its contacts, causing the rotary switch to rotate one step forwards. This operation is repeated automatically, as many times as there are pulses in the transmitted signal. When the rotary switch reaches its end position, the battery supplies current to the selected servomotor. The servomotor is then set in motion, producing the predefined action.

Torres Quevedo discovers that in order to achieve a finite but not limited set of actions based on a binary system, such as the telegraph system (with two connection-disconnection states), it was necessary to create a fixed number of code words by means of a sequence of binary states. It is known that with two binary states, four different code words can be achieved.

The problem, at that time, was the impossibility of having a synchronisation mechanism that was able to detect the end of one symbol or character and the beginning of the next. In this situation, the only way to solve this difficulty was to use an asynchronous synchronisation method, based on the state of the telegraph signal. Torres Quevedo's final proposal was to use a code based on the number of impulses transmitted; thus, one impulse corresponds to action number 1, two impulses to action number 2, and so on.

Fig. 17 Experiment with the telekino embarked on a boat, in the Abra de Bilbao, in 1906. Left, Torres Quevedo and King Alfonso XIII with the telekino. Right, manoeuvres of the remote-controlled boat with the telekino

The first tests with the telekino, incorporated into a tricycle, were carried out in 1904 at the Beti-Jai fronton in Madrid (see Fig. 15). In September and October 1905, tests were carried out with the "Vizcaya" boat on the Nervión estuary in Bilbao. And in 1906, also with a boat, on the pond of the Casa de Campo in Madrid.

The "official test" or main test of the telekino was held in the Abra de Bilbao, in September 1906, in the presence of the King of Spain and in front of a large crowd (see Fig. 17). This test consisted of guiding the manoeuvres of a remote-controlled boat with the telekino from land. The boat had two servomotors, one acting on the propeller and the other on the rudder. The test was a resounding success.

Attempts were also made to drive the manoeuvres of the "Torres Quevedo" airship from the ground with the telekino, during tests in Guadalajara, in the summer of 1907, but these tests were not completed due to problems in the manufacture of the aerostats.

In 2006, the International Institute of Electrical and Electronics Engineers (IEEE) awarded a Milestone to Leonardo Torres Quevedo's Telekino as part of its "Milestones Programme" (Torres Quevedo 1653). The aim of this programme is to highlight the most relevant milestones in the branches of Electrical Engineering, Electronics and Computing, those advances, inventions or discoveries that marked a before and after in the History of Science and Technology.

6 Airships (Balloons)

By 1900 there were two constructive types of steerable balloons (García Santesmases 1980; González de Posada 1992; II and III Symposium 'Leonardo Torres Quevedo: su vida, su tiempo, su obra' 1987; los transbordadores de Torres Quevedo 2012; Torres Quevedo 1710):

- Flexible airships (models promoted by Alberto Santos Dumont). This type of airship consisted of a gas-filled envelope, maintained its shape by the internal pressure of the gas and was regulated by means of compensation chambers and automatic valves. They had one major advantage: they could be easily folded and transported, but had the disadvantage that the suspension of the basket tended to arch the balloon in the centre. Although they were tested with some success, they needed further study to be truly effective.
- Rigid (following the system of Count Ferdinand Von Zeppelin). They had a rigid inner structure (wood or metal) with several compartments containing gas-tight balloons, and a fabric or metal (aluminium or duraluminium) liner supported from the inside by the frame. Their main disadvantage was the impossibility of folding the balloon for easy transport, and they also had a high walking resistance (due to the large volume), were brittle and had transverse oscillations. Their great advantage was their stability.

In 1901, a semi-rigid system appeared, designed by the engineer Julliot for the Lebaudy brothers (businessmen who were aerospace enthusiasts). This system was intended to combine the advantages of the two previous systems. The airship had a rigid keel at the bottom (without an internal structure), from which the basket hung. It was made by the builder Edouard Surcouf, two years later.

Torres Quevedo was committed to semi-rigid airships, involved in the research and scientific path of airships, and in 1902 he presented a preliminary project on airships to the Academy of Sciences of Madrid and Paris: '*Perfeccionamiento de los aerostatos dirigibles*' (Improvement of dirigible aerostats) (see Fig. 18). The favourable report of his Spanish report, by Echegaray, on his studies on the instability of flexible balloons and the gradient of pressures that occurred in them, and the solutions proposed by Torres Quevedo to solve them, earned him the support of the Spanish government, and in 1905 he started the project for his first airship, in collaboration with the captain of engineers Alfredo Kindelán.

The airships of the time were elongated in shape to favour speed, with a circular cross-section and a long basket suspended from the balloon itself by means of a rigid structure. Torres Quevedo paid special attention to the stability achieved by means of an extensible beam, sewn to the fabrics, which was stiffened when the balloon was inflated.

This first Torres Quevedo airship presented important technical innovations: to make the flexible part of the balloon sufficiently rigid, a beam formed by rods that separated the different gas compartments with membranes was implanted inside the balloon. In addition, it had a series of air pockets which could be filled to increase the

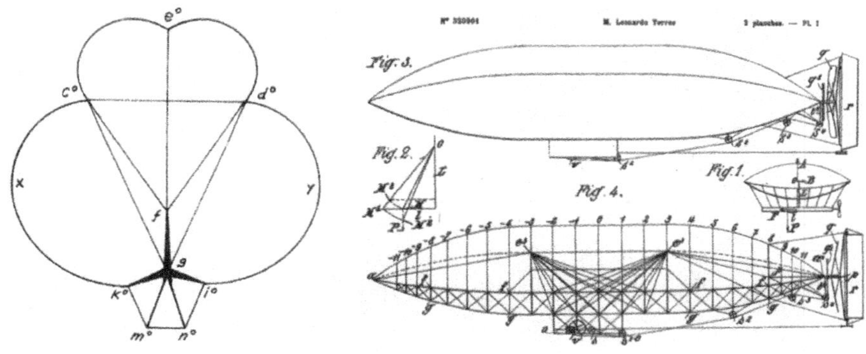

Fig. 18 Lobed cross-section and diagrams of the patent *'Improvement of dirigible aerostats'*, of 1902

weight and thus facilitate the stability of the airship. The cross-section of the airship was reshaped, giving it a trilobed shape (see Fig. 18). The nacelle frame was also placed, like the keel of a ship, alongside and inside the balloon.

On 11 July 1906 Torres Quevedo applied for a new patent for *'A new system of deformable fusiform balloons'* which has the advantages of the previous systems and eliminates the disadvantages because the airship is flexible, deflatable, transportable and rigid, due to the effect of the internal pressure, i.e. self-rigid airships.

Under the supervision of Torres Quevedo himself, the first prototype airship, measuring 630 m3, the so-called 'Torres Quevedo 1', was built in 1906 at the Military Aerostatic Park in Guadalajara (Spain). Tests with this first prototype, carried out in 1907, detected some anomalies. In 1908, these errors were corrected and a more advanced model of 960 m3, the 'Torres Quevedo 2', was manufactured, whose tests were a resounding success (see Fig. 19, left).

In October 1909, with slight modifications, the "Torres Quevedo 2" was tested in the vicinity of Paris (see Fig. 19, right) and, although an accident occurred during one of the flights, the French company *Astra* became interested in the project. As

Fig. 19 Tests of the airship 'Torres-Quevedo n° 2'. Left, in Guadalajara (1908). Right, in the vicinity of Paris (1909)

Fig. 20 Left, the "Astra-Torres No. 1" flying over the Astra hangar, 1911. Right, the world speed record of the "Astra-Torres XIV", 1913

a result of the collaboration between Torres Quevedo and *Astra*, a new airship was built in 1911, already named *Astra-Torres No. 1* (see Fig. 20, left).

In the summer of 1913 the "Astra-Torres XIV" was built, with a three-lobed section and a length of 47.7 m. This airship marked the international recognition of the Torres-Quevedo system by breaking the world speed record for an airship with 83.2 km/h (see Fig. 20-right), recorded during the acceptance tests.

In 1914 the "Astra-Torres XV" was built, also with the Torres-Quevedo system, with a three-lobed shape, two nacelles and dimensions analogous to the German Zeppelin (82 m) and which could reach a speed of 100 km/h. These airship models were acquired by the French and British governments and operated in the First World War in the French and British armies.

Mr. Leonardo presented a new patent on airships on 2 March 1914: *'Deformable fusiform balloons'*. The new system corresponded, as in the case of his first patent of 1902, to a semi-rigid airship, with a straight inner keel, triangular in section, with two metal hulls at the bow and stern.

In 1919 he registered the patent for *'A new type of balloon called Hispania'*, a new airship project, with the engineer Emilio Herrera, the *Hispania,* a large airship for transoceanic journeys with passengers. This project was never realised for funding reasons.

In addition to the four patents on airship technology, Torres Quevedo would contribute three other complementary projects:

- The *'Mooring post'* (1911) (see Fig. 21, left). The mechanical device consisted of a mooring post with a pivoting upper head, specially designed to anchor the self-rigid airships of his system in the open air. For this purpose, at the longitudinal intersections of the three lobes of the airship three cables were placed which ended coinciding at the point where the airship was tied. The airship could rotate

Fig. 21 Left, diagrams of the English patent for the 'Mooring post'. Right, the aircraft carrier 'Daedalus' with seaplanes and an airship anchored to a mooring post

around the axis of the pole by the action of the wind and was self-orienting, i.e. always presenting the least resistance to the wind. The success of the new invention, patented in 1912 in France and the United Kingdom, would be total and it would become the mooring system used for all types of airships, including the "Zeppelins" and the system that is still used today.

- The *'Rotating shed"* (1912). A shelter made of flexible elements that acquired the appropriate shape and rigidity by injecting pressurised air into it. It rotated with the simple action of the wind, facilitating the entry and exit of the airships.
- A new type of ship called a *'Camp-ship'* (1913). A new ship design conceived to accommodate one or more airships of the "Torres Quevedo" system, with all the necessary accessories to inflate them on board, raise them to the mooring post, lower them and accommodate them in the hold-hangar. It was the first aircraft carrier transport project in history. Although it was never built, the Spanish Navy began construction of the "Dédalo" in 1921, following this ship philosophy, the first Spanish aircraft carrier (airships and seaplanes) (see Fig. 21, right).

With the appearance of aeroplanes, airships stopped beingused and, after the 'Hindenburg disaster' in 1937, only the Americans continued to manufacture fleets of airships (before, during and after World War II), but the designs would not be the huge models with an internal metal structure but non-rigid airships that would retain their shape thanks to the pressure of the helium gas that fills the envelope and an interior beam formed by curtains and braces. All the non-rigid airships built from the 1920s to the present are a simplified evolution of the self-rigid system patented by Torres Quevedo in 1906. And the "mooring pole" patented in 1911 has since been the standard mooring for all airships.

7 Other Works and Inventions

Spanish-American Technological Dictionary. (Torres-Quevedo Torres-Quevedo 2007) At the American International Scientific Congress of 1910, in Buenos Aires (Argentina), Torres Quevedo proposed, together with the Argentinean engineer Santiago Barabino, the constitution of a Spanish-American board of scientific technology for the edition of this dictionary, which would eventually become the *'Unión Internacional Hispanoamericana de Bibliografía y Terminología Científicas'*, initially chaired by Torres Quevedo himself, who would address this issue. As a result of the work of this board, in 1930 what was to be Volume I of the *'Diccionario Tecnológico Hispanoamericano'* (Hispanic American Technological Dictionary) appeared. In 1920, he dedicated his speech for admission to the Real Academia Española de la Lengua to this subject (González Redondo 2013).

System of notations and symbols for the description of machinery. (Rodríguez Alcalde 1966; Torres Quevedo 1920) Aimed at facilitating the description of machines. He published this work in the Revista de la Real Academia de Ciencias (April 1906) under the title *'Sobre un sistema de notaciones y símbolos destinados a facilitar la descripción de las máquinas'* (On a system of notations and symbols intended to facilitate the description of machines).

It is an attempt at systematisation and generalisation in the description of mechanisms and their connection in machines. Torres Quevedo developed a new symbolic language (see Fig. 22) for the description of machines with the proposal of some rules for its grammar and some symbols from his dictionary. He builds a table of symbols, a collection of rules and, as usual in his work, applies them to an example.

The dianemologo, (II and III Symposium 'Leonardo Torres Quevedo: su vida, su tiempo, su obra' 1987) a process patented in 1907 for*'copying a speech as it is delivered without the need for shorthand…';'he object of the apparatus is to divide the speech among several copyists, as it is pronounced, so that each of them makes copies of only a small part of it'*.

Automatic weighing scales. (Torres-Quevedo Polanco 1951) He designed and built a weighing scales that was capable of weighing automatically, after some trial and error, without human intervention.

The coordinate indicator. (II and III Symposium 'Leonardo Torres Quevedo: su vida, su tiempo, su obra' 1987) It was a system for finding one's way around a city. He proposed it for Madrid and Paris. It consisted of locating the various points in a city by using rectangular coordinates analogous to geographical longitude and latitude. These coordinates defining a given position would be located (written) on certain indicators (he chose street lamps). A guide would contain the list of streets, squares, etc., relating their position in the coordinate system devised by Torres Quevedo. For example, in Fig. 23a, on the streetlight are the coordinates of a point corresponding to row 26 and column 87 and the weathercock would indicate the north. The coordinate

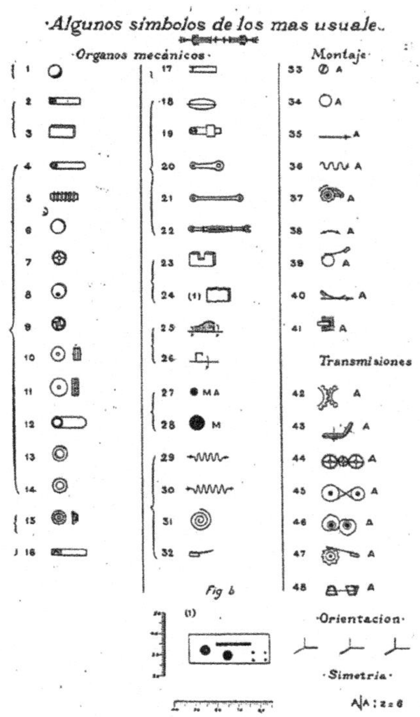

SIGNIFICADO DE LOS SÍMBOLOS

Órganos mecánicos.

Representa el elemento descrito:

1. Esfera.—2. Paralelepipedo rectangular recto (barra).—3. Paralelepipedo rectangular recto (placa).—4. Cilindro de base circular (eje, columna).—5. Tornillo.—6. Cilindro de base circular (disco).—7. Rueda dentada recta.—8. Disco excéntrico.—9. Polea de llanta convexa.—10. Polea de gargantas.—11. Rueda que engrana con un tornillo sin fin.—12. Cilindro de base anular (manguito).—13. Placa anular.—14. Corona dentada.—15. Rueda dentada cónica.—16 y 17. Barras de sección constante.—18 y 19. Barras de sección variable.—20. Manivela.—21. Barra articulada en sus dos extremos.—22. Biela.—23. Placa cuya forma se indica al dibujar el símbolo.—24. Placa cuya forma se dibuja exactamente en una figura aparte (fig. *b*).—25 y 26. Superficies de revolución.—27. Caja *M* labrada en *A*.—28. Plaza hueca *M*.—29 á 32. Resortes.

Montaje.

Expresa el movimiento que puede tener el elemento descrito con relación al que le sirve de soporte; este último se indica por la letra *A* en las fórmulas 33 y 41, que se leen así:

33. Fijo en *A*.—34. Gira libremente alrededor de *A*.—35. Resbala sobre *A*, siguiendo una trayectoria rectilinea (corredera).—36. Resbala sobre *A*, siguiendo una trayectoria helizoidal (tornillo y tuerca).—37. Gira al rededor de *A* sólo en un sentido.—38. Gira al rededor de *A*, oscilando entre dos posiciones extremas que limiten la amplitud de su movimiento.—39. Gira alrededor de *A* con rozamiento suave (suave porque el resorte está representado por una linea fina). 40. Resbala sobre *A* con rozamiento fuerte (fuerte porque el resorte está pintado con linea gruesa).—41. Encaja en *A*.

Transmisiones.

Indica la conexión mecánica que existe entre el elemento descrito y otro mecanismo representado por *A* en las fórmulas 42 á 48.—42. Engranaje cilíndrico.—43. Engranaje cónico.—44. Tren de ruedas dentadas.—45. Correa sin fin.—46. Transmisión que establece una relación de velocidades variables.—47. Rueda de roquetes y trinquetes. 48. Transmisión por medio de husillos cónicos.

Orientación.

Indica que la dimensión singular del elemento descrito es pararela al eje dibujado con linea gruesa ó al plano determinado por los dos ejes que se dibuján con trazo grueso.

Simetría.

Expresa el plano con relación al cual son simétricos los dos elementoa *A A*.

Fig. 22 List of symbols used to describe machines

indicator for orientation in large cities was first presented in 1896 (‘*Orientation in large populations. Coordinated indicators*’ is published in Madrid Científico).

In the last years of his life, Torres Quevedo made contributions in the field of educational disciplines: his last patents related to subjects such as typewriters and their improvement (patents nos. 80121, 82,369, 86,155 and 87,428), the marginal pagination of books (patents nos. 99176 and 99,177) and, especially, the projectable pointer (patent no. 116770) and the didactic projector (patent no. 117853).

The didactic projector. (II and III Symposium ‘Leonardo Torres Quevedo: su vida, su tiempo, su obra’ 1987) The author says on the subject:

> For some time now I have been studying how to facilitate technical teaching by means of the use of luminous projections.... The views that are usually projected are each photographed on a glass slide and all of them are placed one by one, in front of the flashlight, in a frame prepared to receive them.... I have decided to study the way of constructing, projecting and changing the views, taking into account their constitution. It could, in my opinion, be executed more advantageously by the apparatus represented in the figure that I am describing.

The projectable pointer. (II and III Symposium ‘Leonardo Torres Quevedo: su vida, su tiempo, su obra’ 1987) Also known as the laser pointer, the inventor himself points out its purpose:

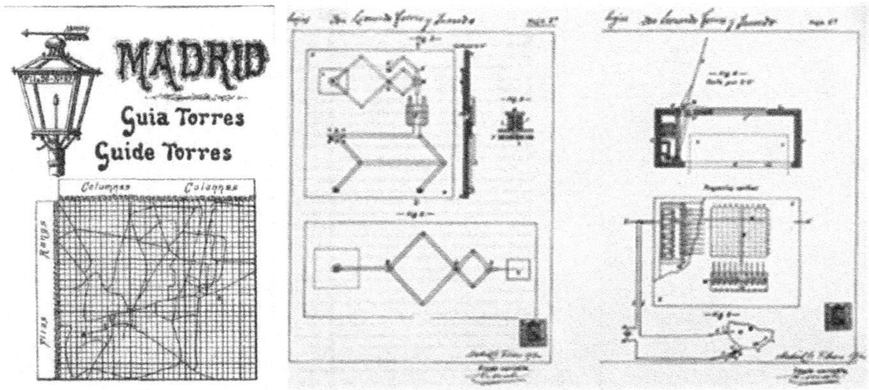

(a) (b)

Fig. 23 **a** Cover of the Torres Guide for Madrid. **b** Detail of the patent of the projectable pointer

> The difficulties encountered by a teacher in illustrating his speech by means of luminous projections are well known. He needs to stand in front of the screen, taking care not to obscure the projected figure, to draw his pupils' attention to the details that interest them most and to show them with a pointer.

It was an invention of the last years of his life (1930). It was based on the shadow produced by an opaque body moving on a screen. The presenter could move the pointer at will on the plate (today we would call it a slide) by operating an articulated system which was actually the invention (see Fig. 23b).

8 Conclusions

It is difficult to find engineers who have researched and worked in so many different areas of engineering as Leonardo Torres Quevedo. The main subjects to which he devoted his life were calculation machines, Automatics and its applications, cableways and airships.

He was not a researcher who limited himself to theoretical approaches but developed them, he wanted to be effective by putting his theories into practice, he was a precursor of what we now call R&D +i.

Faced with any problem posed, either by him or by an outer commission, he was looking for the solution, but at the same time he developed the theory that allowed him to justify and construct the whole process. His solutions were original and, in some works, he was the first to state and solve certain problems; in other works, the fewest, he was the precursor of situations that he was unable to develop but he left left them proposed.

In summary, we can highlight from his huge work:

- Cable cars. Its funicular systems were the first in the world to be used to transport people. The Niagara cable car opened in 1916 and is still in operation.
- Algebraic machines or analogic calculation machines. For the first time, a general and complete solution to the problem of constructing algebraic and transcendental relations through mechanisms implemented in machines is given.
- His work papers on analogic calculation machines, where, among other theoretical contributions, he establishes a new concept of kinematics.
- Telekino. It is the world's first remote control device. In 2006, it was awarded a Milestone by the IEEE.
- Chess player. It is the first programmed automaton to play a chess endgame. The machine plays the white king and rook and gives checkmate to the black king, managed by a human opponent.
- Electromechanical arithmometer. In terms of its design and construction, it can be considered the first computer in history. The arithmometer received its instructions from a typewriter. It performed a sequence of basic operations: addition, subtraction, multiplication and division. The machine itself executes the operations and prints them out with the typewriter.
- His written work in the field of Automatics and its application to automatic electromechanical machines. He is the first and most important work in areas of Science and Engineering such as Automatics, Cybernetics and Artificial Intelligence.
- Airships. He designed and built airships (balloons) that were marketed. He incorporated original innovations into them which, over time, proved to be very effective.
- Camp Ship. A ship design for transporting deflated Torres Quevedo type airships. They could be inflated and launched from the ship itself. On its return it had a fastening system (original to Torres Quevedo) and could be stowed until it could be used again. Similar to today's aircraft carriers.

The main characteristics of his work as a scientist and engineer were: the universality of his achievements; the resolution of problems and the application of appropriate solutions; the technical and scientific accuracy of his approaches; and being the forerunner and pioneer in many topics of engineering.

In the celebration of the centenary of Leonardo Torres Quevedo's birth, Professor Pedro Puig Adam said that "*the best tribute to his memory is to know his work and disseminate it*". May this document serve to increase the knowledge of his person and his works.

Acknowledgements The authors of this work would like to express their special thanks to the Torres Quevedo Museum (E.T.S.I. Caminos, Canales y Puertos, U.P.M.) and in particular to Professor Manuel Romana, for the facilities provided when we visited the museum and for the permission granted to display, in this document, the photographs of many of the original Torres Quevedo prototypes that are kept in the aforementioned museum.

We would also like to thank the biographers of D. Leonardo Torres Quevedo: J. García Santesmases, F. González de la Posada, F. A. González Redondo, L. Rodríguez Alcalde, et al. who have helped us to understand his enormous work and without whose writings this work would not have been possible.

References

G. Torres-Quevedo Polanco. 'Torres Quevedo y la Automática'. Revista de la Real Academia de Ciencias Exactas, Físicas y Naturales. Volume XLV, Cuaderno Primero. 31 p. Madrid (1951)

Industrial Property Register, *Patentes de invención de Don Leonardo Torres Quevedo* (Ministry of Industry and Energy, Madrid, 1988)

J. García Santesmases. 'Obras e Inventos de Torres Quevedo'. Edita el Instituto de España, Colección Cultura y Ciencia. Madrid, (1980). ISBN: 84–85559–07-X

J. A. del Barrio Unquera. Leonardo Torres Quevedo y el esperanto. Conmemoración del sesquicentenario del nacimiento de Leonardo Torres Quevedo (1852), pp. 281–302. (2003). ISBN 84–95486–63–6

L. Torres-Quevedo Torres-Quevedo. 'Leonardo Torres Quevedo'. Ingeniería y Territorio. N° 79. Pp. 54–57 (2007). ISSN 1695–9647

F. García Merayo. 'Leonardo Torres Quevedo'. Revista Digital de ACTA, Número 002. 17 pág. (2013)

F. González de Posada, *Leonardo Torres Quevedo* (Published by the Fundación Banco Exterior de España, Biblioteca de la Ciencia Española, 1992)

L. Rodríguez Alcalde. Biography of Leonardo Torres Quevedo. Santander: Institución Cultural de Cantabria-CSIC. Madrid (1974)

M. de la Fuente Merás. 'Leonardo Torres Quevedo en su 150 aniversario'. 1st Conference on Educational Innovation and Research. Government of Cantabria. Santander (2003)

Proceedings of the I, II and III Symposium 'Leonardo Torres Quevedo: su vida, su tiempo, su obra'. Amigos de la Cultura Científica (1987, 1991 and 1995)

Proceedings of the I, II, III and IV Symposium 'Ciencia y Técnica en España de 1898 a 1945: Cabrera, Cajal y Torres Quevedo' (1999, 2000, 2001 y 2002). I.S.B.N.: 84–87635–35–0

The **Proceedings**, lectures and papers in digital format (more than 2,000 pages) presented at the seven Symposia, held between 1987 and 2002, can be found at: http://www.torresquevedo.org/libros/

F. González de Posada, "Leonardo Torres Quevedo. Investigación y Ciencia, No **166**(July), 80–87 (1990)

J. Rui-Wamba, F. Sáenz Ridruejo. En torno a Leonardo Torres Quevedo y el Transbordador del Niágara. Edited by: Esteyco Foundation. Madrid (1995)

los transbordadores de Torres Quevedo, J. Aramberri, F. A. González Redondo. Innovación y tecnología. Fabrikart N° **10**, 26–45 (2012)

Leonardo Torres Quevedo y la conquista del aire. Centenario de la botadura del dirigible 'Torres Quevedo'. 1907-Guadalajara-2007. (Annals of the Exhibition. Curator: F. Gónzalez de Posada). Published by Amigos de la Cultura Científica (2007). ISBN: 978–84–87635–37–3

F. González de Posada, F. A. González Redondo. "Leonardo Torres Quevedo (1852–1936). Part 1. Las máquinas algébricas", La Gaceta de la RSME. Vol 7.3, pp. 787–810. Madrid (2004)

P. Puig Adam. 'Torres Quevedo. El cálculo mecánico y la Automática'. Revista de la Real Academia de Ciencias Exactas, Físicas y Naturales de Madrid, XLVII, pp. 11–28 (1953)

E. Huélamo Martínez. Algunos recuerdos del cálculo mecánico. DYNA. N° 7. October, 1995. Pp. 14–24

B. Randell, From Analytical Engine to Electronic Digital Computer: The Contributions of Ludgate, Torres and Bush. Annals of History of Computing. **4**(4), 327–341 (1982a)

B. Randell. Origins of Digital Computers: Selected Papers. 3rd edition. Springer-Verlag. Berlin (1982b) (Includes: "Essays on Automatics. Its Definition. Theoretical Extent of its Applications" (original 1914) and "Electromechanical Calculating Machine", 1920)

L. Torres Quevedo. Memoria sobre las máquinas algébricas. Revista de Obras Públicas, ISSN 0034–8619, 43, volume I (26): pp. 202–205. (1895a)

L. Torres Quevedo. Memoria sobre las máquinas algébricas. R. O. P., 43, volume I (27): pp. 209–215. (1895b).

L. Torres Quevedo. Memoria sobre las máquinas algébricas. R. O. P., 43, volume I (28): pp. 217–222. (1895c)

L. Torres Quevedo. Memoria sobre las máquinas algébricas. R. O. P., 43, volume I (29): pp. 225–227. (1895d)

L. Torres Quevedo. Memoria sobre las máquinas algébricas. R. O. P., 43, volume I (30): pp. 233–240. (1895e)

L. Torres Quevedo. Memoria sobre las máquinas algébricas. R. O. P., 43, volume I (31): pp. 241–246. (1895f)

L. Torres Quevedo. Memoria sobre las máquinas algébricas. R. O. P., 43, volume I (32): pp. 249–254. (1895g)

L. Torres Quevedo. Memoria sobre las máquinas algébricas. R. O. P., 43, volume I (33): pp. 257–262. (1895h)

L. Torres Quevedo. Máquinas algébricas. R. O. P., 48, volume I (1339): pp. 195–196. (1901)

L. Torres Quevedo. Máquinas algébricas. R. O. P., 48, volume I (1341): pp. 205–209. (1901)

F. Thomas, A short account on Leonardo Torres' endless spindle. Mech. Mach. Theory **43**(8), 1055–1063 (2008)

L. Rodríguez Alcalde. Torres Quevedo y la Cibernética. Ediciones Cid. Madrid (1966)

F. González de Posada, F. A. González Redondo. 'Leonardo Torres Quevedo (1852–1936). 2ª Parte. Automática, máquinas analíticas', La Gaceta de la RSME. Vol 8.1, pp. 267–293. Madrid (2005)

L. Torres Quevedo. Sobre un nuevo sistema de máquinas de calcular electro-mecánicas. R. O. P., 59, volume I (1857): pp. 227–233 (1911)

L. Torres Quevedo. Sobre un nuevo sistema de máquinas de calcular electro-mecánicas. R. O. P., 59, volume I (1860): pp. 274–278. (1911)

L. Torres Quevedo. El telekino. R. O. P. 55, volume I (1653): pp. 325–328 (1907)

M. E. Andrés Hernández. El Telekino de Leonardo Torres Quevedo. ETSI de Telecomunicación, University of Valladolid (2006)

A. Pérez Yuste, M. Salzar Palma. "The first wireless remote-control: the telekine of Torres Quevedo", Conference on the History of Electronics (CHE), United Kingdom (2004)

L. Torres Quevedo. Speech by Mr. Leonardo Torres Quevedo, in reply to that of Mr. Garcini. Revista de Obras Públicas, 56, volume I (1710): pp. 315–317. (1908)

F. A. González Redondo. La contribución de Leonardo Torres Quevedo a la historia mundial de la aeronáutica. Aeroplano: Revista de Historia Aeronáutica.Nº 31, pp. 22–37 (2013)

L. Torres Quevedo. Discurso de aceptación en la Real Academia Española (31 de Octubre de 1920). Revista de Archivos, Bibliotecas y Museos. Madrid (1920)

L. Torres Quevedo. Sobre un sistema de notaciones y símbolos destinados a facilitar la descripción de las máquinas. R. O. P. 55, volume I (1634): pp. 25–30 (1907)

The Yeregui Family (18th–Twentieth Century)

J. Aginaga, A. Claver, J. M. Pintor, and X. Iriarte

Abstract Since their appearance in the Middle Ages, mechanical clocks and watches have been ingenious devices with complex assemblies of gears, cams and actuators, some of which have survived to the present day. Over the centuries, advances in mechanical engineering were applied to clockmaking, achieving clocks with ever-increasing precision. In the north of Navarre, the Yeregui family formed an outstanding family of mechanical clockmakers who, from the end of the eighteenth century until the middle of the twentieth century, designed, built, assembled and maintained tower clocks for five generations. Their first recorded clock was designed and assembled by José Francisco Yeregui Zabaleta for the village of Betelu and its deed dates from 15th April 1796. Subsequently, different members of the family manufactured numerous clocks for towns and cities in and around Navarre, including the old clock of Pamplona Town Hall, in operation for more than 150 years until 1991 and recently restored. This chapter briefly describes the most significant advances in mechanical clockmaking and delves into the history and work of the Yeregui clockmaking lineage.

Keywords History of MMS · Clockmaking · Tower clock · Yeregui family

J. Aginaga (✉) · A. Claver · J. M. Pintor · X. Iriarte
Public University of Navarra, Pamplona, Spain
e-mail: jokin.aginaga@unavarra.es

A. Claver
e-mail: adrian.claver@unavarra.es

J. M. Pintor
e-mail: txma@unavarra.es

X. Iriarte
e-mail: xabier.iriarte@unavarra.es

R. López-García and M. Ceccarelli (eds.), *Distinguished Figures in Mechanical Engineering in Spain and Ibero-America*, History of Mechanism and Machine Science 43, https://doi.org/10.1007/978-3-031-31075-1_15

1 Introduction

Mechanical clocks, although they have been largely replaced by electronic clocks since the mid-twentieth century, continue to arouse great interest and are considered by many to be mechanical works of art. Although in the twenty-first century they have become museum exhibits or the property of collectors rather than instruments for telling the time accurately, many of the mechanical clocks built in previous centuries are still in use today. This fact, as well as the complexity of their operating mechanism, means that mechanical clockmaking remains a relevant discipline or field of knowledge from the point of view of machine mechanics and its history. It should also be noted that, like most machines prior to the Industrial Revolution, mechanical clocks are mechanisms powered by renewable energies, the source of energy being the person who once or twice a week is responsible for winding up the clock. In the case of tower clocks, the energy was provided by raising the weights whose fall due to the effect of gravity acts as the clock's engine.

Despite having the same working principle, mechanical clocks can be divided into two main groups: domestic or portable clocks and monumental tower clocks. While the former were often small devices in the possession of the wealthy social classes used to create admiration in society, the latter played a public role in ordering the religious and civil life of populations.

As far as tower clocks are concerned, mechanical clockmaking is arguably a discipline that is on the borderline between blacksmithing and mechanical engineering. Generally, clockmakers are not considered mechanical engineers because, in fact, they do not have or historically have not had formal education as engineers. This fact is logical since the origins of the engineering profession date back to the eighteenth century, whereas mechanical clockmaking may have started towards the end of the thirteenth century (Pérez Álvarez 2015). Thus, mechanical clockmaking has historically been linked to trades such as blacksmithing, locksmithing and even gunsmithing (Ramírez Martínez 2002). However, in addition to working the metal, the clockmaker had to design complex mechanisms of gears and cams, as well as adjusting its movement for correct time measurement by means of the escapement mechanism, a set of elements which, by transforming the rotary movement into oscillation, makes the clock move regularly and determines the time or frequency of the clock. The clockmaker also had to think about storing the energy needed to run the clock, which was solved by compressing and decompressing springs or by means of weights which, after being raised, were lowered by the action of gravity, thus driving the movement of the clock. Finally, the work of the clockmakers did not end once the clock had been built and was running; they also had to define the maintenance operations necessary to ensure that the clock continued to run in proper condition.

The education of the clockmakers was therefore not highly qualified, which today would correspond to university studies, but they did have a high component of mechanism design. In fact, apprentice aspiring clockmakers were usually trained for 5 or 6 years with a master clockmaker who undertook to teach them all the secrets of the trade and to feed them during this period (Ramírez Martínez 2002). In addition to acquiring the necessary metalworking skills, the apprentices were also required to develop the overall idea of how to fit the various components of the clock. Given the skill and ingenuity required for clockmaking, it could be said that the master clockmaker, along with the master gunsmith and the locksmith, had a higher level of knowledge than the master blacksmiths, a fact that gave them a certain respect and admiration from the common people.

In this context of master clockmakers is the Yeregui family, a family of Navarrese origin with as many as 7 clockmakers over 5 generations. The fact that the trade extended over 5 generations gives us an idea of the relevance they had in Navarre and other neighbouring regions, where they installed clocks in numerous localities.

2 Historical Remarks of Mechanical Clocks

Prior to turning to the life and work of the clockmakers of the Yeregui family, it is necessary to briefly review the main landmarks in the development of mechanical clockmaking. This is not intended to be an exhaustive review, but rather to highlight the most important advances in the technique and operation of mechanical clocks and watches. It should be said that, prior to the invention of mechanical clocks, there were other types of instruments for measuring time, such as sundials, hourglasses, clepsydra, etc. But it was mechanical clockmaking that reached the highest levels of precision for several centuries, until mechanical clocks were surpassed by electric clocks in the middle of the twentieth century.

First of all, the importance of the escapement mechanism must be emphasised. The escapement mechanism is the key element that distinguishes mechanical clocks and watches from clepsydras and other earlier time-measuring instruments. It is a mechanism that regulates the speed of the clock, making it constant. The exact date of its invention is unknown, but it is accepted that it may have been invented towards the end of the thirteenth century. There may have been two reasons for the need to measure time: on the one hand, monastic needs, meaning the calls of the church to the parishioners for praying; on the other hand, scientific needs, since astronomers needed an instrument capable of measuring time accurately in order to describe the movement of the planets and model the universe. It was probably in this context that the escapement mechanism was invented and the first steps towards more precise clockmaking were taken (Pérez Álvarez 2015).

There existed, therefore, people in the scientific field, mainly physicists or astronomers, interested in the measurement of time who have gone down in history as clockmakers. One of the earliest may have been Richard of Wallingford (1292–1336). Being the son of a blacksmith, he combined his metalworking skills with his knowledge of astronomy to build the clock at St Albans Abbey in Oxford. Also in the fourteenth century, Giovanni di Dondi (1330–1388), the son of a clockmaker, designed the Astrario, an extraordinary astronomical clock that showed the position of the sun, moon and various planets known at the time. It is remarkable that, additionally, the Astrario is also considered to be one of the earliest known mechanisms to use non-circular gears (Addomine et al. 2018).

Beyond astronomical purposes, in the fourteenth and fifteenth centuries, mechanical clocks were considered to be exclusive items that the aristocracy displayed in society to arouse their admiration. Similarly, the monastic and governmental needs of populations gave rise to another type of clock with a completely different function: to regulate the functioning of towns and villages by telling the time. In this respect, curiously enough, it could be said that the invention of mechanical clocks preceded the need for civil society to know the time. This is how tower clocks were born, which often did not even have a dial to show the time, but transmitted the time by ringing a bell located in a tower to be heard by as many people as possible.

In terms of technology, regardless of using them as tower clocks or for astronomical purposes, the development of mechanical clocks is linked to the invention of the escapement mechanism. It is believed that the first type of escapement was known as the verge and foliot escapement. This escapement mechanism consists of a saw-toothed or crown wheel, a vertical rod with two pallets and a horizontal rod attached to the first with two small weights. As can be deduced from Fig. 1, the rotation of the crown wheel, always in the same direction, alternately collides with the pallets of the vertical rod, converting the rotating movement into oscillation. The location of the weights on the horizontal rod regulates the frequency of the oscillation. Thus, the clock provides the time depending on the frequency of these oscillations.

Fig. 1 Verge and foliot escapement (Pérez Álvarez 2018)

In the mid-seventeenth century, the verge and foliot escapement mechanism was replaced by the anchor and pendulum escapement. The authorship of this invention is not clear, as some references consider it to be the work of Robert Hooke (1635–1703), while others attribute it to Cristian Huygens (1629–1695), a Dutch astronomer, physicist and mathematician (Bennett et al. 2002). This type of escapement mechanism consists of a saw-toothed wheel called the escape wheel, the anchor and the pendulum (Fig. 2).

The anchor has two protruding pallets which are alternately pushed by the escape wheel, similar to the way the pallets of the vertical rod did in the verge and foliot escapement. Thus, the escape wheel always rotates in the same direction while the anchor oscillates, converting the rotation into oscillation. This new type of escapement mechanism greatly improved the accuracy of clocks from an error of a few hours a day to an error of minutes (Stoimenov et al. 2012). Huygens himself showed that reducing the amplitude of the oscillation reduced the effect of non-linearities present when the angle rotated was greater, which would undoubtedly help the accuracy of clocks (Du and Xie 2013).

Fig. 2 Anchor and pendulum escapement (Du and Xie 2013)

The frequency of oscillation of the pendulum depends, as is well known, on the location of its centre of gravity. Nevertheless, the expansion of metals with temperature and the consequent increase in the length of the pendulum caused clocks to lose accuracy depending on the ambient temperature. George Graham (1673–1751) conceived the idea of using mercury to compensate for the loss of accuracy due to the expansion of the pendulum; by placing a small container of mercury in the pendulum, the expansion of the mercury compensated for the expansion of the pendulum. The mercury pendulum reduced the error of clocks to a few seconds a day.

Clocks were also a breakthrough in navigation, as in the mid-eighteenth century they made it possible to measure longitude on the open ocean. This required greater accuracy, as the clock had to run for months at a time, keeping the time at the point of departure. By measuring the local time and latitude from the sun and knowing the time at the point of departure, it was possible to know the longitude where a ship was. It was John Harrison (1693–1776) who made the necessary advances to design clocks that were capable of measuring time and were not affected by waves. Temperature compensation was achieved by means of two metal strips, one of brass and the other of steel, riveted together and attached to a spring. As the two metals expanded differently, the metal strip curved and shortened the spring, increasing its rigidity and readjusting the frequency of the clock. With this and other advances, the accuracy of his H4 clock was computed as an error of 39.2 s after 47 days of navigation (Betts 1993).

Most of the aforementioned developments in clockmaking took place in England. In Spain, the first mechanical clocks may have been the one in Perpignan, at that time part of the Kingdom of Aragon, and the one in Toledo Cathedral, both dating from the mid-fourteenth century (Pérez Álvarez 2018). Gradually, from the fifteenth century onwards, the number of public clocks in the main cities began to increase, and by the seventeenth and eighteenth centuries there were tower clocks in almost every town. Usually, these clocks were troublesome and needed a lot of maintenance and repairs. Contracts were usually made not only to build the clocks but also to keep them in good working order for 5 or 6 years. In spite of this, often other people with no clockmaking knowledge would make adjustments or try to maintain the clock and, as a result, make it run worse. Then, clocks were frequently replaced after a few decades of operation. Thus, contracts went so far as to explicitly state that the responsibility of the clockmakers to carry out repairs excluded cases where the clock had failed "due to the fault of the person who manages it".

3 Biographical Notes

Most of people usually considered as a distinguished figure in a knowledge field come from whealthy or, at least, well-off families. They lived in major cities, they had a formal education and they had studied in renowned universities and often with distinguished menthors. The history of the Yeregui family is quite different. They

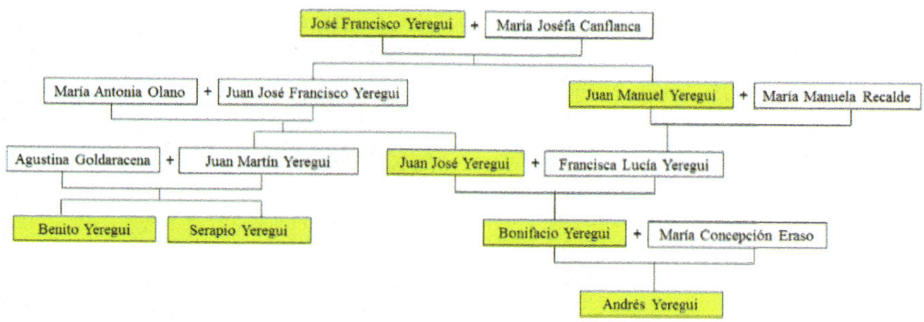

Fig. 3 Family tree of the Yeregui clockmakers (courtesy of José Ignacio Yeregui Arlazón, Yeregui Elkatea)

come from small villages in the mountains of Navarre and it is there where they carried out their activity.

The family clockmakers are divided into two branches, those from Betelu (Navarre) and those from Zumaia (Gipuzkoa). The former comprise all five generations, from José Francisco to Andrés, while those from Zumaia are two clockmakers, Benito and Serapio, brothers to each other and cousins of the fourth generation of those from Betelu. Figure 3 shows part of the Yeregui family tree in which we can see the 7 clockmakers in the dynasty, not including the family members not dedicated to clockmaking.

As shown in Fig. 3, Jose Francisco Yeregui was the first clockmaker of the family. He was born in 1760 in Leitza and initially worked as a carpenter. However, his curiosity about how the clock in his village's parish worked led him to build a similar clock but made of wood. After exhibiting the clock at a fair in Pamplona, and given the success it had at that exhibition and the recommendations made to him, he decided to give up carpentry and devote himself to clockmaking (Garmendia Larrañaga 1970). There is a curious coincidence between José Francisco Yeregui and the eminent John Harrison: anecdotally, and without wishing to place them on the same level in the history of clockmaking, both built their first clocks in wood and in a self-taught way.

After José Francisco, four generations of his descendants were also involved in clockmaking. The first of these was Juan Manuel (1798–1848), who continued his father's work in the Betelu workshop. He was followed by his son Juan José (1819–1887) and his grandson Bonifacio (1850–1911), also in Betelu. Benito (1843–1912) and Serapio (1859–1923), Bonifacio's cousins, learned their trade in the same workshop. After training in Betelu, they developed their activity in the Gipuzkoan towns of Usurbil and Zumaia. As industrialisation progressed, Benito and Serapio began to manufacture clocks with imported parts and expanded their activity in the metal sector. Meanwhile, in Betelu the other branch of the family continued to design clocks in a more artisan manner, with Andrés (1884–1975), Bonifacio's son, being the last clockmaker in the Yeregui family. Figure 4 shows a photograph of Benito Yeregui Goldaracena.

Fig. 4 Benito Yeregui Goldaracena (1843–1912), the most prolific clockmaker of the Yeregui dynasty

After the clockmaking dynasty died out, their descendants created the Yeregui Elkartea Foundation in the twenty-first century. One of its objectives is to try to recover the memory of the clockmaking dynasty, as well as to restore and promote the restoration of mechanical clocks in order to put them back into operation, whether they were built by their predecessors or by other clockmakers.

Although there is little documentation concerning their work and legacy, the activity of each of the clockmakers in the Yeregui family is described below.

4 Review of Their Main Works

As previously mentioned, the Yeregui family consisted of 7 clockmakers over 5 generations. From the first clocks made by José Francisco Yeregui Zabaleta at the end of the eighteenth century to the last clocks made by Andrés Yeregui Eraso well into the twentieth century, they made numerous clocks for different towns and cities in Navarre and the surrounding area. Documenting the activity and production of each of them is an arduous task and the information available is often closer to hypothesis than to certainty. This is due to two fundamental reasons: on the one hand, clockmakers did not usually sign their works until the middle of the nineteenth century (Ramírez Martínez 2002). On the other hand, when the clocks had problems or stopped working, they were replaced by new ones and often the old clock was part of the payment made to the clockmaker in charge of building the new one, so that he could use the pieces for other clocks or other ironwork, leaving the old clock dismembered and leaving no trace of its existence.

4.1 *José Francisco Yeregui Zabaleta (1760–1834)*

José Francisco was the first clockmaker in the Yeregui dynasty. The first Yeregui clock recorded in notarised documents is the one made for the town of Betelu in 1796. However, it was not the first clock he built, as the 1796 contract mentions that "*they are well aware of his expertise and that he has made several clocks in the towns of Burguete, Berastegui, Lugar de Oreja and other parts*" (Royal and General Archive of Navarre 1796). The clock was to be built for 86 ducats, to be paid half by the town and half by the church. José Francisco also acquired the responsibility of conserving and repairing it for 6 years and the contracting party had to provide him "*wage and free forge*". Figure 5 shows part of the deed of obligation to build the Betelu clock.

This clock had to replace another one that was built in 1748 because "*it was entirely out of order*" and "*it was so necessary for the government of the people and the attendance at sacred functions*". The contract specified "*that the said clock should be with its hour and half hours*" and gave indications regarding the diameters and thicknesses that some of its wheels should have, including the Santa Catalina or escapement wheel. The clock was placed in the parish tower of Betelu. In addition to the specifications of the contract, at three o'clock in the afternoon the clock rang 33 chimes, with which it was intended to commemorate the death of Jesus. This clock was in operation until it was replaced in 1962 by another clock made by the Murua company of Vitoria (Garmendia Larrañaga 1970).

In 1804 there is evidence of two other contracts for the construction of clocks in the villages of Egiarreta and Ihabar, the first for the construction of a new clock and the second for the replacement of the old one because it was "*quite worn out*" (Royal and General Archive of Navarre 1804). The former is still in the parish church of

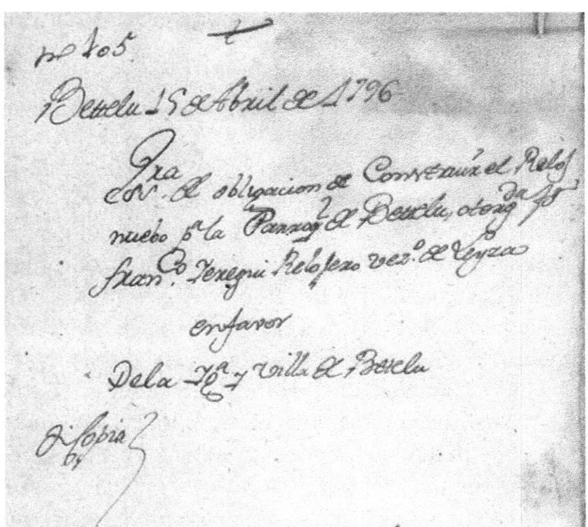

Fig. 5 Extract from the Betelu contract deed of 1796 (Royal and General Archive of Navarre 1796). Translation: "*Betelu 19th April 1796. Deed of obligation to Build the new Clock for the Parish of Betelu, granted to Francisco Yeregui Clockmaker, Neighbour of Leitza. In favour of the Church and Town of Betelu. Copy*"

Fig. 6 Current state of the Egiarreta clock (photograph taken in 2019)

Santa María de Egiarreta, although it has been derelict for decades. Figure 6 shows its current state, dirty and deteriorated. However, the movement train with its escapement wheel and anchor can be easily identified, as well as the chiming train with its 2-blade regulator or fly fan.

The Egiarreta and Ihabar contracts are similar and describe in detail some of the characteristics that the clocks and their components must have, which help us to get an idea of the technology of that time. One example is the use of bronze bearings. The contract does not mention them as such, but it does specify that the clock must have "*bronze pieces riveted to the bars of another frame, where the shaft of the wheels play, and these are to make the wheels run more smoothly*" (Royal and General Archive of Navarre, Pamplona Notary's Office 1804). Other aspects mentioned include, among others, the "*pinions of solid piece well hardened and polished so that they run more smoothly*", the "*anchor with its pallets well hardened and tempered*" and "*at the end of the pendulum a male and female lenticular bob to fix its movement*". The latter probably refers to the pendulum being made up of two tongue-and-groove parts, so that by adjusting them the length of the pendulum could be regulated to compensate for seasonal expansion. In general, the level of technical detail that appears in these contracts suggests that the clockmaker himself would have helped in the redaction of the contract, as this level of knowledge of clockmaking would not be expected from the notary or the contracting party, consisting of neighbours and the church.

The next known clock is the one made for the town of Olazagutia. The contract was signed on 6 December 1805. While the technical aspects are similar to those described in the two previous contracts, this one adds a new requirement: "*that the aforementioned clock should have hours and half-hours and that the bell should chime at 12 noon, striking the first three strokes with a pause and the following six in the manner of the clock and the tolls for the Animas in summer time should strike at nine o'clock at night and in winter at eight o'clock, striking from twenty-seven to thirty-three*" (Royal and General Archive of Navarre 1805). This is a new requirement that José Francisco would probably have solved by means of actuators, notches in certain wheels and stops to make the necessary pauses. Something similar was required for a considerably later clock for the town of Burlada, the contract for which dates from 12 March 1821 (Garmendia Larrañaga 1970). Considering that in all the above contracts, except the first, José Francisco is described as the clockmaker of Betelu, it can be deduced that after the construction of the clock in that town in 1796, José Francisco and his family moved to Betelu, where another four generations of Yeregui clockmakers worked.

Although there is documentary evidence of only the five clocks mentioned above, it is very likely that he had built many more. From the technical details found in the documentation, it can be assumed that they were clocks with two trains, the movement train and the chiming train. This can be seen in the Egiarreta clock, in which the two trains can be distinguished (Fig. 6). This system was the usual one for clocks that struck the hours and half-hours.

4.2 *Juan Manuel Yeregui Canflanca (1795–1848)*

Juan Manuel was the youngest of José Francisco's six children and the first to continue his father's work in the Betelu workshop. There is evidence of four clocks built by him, although, as in the case of his father, there were probably more.

The first clock made by Juan Manuel for which there is evidence is the one in Pamplona. There is no documentary record of the construction contract, but the clock still exists and has a nameplate indicating its author, place and year of manufacture, 1827 (Fig. 7). The fact that it has an identifying plaque is noteworthy, since until this period tower clocks were usually anonymous works, and it is from the mid-nineteenth century that the signatures of the builders began to appear (Ramírez Martínez 2002).

The clock must have been initially installed in the church of San Lorenzo in Pamplona and taken to the City Hall in 1849 (Garmendia Larrañaga 1970). It remained in the City Hall until 1991, when it was replaced by an electronic clock. The withdrawn clock kept most of the original parts, although it also had components incorporated later. One example is the fact that it has ball bearings instead of the bronze bearings that were used at the time. Presumably, the wear of these bearings would have produced an excessive clearance in the shafts, which would have disturbed the clock's operation. Thus, at some point in the twentieth century they would have been replaced by ball bearings. Similarly, the anchor may also have been replaced or modified, as it has screwed pallets instead of being a single piece.

Recently, the clock has been repaired and restored to be exhibited in operation in the Pamplona Planetarium (Aginaga et al. 2021). Figure 8 shows the clock with its three trains, the movement train and the two chiming trains for the hours and quarter

Fig. 7 Nameplate of the Pamplona City Hall clock. It can be read: "*I was made by Juan Manuel Yeregui in Betelu in 1827*"

hours. In order for it to be fully operational, including the ringing of bells, a small bell has been installed, as the original bells obviously remain in the Town Hall.

Fig. 8 Pamplona City Hall clock (1827)

Another clock by Juan Manuel for which there is evidence is the Arbizu clock, built to replace the previous one according to an account book of the aforementioned town to which we have had access. This account book places the construction of the clock in 1831 and, although the name of Juan Manuel Yeregui does not appear, it indicates the costs of moving the old clock to Betelu as well as those of driving the new clock from Betelu. It is plausible to believe that the clock in the account book is the one that currently stands in the church of Arbizu, as its structure is very similar to that of the clock in Pamplona.

According to the historian Juan Garmendia, he also built the clocks for Lizarraga in 1843 and Aránzazu in 1853. It is curious in the case of the second clock, as the date is after his death. However, Juan Garmendia had the opportunity to meet Andrés Yeregui, the last clockmaker in the dynasty, and it was he who cited the Aránzazu clock as having been built by Juan Manuel, although he was probably helped by his nephew Juan José, the next clockmaker in the family (Garmendia Larrañaga 1970).

From the available information, two changes or advances can be deduced with respect to the clocks built by his father, José Francisco. On the one hand, Juan Manuel's clocks incorporate two chiming trains: one for the hours and the other for the quarter hours. On the other hand, the arrangement of the trains is different, since in the Egiarreta clock the shafts are arranged horizontally, whereas in the Pamplona and Arbizu clocks the shafts of each gear train are arranged in vertical planes.

4.3 *Juan José Yeregui Olano (1819–1887)*

The next clockmaker in the family was Juan José, Juan Manuel's nephew, who took over the Betelu clockmaking workshop after the death of his uncle. Little is known of the clocks built by Juan José, except for the two clocks mentioned by the historian Juan Garmendia.

On 12 April 1877, the contract between Juan José and the town of Altzaga (Gipuzkoa) was formalised for the construction of a clock for the parish church. Among the requirements for this clock, the request to "*strike the hours repeatedly and the half hours*" (Garmendia Larrañaga 1982) stands out. The repetition of the hours is something that has not been seen in the previous contracts of the Yeregui clockmakers to which we have had access. In order for the clock to repeat the hours, a snail-shaped cam was used, similar to the one shown in Fig. 9. This is a disc cam —located in the image behind the gearing— which releases the actuators of the bells. At the bottom left, it can be seen that the disc has a double notch, so that the chime was activated once after the first jump and a second time after the second jump. Note that on the right side of the disc there is only one jump, so the half-hour chimes would not be repeated.

The other known clock by Juan José is the one built for Espinal (Navarre) in 1884. The only details we have are the cost of "*three thousand five hundred vellon reals (an ancient Spanish coin)*" and how it was paid for. We do not know how long it was in operation, but there is no trace of it in Espinal today.

Fig. 9 Snail-shape cam for repeating hour chimes. *Source* Yeregui Elkartea

Although there is only record of this clock, it is more than likely that Juan José built many more, since in 1856 he had built a large house in Betelu which is still preserved and whose ground floor would have been designed as a clockmaking and blacksmith's workshop. He probably made the ironwork for the house himself, given that the surname "Yeregui" can be read on the highest balcony of the house, as shown in Fig. 10.

Fig. 10 Balcony of the Yeregui family house in Betelu

4.4 *Bonifacio Yeregui Yeregui (1850–1911)*

Bonifacio, the third of the six children of the marriage between Juan José and his cousin Francisca Lucía, daughter of Juan Manuel, was the one who continued with the Yeregui's clockmaking workshop. Among the clocks built by Bonifacio were those in the villages of Marcilla, Saldias, Gaintza and Villanueva. It is not known what became of the first three clocks. Regarding the one in Villanueva, in the Arakil valley in Navarre, the clock must have stopped working a few decades ago and, after a voluntary but unsuccessful attempt to repair it, in which the anchor was replaced by an electric motor, the clock is still standing -stopped, obviously— in the church of Villanueva but without some of its essential components.

Bonifacio, like probably some of his predecessors, in addition to clockmaker, was also a blacksmith. He is also mentioned as the teacher of Ignacio Zubillaga (1868–1948), who, after training in the Betelu workshop, was also an outstanding blacksmith and clockmaker (Gil and San Vicente 2010). Zubillaga built numerous tower clocks for different towns in Gipuzkoa, among them the one in Asteasu in 1896, the restoration of which has recently been promoted by the Yeregui Elkartea Foundation.

4.5 *Benito Yeregui Goldaracena (1843–1912)*

Cousin of Bonifacio, after learning the trade in the Betelu workshop, he worked as a clockmaker first in his workshop in the Aginaga neighbourhood of Usurbil and then in Zumaia. Benito is the most prolific of the Yeregui dynasty, or at least the one with the most documented clocks.

According to Yeregui Elkartea, more than 80 clocks made by Benito have been counted. Among them is the clock of the Cathedral of the Good Shepherd in San Sebastian. The cathedral was built between 1888 and 1897 and its clock was entrusted to Benito Yeregui for 3,970 pesetas (Spanish currency before the euro) (Murugarren 1996). It functioned correctly until 1947, when it had to be repaired by the Viuda de Perea company. Then, it was in operation until the end of the 70 s of the twentieth century. Now, it is currently being restored by the Yeregui Elkartea foundation (Fig. 11).

Although he started out building handcrafted clocks like his predecessors, he later moved on to industrialised clock making. Thus, most of Benito's known clocks have the same format or structure and are of the type known as Morez. This name comes from the factory in Morez (France), founded in 1858 by Louis-Delphin Odobey. This factory was one of the most important in Europe and was in operation until 1964. From the architecture of the clocks, it can be assumed that Benito Yeregui's clocks were not of his own design, but that he bought the parts from the Morez factory and his job was to assemble and attach them to the dial and exterior bells, as well as possible modifications to the chimes requested by churches and town halls.

Fig. 11 Antique clock from the Cathedral of the Good Shepherd in San Sebastian in the workshop where it is being restored. *Source* Yeregui Elkartea

Thus, Benito may be considered as the clockmaker of the dynasty that went from handcrafted clocks to the industrialisation or almost mass production of clocks. He himself offered to place his clocks in town halls and churches, as in the case of Zizurkil (Hernandorena Kultur Elkartea 2016). Likewise, Benito's clocks bore a plate on which the serial number was identified. By way of example, the Catalogue of clocks in La Rioja shows a clock in the Torremúzquiz Palace in Ezcaray (La Rioja) made by Benito Yeregui in 1891 with the serial number 32 (Ramírez Martínez 2002).

Many of the clocks made by Benito are still in use, such as those located in the parish church of the Natividad de María Santísima de Zestoa (1879), that of Nuestra Señora de la Ascensión de Urrexola (1884), the church of San Lorenzo de Ikaztegieta (1887), the church of San Millán de Zizurkil (1890), the parish church of Santísima Trinidad de Nuarbe in Azpeitia (1895), the church of San Juan Bautista de Alegi (1904) or the church of La Anunciación de Urrestilla (1905).

At the beginning of the twentieth century, Benito and his family moved to Zumaia, where Benito and his eldest son Calixto founded the company "Yeregui e hijos". Shortly afterwards, Calixto would found "Yeregui y Compañía" together with his brother-in-law Ángel Galardi, which would not only be dedicated to clockmaking but also to the manufacture of naval engines. Figure 12 shows a page from the catalogue of Yeregui y Compañía, in which we can appreciate that the structure of the clock is of the Morez type. The rest of Benito's sons, who in principle would have joined their brother Calixto's company, would later create the company "Yeregui Hermanos", dedicated to the construction of marine engines, becoming one of the most important companies in the sector at state level in the 1920s (Herreras Moratinos 1998).

Fig. 12 Catalogue of the company "Yeregui y Compañía", first page

4.6 Serapio Yeregui Goldaracena (1859–1926)

Similar to his brother Benito, Serapio was also a businessman in addition to being a clockmaker. The information we have about Serapio's clocks indicates that he brought the parts from Morez for subsequent assembly in his workshop in San Sebastián. There is evidence of the manufacture of four of his clocks in the villages of Arano, Irun, Elizondo and Goizueta. It is worth noting that in the parish of the Assumption in Arano (Navarre), the 1909 clock still works. Like other previously mentioned clocks, the hours strike twice while the half hours strike only once. Figure 13 shows a photograph of Serapio together with another of the Arano clock.

Fig. 13 Photograph of Serapio Yeregui and the machinery of the Arano clock

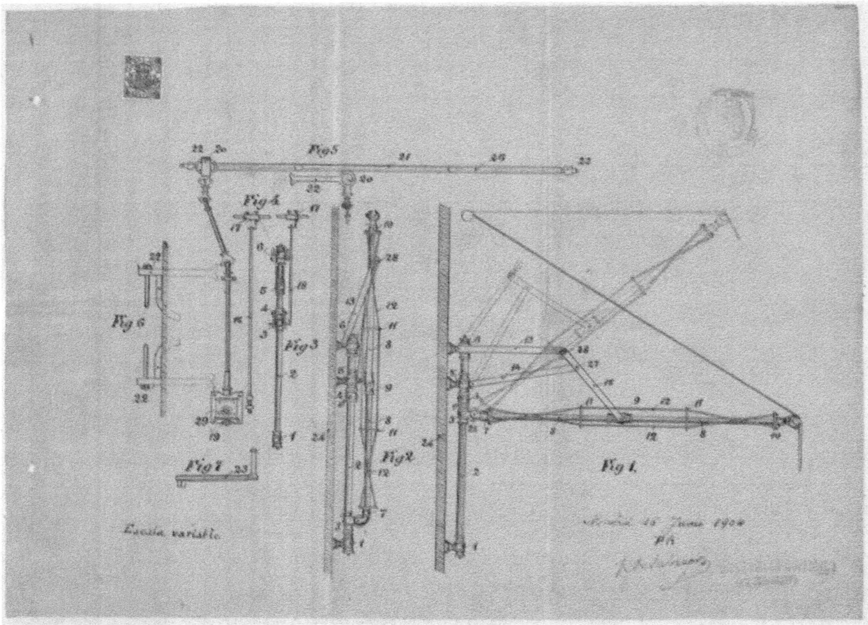

Fig. 14 Awning mechanism patented by Serapio Yeregui in 1904 (Industria 1904)

Serapio diversified the business by also offering other blacksmithing products in addition to tower clocks. He advertised products such as winch machines, apple crushers, the installation of lightning rods and the manufacture of awnings, for which he had a patent for invention in Spain. Figure 14 shows a draw of the awning mechanism patented in 1904 (Industria 1904).

4.7 *Andrés Yeregui Eraso (1884–1975)*

Son of Bonifacio, he was the last clockmaker in the family. He trained as a clockmaker in the Betelu workshop and there he continued the family work that his great-great-grandfather had begun at the end of the eighteenth century. It is worth mentioning that while his cousins in Zumaia had moved on to industrialisation, Bonifacio had continued with the family tradition of making handcrafted clocks. Andrés continued his father's work, keeping the forge active until the mid-1960s. However, it seems that his initial idea was for someone else to continue with his clocks, as until recently there was a sign on the Betelu mansion that read "*Factory of tower clocks. Andrés Yeregui and successors*" (Ramírez Martínez 2002).

According to Juan Garmendia, "*Andrés Yeregui has been a consummate craftsman. Both in the most perfect finishing of works considered to be routine and*

Fig. 15 Clock built by Andrés Yeregui for the town of Latasa in 1914

in those that lent themselves to innovation, he has given continuous demonstrations of talent". Andrés built clocks for different towns in Navarre, including those of Izurdiaga, Igoa and Latasa. The latter, built in 1914, is still in the church in that town, although it has been stationary for a long time. Figure 15 shows a picture of the clock and a detail of its plate. A glance at its components shows that some of the original parts have been clumsily replaced.

With Andrés concluded the tradition of five generations who built tower clocks for more than 150 years. The Yeregui family's small artisan clockmaking workshop in Betelu, which produced many different clocks, was unable to compete with the industrialisation and importation of clocks from European factories such as Morez.

5 Legacy and Today Interpretation of Contributions

The Yeregui family was a prolific family of clockmakers who built numerous tower clocks that were installed in different locations in and around Navarre. As clockmaking techniques advanced, they introduced improvements in their designs in order to make more complete, complex and precise clocks. Many of their clocks worked for many years, since the mere fact that they still exist, despite not working, indicates that they were not replaced by other mechanical clocks, meaning that they stopped working well into the twentieth century when it was no longer necessary for the church or town clock to govern the running of the town.

José Francisco was the first of the dynasty and, in addition to making numerous clocks, he introduced technical improvements to his designs. It is noteworthy that, in the successive contracts found, we can observe an increasing complexity in the requirements for the chiming of the clock. Juan Manuel probably took advantage of his father's fame to continue with clockmaking, even building a clock for Pamplona when it is more than likely that there were other clockmakers in the capital of Navarre. Juan José was also very active as a clockmaker, as he ordered the construction of a large house, the ground floor of which was used as a clockmaking workshop.

Bonifacio worked in the second half of the nineteenth century, when the level of industrialisation was beginning to be considerable and great clock factories opened, with which a small workshop could hardly compete. In fact, his cousins Benito and Serapio took advantage of this industrialisation to import parts from the large factories and assemble a large number of clocks, some of which are still in use today. Andrés, the last of the dynasty, would be overtaken by industrialisation in the twentieth century and, although he designed some clocks, it is possible that he was largely dedicated to the maintenance and repair of others, as well as other blacksmith work.

The Yeregui family were not the only clockmaker family of their time and perhaps their legacy could be considered similar to that of other families of clockmakers who made clocks in other regions of Spain. Nevertheless, they made numerous clocks for a multitude of towns in the area, including works in the two nearest large cities, Pamplona and San Sebastián. It is worth mentioning the figure of José Francisco, the first clockmaker in the dynasty. He made his first clock in a self-taught way and with wooden parts, as he had not yet learnt to work with metal. Subsequently, he designed and manufactured clocks with different chiming characteristics and increasing complexity. For example, he solved the need to strike a certain chiming melody at different times in winter and summer in the Olazagutia clock (1805) by placing a screwable sheet at different points of the counting wheel. Considering that this was the early nineteenth century and that he lived in a mountain village, it is very likely that such ingenious solution was his own invention. Continuing with the rest of the dynasty, they were designing and manufacturing clocks for about 150 years. Thus, the fact that as many as 7 clockmakers passed through the Betelu workshop over 5 generations, together with the certainty that some of their clocks are still working and many others exceeded 100 years of life, gives this dynasty of clockmakers a certain notoriety.

Acknowledgements The authors of this work would like to thank the Yeregui Elkartea Foundation, and in particular Xabier Álvarez Yeregui, for the information provided about the numerous clocks built by the Yeregui family, as well as his willingness to accompany and guide the authors on their trips to the villages where Yeregui clocks still exist.

References

V. Pérez Álvarez, The role of the mechanical clock in medieval science. Endeavour **39**(1), 63–68 (2015)

J.M. Ramírez Martínez, J.L. Tomás San Román, El discurrir del tiempo en la Rioja. Relojes y Relojeros. Fundación Caja Rioja, Logroño (2002)

M. Addomine, G. Figliolini, E. Pennestri, A landmark in the history of non-circular gears design: The mechanical masterpiece of Dondi's astrarium. Mech. Mach. Theory **122**, 219–232 (2018)

V. Pérez Álvarez, *Técnica y fe: el reloj medieval de la catedral de Toledo* (Fundación Juanelo Turriano, Madrid, 2018)

M. Bennett, M.F. Schatz, H. Rockwood, K. Wiesenfeld, Huygen's Clocks. Proceedings: Mathematical, Physical and Engineering Siences 458, 563–579 (2002)
M. Stoimenov, B. Popkonstantinovic, L. Miladinovic, D. Petrovic, Evolution of Clock Escapement Mechanisms. FME Transactions **40**, 17–23 (2012)
R. Du, L. Xie, The Mechanics of Mechanical Watches and Clocks. History of Mechanism and Machine Science 21. Springer, Berlin, Heidelberg (2013)
J. Betts, John Harrison: inventor of the precision timekeeper. Endeavour **17**(4), 160–167 (1993)
J. Garmendia Larrañaga, Artesania Vasca – Euskal esku-langintza, Auñamendi, San Sebastián (1970)
Royal and General Archive of Navarre, Lekunberri Notary's Office, Juan José Peralta Protocol, no. 105, 15th April 1796
Royal and General Archive of Navarre, Lekunberri Notary's Office, Matías Antonio Goicoa Protocol, nº 158, 23th December 1804
Royal and General Archive of Navarre, Pamplona Notary's Office, Matías Antonio Goicoa Protocol, no. 26, 3rd February 1804
Royal and General Archive of Navarre, Altsasu Notary's Office, Gabriel Albizu Protocol, no. 146, 6th December 1805
J. Aginaga, A. Claver, N. de Bilbao, M. Benito, Restauración, reconstrucción virtual e impresión 3D del antiguo reloj del Ayuntamiento de Pamplona. In: XXIII Congreso Nacional de Ingeniería Mecánica, Jaén (2021)
Garmendia Larrañaga, J.: La forja del hierro y el labrado de la madera en Berastegui. Caja de Ahorros Provincial de Guipuzcoa, San Sebastián (1982)
Prieto Gil de San Vicente, M.: Muestrario de bisagras del Albiztur Kobie Antropología Cultural 14, 171–188 (2010)
L. Murugarren, Catedral de el Buen Pastor: Donostia-San Sebastián 1897–1997. Kutxa Fundazioa, San Sebastián (1996)
Hernandorena Kultur Elkartea: Kanpandorreko ordularia, 125 urte lanean. Sustraiak 19 (2016)
B. Herreras Moratinos, Patrimonio y construcción naval en Gipuzkoa. Revista De Estudios Marítimos Del País Vasco **2**, 431–471 (1998)
Ministerio de Industria, Comercio y Turismo. Oficina Española de Patentes y Marcas, O.A. Archivo. Fondo Histórico, P_34185 (1904)

The manufacturer's authorised representative in the EU is Springer Nature Customer Service Centre GmbH, Europaplatz 3, 69115 Heidelberg, Germany. If you have any concerns regarding our products, please contact ProductSafety@springernature.com

Printed and bound by CPI Group (UK) Ltd, Croydon, CR0 4YY
28/09/2026
02224085-0012